Towards a Data-Driven Military

TOWARDS A
DATA-DRIVEN MILITARY

A multidisciplinary perspective

Edited by

Peter B.M.J. Pijpers, Mark Voskuijl, Robert J.M. Beeres

LEIDEN UNIVERSITY PRESS

The Open Access edition was made possible by a contribution from The Netherlands Defence Academy

Netherlands Annual Review of Military Studies 2022

Cover design: Andre Klijsen
Cover illustration: Media Centre of Defence
Lay-out: Crius Group

ISBN 9789087284084
e-ISBN 9789400604537
https://doi.org/10.24415/9789087284084
NUR 805

Table of Contents

List of Figures

List of Tables

Foreword by the Dean of the Faculty, Prof. Dr. Patrick Oonincx

Every year the Faculty of Military Sciences at the Netherlands Defence Academy publishes the Netherlands Annual Review of Military Sciences. This year marks the starting point on a somewhat different approach. As of this year the series will address themes within society and the Ministry of Defence in particular from a multidisciplinary scientific point of view with contributions from all departments of our faculty.

This year we have chosen to study the use of data in the military, one of the keystones in the Defence Vision 2035. The current importance of data in the military cannot be overstated. Good intelligence in the Ukraine-Russian war will provide an advantageous position and satellite communications is available at places where regular means of communications are absent.

As a mathematician this topic reminds me of Alan Turing, one of the founding fathers of modern numerical mathematics and artificial intelligence (AI). Already in 1935 Turing described the contours of an abstract computing machine using a moving scanner to read and write symbols. During World War II he acted as a leading cryptanalyst at Bletchley Park improving issues of machine intelligence. Amongst others this led to development of an electromechanical machine called The Bombe for enciphering the German Enigma code. While this is a classical and early example of data in the military, nowadays data science and AI give rise to many new capabilities that need to be explored scientifically to properly understand their capabilities. This volume is a starting point on that scientific journey. Where Turing made the first steps into the world of data science, we embrace the topic and take it to a new level in the military with several studies, from technological fields including autonomous systems, to the fields of military management and war studies including addressing topical legal challenges related to the use of data.

The editors and authors were really challenged this year, but they succeeded in writing a volume that gives many answers to issues in relation to the topic of a data-driven military. I wish you interesting hours of reading the volume and do not hesitate to contact us for a deeper dive into any of the studies after reading about them.

Towards a data-driven military
– an introduction

Peter B.M.J. Pijpers, Mark Voskuijl & Robert J.M. Beeres

Introduction

In October 2020, the Netherlands Ministry of Defence (MoD) published its Defence Vision 2035 (DV35) setting out guidelines for future doctrine, policy, innovation and acquisition.[1] This so-called White Paper envisages the 2035 defence organisation as a reliable partner and protector that is technologically advanced and able to execute information-driven operations. The future Netherlands defence organisation will avail of strong innovative capacities and focus on specialisation within the North Atlantic Treaty Organization (NATO) and the European Union (EU). Moreover, it will be an effective military actor in the information environment, making use of available data and information. In the near future, the Netherlands defence organisation will obtain an authoritative information position, be able to execute multi-domain and integrated information-driven operations, and to use information as a weapon.[2]

The use of information and data during war, armed conflict or interstate competition is as old as war itself. The Cold War era, where the military instrument was sometimes rendered obsolete due to the overwhelming strategic nuclear stalemate, proved a cornucopia for activities other than the use of force. Both the Soviet Union and its allies,[3] and the United States and its allies were very active in influencing each other by means of persuasion, coercion or manipulation. These were the heydays of both the Soviet doctrine of *Active Measures* as well as American *Political Warfare*.[4]

Recent technological advances in the field of computer science have impacted the ability to use information and data profoundly. Not only in business, but also in the military. Microelectronics and miniaturisation have enabled the development of a wide range of new products and capabilities for military systems. A relevant example worth mentioning in this context is the use of large remotely piloted aircraft systems (RPAS) such as the General Atomics MQ-1 Predator and MQ-9 Reaper in Iraq and Afghanistan over the last fifteen years.[5] Unmanned aircraft, equipped with impressive sensor suites and precision-guided munitions can be piloted from the other side of the world by means of satellite communication. At present, small low-cost unmanned aircraft systems are used world-wide for a variety of military applications such as information, surveillance and reconnaissance missions. In the future, swarms of weaponised unmanned aircraft with high levels of autonomy may be expected.

Meanwhile, information technology (IT) interconnectedness serves to enable cyberspace. This novel man-made domain unlocked the information environment, thereby providing new opportunities for trade and communication.[6] Customers anywhere in the world can easily buy products in China without Mandarin proficiency, and the Internet supports upcoming firms in their search for global markets instantly. As cyberspace will also serve as a platform in competition and conflict,[7] malign actors will be provided similar opportunities.

On 24 February 2022, the Russian Federation, conducting a 'special military operation', invaded Ukraine, thereby flagrantly violating international law and alarming international society.[8] Though military and kinetic activities gained most public and media attention – largely due to the damage and destruction they cause – the Ukraine-Russian war and, specifically, the run-up to the war, illustrates the use of data in modern warfare.

In hindsight, the 2014 Maidan revolution, the annexation of the Crimean Peninsula in February and March 2014, the subsequent pro-Russian revolts in the Donbass region, can be seen as triggers shaping up to the 2022 Ukraine-Russian war. Since 2014, the Russian Federation has executed numerous operations in or via cyberspace to undermine, compel or deter Ukraine, most prominently, the 2015 cyber-attack on the Ukrainian electricity network, leaving more than 200,000 people without power for over four hours.[9] In the direct run-up to the 2022 Ukraine-Russian war, Ukraine faced attacks by numerous cyber operations, ranging from the instalment of 'wiper-malware', destroying computer software, blocking government websites via distributed denial-of-service (DDoS) attacks, sabotaging the Viasat satellite system and leaking personal data of more than 2 million Ukrainians.[10]

The attributes of cyberspace, and the ubiquitous access to data changed the character of conflict. States are no longer the sole actors involved, and physical borders have lost relevance in cyberspace. In modern conflict numerous non-state actors, sympathisers (hacker groups such as Anonymous) have sided with one of the warring states without necessarily being belligerent entities; ICT businesses are more outspoken about intrusions in the ICT infrastructure; citizen journalism (for example Bellingcat) is involved in debunking disinformation disseminated on social media platforms; and even the traditionally covertly operating intelligence services are now sharing data via Twitter hoping to expose Russian plans and intentions.[11]

Increasingly, data has gained importance, not only in society at large but *ipso facto* in modern warfare,[12] the core theme of the Netherlands Annual Review of Military Studies (NL ARMS) 2022.

NL ARMS 2022 assesses the use of data and information on modern conflict from different scientific and methodological disciplines, aiming to generate valuable contributions to the on-going discourse on data, the military and modern warfare. Military Systems and Technology approaches the theme empirically by researching how data can be used to enhance the efficiency and effectiveness of military materiel and equipment, thereby generating valuable data to enhance and accelerate the decision-making process. War Studies takes a multidisciplinary approach on the evolution of warfare, while Military Management Studies takes a holistic organisational and procedural approach. Based on their scientific protocols and methods of research, the three domains put forward different research questions and perspectives, providing the unique character of NL ARMS 2022.

The next section provides an overview of the upcoming chapters. To this end, the editors selected a categorisation with reference to a data-driven military or defence organisation. Case-studies focus on the Netherlands Defence organisation. This volume's first part elaborates on how the use of data impacts organisation, focusing on the logistical, personnel and material aspects of the Defence organisation. The second part assesses how data-driven techniques can enhance or accelerate decisions made within the Defence organisation. The last part discusses how data affect the planning and conducting of operations.

Overview of the Chapters

The first part of the book discusses the organisational aspects of a data-driven military Defence organisation. In Chapter 1, Kramer and van Os sketch a sociotechnical perspective on digital transformation. In general, there is consensus that digital technology will hold profound transformative effects on society, organisations, and human beings. Notwithstanding this broad consensus, these transformative effects of digital technology are also controversial, notably, because the transformative impact of digital technology cannot be straightforwardly deduced from functional specifications of the technology itself. Op den Buijs, in Chapter 2, highlights one of the organisational aspects; human resource management. The use of big data analytics in the Human Resource Management (HRM) field has become enormously popular. It can therefore also be beneficial for the armed forces to cope with changes in the HRM environment related to technology, labour market, aging population, personnel recruitment. However, HRM data analytics do not only offer opportunities, they also pose challenges. The third and fourth chapters focus on maintenance. While data may be considered an important "weapon", data collection and analysis are also crucial in reducing the number of system failures, and thus, potentially, may increase systems availability and military performance considerably. In Chapter 3,

Tinga, Homborg and Rijsdijk introduce the concept of data-driven maintenance using various maturity levels, ranging from detection of failures and automated diagnostics to advanced condition monitoring and predictive maintenance that are tested against practical cases to demonstrate the benefits and discuss the challenges that are encountered. In Chapter 4, Vriend, Tiddens and Jurrius argue that while machine learning is used successfully in many applications, challenges remain; data is often stored in separate places, and data used for training purposes ought to respect privacy. Federated learning circumvents the challenges and allows machine learning models to be trained based on privacy-sensitive data sets of multiple parties without having to share raw data. This promising technique is especially valuable in case of collaborative activities with external parties. De Gooijer, Hoogstrate, Schijvenaars, van Fenema and van Kampen, in Chapter 5, focus on the sustainment organisation of the MoD and explore the usability of data-driven maturity models to explore whether the Netherlands defence organisation can become an information and data-driven organisation.

The second part of this volume assesses the extent to which data can be used to support the decision-making process. Both during armed conflicts as well as while preparing for deployments, data can enhance the effectiveness and efficiency of decision-making processes. In Chapter 6, Hoogstrate analyses the effects of Big Data and Artificial Intelligence (AI) on the practice of forecasting in defence and military applications. By combining Big Data and AI, he expects that forecasting and foresight development will be greatly influenced and will impact on applications at the strategic, operational as well as tactical level. In Chapter 7, Lindelauf, Monsuur and Voskuijl investigate whether algorithmic techniques from the fields of operations research, data science and aircraft trajectory optimisation can aid military flight mission planning. Optimising military helicopter missions relates to aspects including instance route selection, helicopter configuration design, opponent modelling to personnel to platform allocations. In Chapter 8, van Ee, de Lima Filho and Monsuur research how maritime patrols conducted with multiple unmanned aerial vehicles can optimise their objectives to detect, locate and identify (opposing) vessels. The authors make clear that the routing problem including mutual support, can be modelled as a generalised travelling salesman problem (GTSP), thereby investigating the costs of requiring mutual support and comparing it to the costs of using separate drones that detect and identify vessels in the area of operations. Chapter 9 by Theunissen provides an overview of current trends in the development of Detect- and Avoid (DAA) systems required for the integration of remotely operated aircraft into non-segregated airspace. A DAA system provides the pilot with actionable information derived from real-time data about cooperative and non-cooperative traffic. Theunissen discusses the potential AI and Machine Learning techniques for the purpose of DAA and several associated legal, ethical, integration and certification issues are addressed. In Chapter 10, the final chapter of Part II, Horlings, Lindelauf and Rietjens describe how in the current information age, military intelligence and security organisations are confronted with information overload – a situation in which decision-makers face a level of information that is greater than their information processing capacity. Information overload is not only the result of the continuously increasing amount of the data, but also of the high levels of uncertainty of the data. Information overload has serious consequences hampering the effectiveness and efficiency of military intelligence and security organisations. In order to improve decision-making accuracy, organisations need to find ways to process more information without increasing the experienced information load.

The third part of the book is geared towards the use of data during operations. What legal framework applies when military units use, collect and process data in military operations, as well as while preparing for operations. Moreover, the question of how data serves as a weapon of influence is studied. In Chapter 11, Timmermans and Lindelauf elaborate on the advantages of data-driven methodologies regarding their beneficial impact on optimising solutions for decision problems, whilst, on the flipside, causing ethical risks, both to society at large and defence and military organisations in particular. Timmermans and Lindelauf conceptually investigate the trade-off between privacy on the one hand and algorithmic performance on the other, concerning the use of MoD relevant (bulk) datasets from a technical, moral and socio-political view. In Chapter 12, Ducheine, Pijpers and Pouw investigate the legal framework to execute 'information-driven operations', as depicted in the Defence White Paper "Defence Vision 2035". Cyberspace has unlocked the information environment, raising obvious concerns about the use of data and potential infringements of privacy since it simultaneously gives new impetus to use data to improve military intelligence and understanding, as well as to enhance decision making, but also to use information as a "weapon of influence". Deploying armed forces in the information environment is challenging since the current legal framework applicable to information manoeuvre hampers training and preparing for operations. In Chapter 13, Zwanenburg and van de Put analyse the use of biometrics during military operations extraterritorially from the perspective of the right to private life in Article 8 of the European Convention on Human Rights (ECHR). The authors argue that the ECHR is applicable to certain conduct of armed forces outside their own state's territory, and that this includes situations involving the use of biometrics. Therefore, although states have a certain margin of appreciation, compliance with the right to private life during extraterritorial military operations appears to be a tall order. In Chapter 14, the final chapter of Part III, de Jong, de Werd and Bouwmeester argue that the role of information as a source of power in Russia's foreign policy and military actions has received increasing attention of

Western scholars and policymakers, whilst focusing on Russian foreign policies and military operations in Georgia and Ukraine as typical case studies. This chapter aims to retrieve more insight in the nature of information operations by studying the atypical case of the 2020 Nagorno-Karabakh War between Armenia and Azerbaijan, in which Russia proliferated herself as a mediator. De Jong, de Werd and Bouwmeester argue that the Russian narrative is tailored to various national and international audiences and fits with Russian interests.

Finally, NL ARMS 2022 offers an epilogue. In Chapter 15, Baudet and de Jong provide a historical overview of the quantitative use of data, especially in measuring effects, or even success, during warfare. Baudet and de Jong discuss the idea that quantitative data can help manage and predict the course of a war, elaborating on the case of Robert McNamara whose technocratic statistical approach guided the war effort during the Vietnam War. In spite of the fact that the United States lost that war, the underlying idea has had a lasting influence that can be traced to the conduct of contemporary wars. The authors argue that technocratic approaches often disregard the complexity and imponderabilia of unique historical wartime contexts and advocate the integration of quantitative-generalising and qualitative-historicising approaches to understand past and contemporary warfare.

Notes

[1]   Netherlands Ministry of Defence, "Defence Vision 2035: Fighting for a Safer Future," 2020.
[2]   Netherlands Ministry of Defence. Annex p. XII
[3]   Thomas Rid, *Active Measures: The Secret History of Disinformation and Political Warfare* (London: Profile Books, 2020).
[4]   Linda Robinson et al., *Modern Political Warfare: Current Practices and Possible Responses*, 2018.
[5]   Ann Rogers and John Hill, *Drone Warfare and Global Security*, 2014.
[6]   Daniel Susser, Beate Roessler, and Helen Nissenbaum, "Online manipulation: hidden influences in a digital world," *Georgetown Law Technology Review* 4, no. 1 (2019): 1–52.
[7]   Thomas Paterson and Lauren Hanley, "Political warfare in the digital age: cyber subversion, information operations and 'deep fakes,'" *Australian Journal of International Affairs* 74, no. 4 (2020): 439–54.
[8]   James A Green, Christian Henderson, and Tom Ruys, "Russia's attack on Ukraine and the jus ad bellum," *Journal on the Use of Force and International Law*, 2022.
[9]   Robert Lee, Michael Assante, and Tim Conway, "Analysis of the Cyber Attack on the Ukrainian Power Grid," *SANS Industrial Control Systems Security Blog*, 2016.
[10]   United Kingdom Government, "UK Assess Russian Involvement in Cyber Attacks on Ukraine," 2022.
[11]   United Kingdom Ministry of Defence, "Latest Defence Intelligence Update on the Situation in Ukraine – 8 April 2022," Twitter, 2022, https://twitter.com/DefenceHQ/status/1512284278813597702.
[12]   Holger Möldera and Vladimir Sazonovb, "Information warfare as the Hobbesian concept of modern times — the principles, techniques, and tools of Russian information operations in the Donbass," *Journal of Slavic Military Studies* 31, no. 3 (2018): 308–28.

References

Green, James A, Christian Henderson, and Tom Ruys. "Russia's attack on Ukraine and the jus ad bellum." *Journal on the Use of Force and International Law*, 2022.

Lee, Robert, Michael Assante, and Tim Conway. "Analysis of the Cyber Attack on the Ukrainian Power Grid." *SANS Industrial Control Systems Security Blog*, 2016.

Möldera, Holger, and Vladimir Sazonovb. "Information warfare as the Hobbesian concept of modern times — the principles, techniques, and tools of Russian information operations in the Donbass." *Journal of Slavic Military Studies* 31, no. 3 (2018): 308–28.

Netherlands Ministry of Defence. "Defence Vision 2035: Fighting for a Safer Future," 2020.

Paterson, Thomas, and Lauren Hanley. "Political warfare in the digital age: cyber subversion, information operations and 'deep fakes.'" *Australian Journal of International Affairs* 74, no. 4 (2020): 439–54.

Rid, Thomas. *Active Measures: The Secret History of Disinformation and Political Warfare*. London: Profile Books, 2020.

Robinson, Linda, Todd C Helmus, Raphael S Cohen, Alireza Nader, Andrew Radin, Madeline Magnuson, and Katya Migacheva. *Modern Political Warfare: Current Practices and Possible Responses*, 2018.

Rogers, Ann, and John Hill. *Unmanned. Drone Warfare and Global Security*, Pluto Press, London, 2014.

Susser, Daniel, Beate Roessler, and Helen Nissenbaum. "Online manipulation: hidden influences in a digital world." *Georgetown Law Technology Review* 4, no. 1 (2019): 1–52.

United Kingdom Government. "UK Assess Russian Involvement in Cyber Attacks on Ukraine," 2022.

United Kingdom Ministry of Defence. "Latest Defence Intelligence Update on the Situation in Ukraine -8 April 2022." Twitter, 2022. https://twitter.com/DefenceHQ/status/1512284278813597702.

PART I

Data-Driven Organisation

CHAPTER 1

Digitalisation, organising and organisational choice

Exploring the challenges of digital transformation using five applied sociotechnical lenses

Eric-Hans Kramer & Guido van Os

Abstract

One of the main challenges in the process of digital transformation is that the impact of new technology cannot be straightforwardly deduced from the functional specifications of technology itself. Digital technology affords new possibilities which affect organisational form and function which means that to understand the transformational impact of digital technology, human practices and systemic features of organisations need to be viewed in mutual entanglement. We therefore consider the process of digital transformation to be an applied sociotechnical challenge. This chapter explores the sociotechnical challenges of digital transformation using a framework derived from the applied sociotechnical approach in organisation science. This framework consists of five lenses which focus on a particular aspect of this sociotechnical challenge. While this framework could be used to reflect on the challenges of digital transformation in general, we place this discussion in the context of the military organisation. The resulting exploration shows the relevance of this sociotechnical perspective for the military organisation and could be used as a guide for the further developments of an (action) research program.

Keywords: Sociotechnical, Digital transformation, Innovation, Organisational choice, Action research

1.1. Introduction

Across domains and disciplines, there is a broad consensus that digital technology will have profoundly transformative effects on society, organisations, and human beings[1]. Digital technology plays an important role in wide ranging fields such as artificial intelligence, communication technology, machine learning, robotics, 3D printing technology, and some argue that the combined effect of these develop-ments will lead to a fourth industrial revolution[2]. Such a revolution is considered to cause shifts across industries by the emergence of new strategies, business models, reshaping of production, consumption, transportation and delivery systems and organisation design. Digital technology is an example of what the *Netherlands Scientific Council for Government Policy* calls a "systems technology"[3], whose applicability is pervasive rather than focused on a single domain, with significant potential for technological improvements and complementary innovations. There is, however, controversy about the specific transformative effects of digital technol-ogy. Some argue that digitalisation will replace half of the occupations and related jobs[4], also higher skilled jobs and knowledge work[5], while others argue that new technology has the potential to impoverish jobs[6], and enable far-reaching forms of surveillance and control[7]. Against this background, some warn against (implicit) technological determinism[8] while others warn against over-hyped claims about the potential of new technology[9].

Such controversy indicates that transformative impacts of digital technology cannot be straightforwardly deduced from functional specifications of technology itself. Digitalisation affords new possibilities and potentially has a profound effect on work practices, structural design, and the social fabric of organisations, but these are not a given from the outset. New combinations of technology and organisational features afford new possibilities that affect organisational form and function[10]. While frequently technological innovation is viewed from a perspective which treats technological adaption, diffusion and use as separate phenomena[11], digital transformation requires the ability to view the social and technological systems of organisations in concert[12], with an open view towards integral effects on different aspects of an organisation's functioning. We will refer to this as a sociotechnical challenge[13].

The previous indicates that shaping of future organisations takes place against the background of the transformative potential of new technology. In this chapter, we start from the assertion that this challenge requires an applied perspective. The aim of this chapter is to sketch the challenges of digital transformation using insights from the applied sociotechnical approach in organisation science. This approach offers an integrated perspective to analyse organisations and to design interventions[14]. We believe that such a perspective has particular relevance when it comes to understanding the ways in which technological artefacts, systemic organisational features and social practices are entangled. As such, in this chapter the applied sociotechnical approach functions as a compass to explore the different related challenges that constitute digital transformation.

While we believe that the sociotechnical perspective is of general relevance to organisations, we place the discussion in the context of the military organisation.

Evidently, developments in the realms of digital technology are also of great significance to military organisations. Not only will digitalisation provide new possibilities for existing weapons systems across domains (land, sea, air), and create new possibilities for strategy and organisational design. It has also already afforded a cyber domain which not only cuts across the other domains but also has distinct security effects. Therefore, the military organisation is confronted with the sociotechnical challenge mentioned above as well. Below, we start by discussing the topic of digital transformation and organisational change in a general sense to develop an idea of the potential affordances of digital technology (section 1.2). Subsequently, we sketch the framework with the five lenses and apply those to the topic of digital transformation (sections 1.3 to 1.8).

1.2. Digital transformation and organisational change

Digital transformation is a heavily contested concept[15]. There is a strong and rich – but fragmented – body of knowledge with very interesting insights, frameworks and explanations on technology and organisational change[16]. The focus is on the opportunities – affordances – of technology to, among other things, enforce business transformation, implement IT strategies, create digital business ecosystems, optimise customer engagement, and digitise business processes[17]. Hanelt et al.[18], agree that the literature on digital transformation spawns in multiple directions. However, it is possible to structure the extensive literature along three conceptual categories:

1) "contextual condition" as input for digital transformation (open and emerging technologies, data availability, organisational strategy),
2) "mechanisms" to link input conditions to perceived outcomes of digital transformation (innovation and integration of technology in existing processes), and
3) "outcome" of digital transformation (organisational setups and improved performance)[19].

Combining the former conceptual categories Hanelt et al[20] as well as Kraus et al[21] show two dominant patterns in current literature on digital transformation. First, the influence of technology on the organisational design and the aim to create a malleable organisational design. This type of organisational design is characterised by– what's referred to as – an agile structure driven by opportunities of technology, to constantly adapt to competitive and social demands from their environment by using technology[22]. For example, Amazon is depending on technology and user data to adjust their supply-chain on the demand in certain regions and optimise logistic processes within their warehouse by using current information (collected from customers). However, changing organisational processes is a complex undertaking[23]. It demands a constant and explicit interaction between technology and structural design during the process of developing and deploying digital tools[24]. In this process there is a strong emphasis on who within the organisations can innovate, change, and socially influence the quality of work and quality of working relations all within a subset of organisation formal rules and organisational values.

Second, and inherently connected to the idea of organisational design, is how technology opens the path of interorganisational cooperation. For example, emerging technologies (AI, Internet of Things) are open and easily applicable and can simply bring previously disconnected markets together (for example internet, phone, video). Therefore, organisations move to participate in business ecosystems – because of the competitive environment – to maintain their accessibility to these open emerging tecnologies. Although external access does not necessarily change the organisational structure. The move to fragmented networks of strategic parentships, however, blurs the boundaries of an organisation.

Important to take from the former argument is that digital transformation refers to an entangled interaction between technology, structural design, environmental characteristics, and the users of digital solutions[25]. Technology will offer users an opportunity to redesign organisational processes and use it as a tool to react to a competitive environment. However, users will use the technology within the boundaries of the organisational structure and will make conscious choices of how to make use of technology. Digital transformation will set in motion a process within an organisation to discover affordance of technology in a specific setting[26]. In this sense technology is not a deterministic force, but an artefact that will get social and technical significance when it is used within an organisational context (structure, processes, and values)[27]. Organisational context is shaped by its structural design and social values; however, technology will also (re)shape the context. Moreover, social values and organisational structure will enable or constrain users to engage with the technology and use it to its full potential. That is, users will eventually use technology, discover, and maximise what a digital solution can offer them from a technical point of view. However, technology will be implemented within the boundaries of social rules and conventions, which can constrain digital possibilities[28].

Although the former conceptualisation of digital transformation has a strong business-oriented focus, it offers very interesting insights for the military organisation. Digital transformation goes way beyond the conversion of information from analogue to digital form and the use of information technologies. It is a complex transformation – with a vast number of digital tools as the main driver of change – of organisational processes to discover new technological affordances to, among other things, perform a task and increase tactical performance[29]. Introducing technology within an organisation (a military organisation as well), potentially demands a fundamental change in the organisational structure and redefinition of social values[30]. It is a complex multi-dimensional – and therefore interdisciplinary – transformation of exciting organisational conditions – structural and process-design, managerial and legal frames – enforced by (external) disruptive technologies (agile organisational structuring, automation of processes, and smart solutions)[31]. For example, big data in combination with artificial intelligence offers "law enforcement," an extensive digital tool to analyse expected behavioural patterns within large crowds during sports events or demonstrations. However, law enforcement should interact with this specific technological solution to make full use of it. Consequently, they must integrate the technology in current processes, create a structure that allows them to quickly analyse data, and act upon the result of the data-analyses. All within the frame of a legal structure and the social values of society.

The main challenge of the adaptation of technologies – digital transformation – by an organisation (military organisation alike) is influenced by how users of digital tools explore and discover "new" affordances of technology. Although technology can be introduced within an organisation to rethink processes and day-to-day tasks, users enact technology as it will fit their ability to use digital tools to their potential all within the setting of structural design, social values, and norms of an organisation[32]. It is this complex interdisciplinary challenge that requires a "compass" rather than a map, which we will turn to below.

1.3. An applied sociotechnical perspective on digital transformation

The previous indicates that the implications of digitalisation are difficult to capture (because of the many different available digital tools) and difficult to demarcate (because of the sprawling effects in all possible directions)[33]. Traditionally, the applied sociotechnical approach has focused on developing flexible and participative organisations with a distinct emphasis on humane forms of organising. While the first sociotechnical concepts were developed in the era of mechanisation[34], the perspective is continuing to develop in the context of emerging technology[35]. As an applied approach, the sociotechnical tradition is focused on intervention, using different insights on the dynamics that are to be influenced and a tradition of empirical research in which these insights are related to the reality of organisations.

The introduction of this chapter emphasised the use of a compass rather than a map to navigate the interrelated dynamics of a sociotechnical transformation process. While a compass is more generic than a map, it can offer a perspective on a terrain that is characterised by many uncertainties and unknowns, and it can offer a direction on how to navigate that terrain. In this fashion, this chapter will use five sociotechnical lenses to discuss related challenges of digital transformation in organisations. These lenses are built on core insights in the applied sociotechnical approach and explicate ways of looking at different aspects of a transformation process.

Taking this into account, the five lenses are based on (1) the core assumptions of the sociotechnical approach concerning organisational environment, transformation, and development, and on (2) the conceptualisation of the nature of sociotechnical relations. Since applied approaches inevitably are related to questions about whether certain situations are desirable or not[36], (3) the third lens perceives the process of digital transformation from the normative aim of the sociotechnical approach, (4) followed by a lens that focuses on the architecture of systems. The final lens, (5) sketches the basis of an intervention strategy that can be used to take up the topic of digital transformation in organisations. These different lenses are summarised in table 1.1.

Table 1.1. Five applied sociotechnical lenses

	Topic	Sociotechnical lens	
1.	Core assumptions	The principle of organisational choice	§ 1.4
2.	Unit of analysis	The conceptualisation of sociotechnical relations	§ 1.5
3.	Normative aim	Using creative, problem-solving potential	§ 1.6
4.	Architecture	Multilevel dynamics	§ 1.7
5.	Intervention strategy	Sociotechnical action research	§ 1.8

In the subsequent sections, we sketch these lenses and use them to look at the challenges of digital transformation in organisations. The lenses are not specific for the military organisation, but in discussing them we aim to connect them to contemporary discussions in this domain.

1.4. The principle of organisational choice

The first lens explicates the core assumptions with which the applied sociotechnical approach perceives organisations. The principle of organisational choice is based on the contention that the inner workings of organisations are not fully determined by outside forces: they possess certain discretionary elbowroom – choice – in their response to environmental developments[37], whether these are economical, technological, or otherwise. This discretionary space can be used by organisations to develop into a direction that they consider desirable. The principle of organisational choice refers to a fundamental notion of systems theory. If systems did not have "choice" they would have to respond passively to environmental pressures[38], in which case it wouldn't be relevant to study them in the first place. From a socio-technical perspective, the discretionary space defined by organisational choice defines the domain the applied approach can focus on as it is this (metaphorical) space that organisations may use to influence their development.

The principle of organisational choice was introduced by researchers from the Tavistock Institute of Human Relations in the early 1950s, in the context of research into the effects of technological innovations in the mining industry[39]. What initiated this research was the discovery that mechanisation in the mines (conveyer belt systems, power tools, etc) hardly improved productivity compared to traditional work systems.[40] According to the Tavistock researchers, the technological system as it was implemented in the mines caused different negative secondary effects, ranging from stress and alienation in miners to ineffective coordination between parts of the mining system. Furthermore, the case indicated that there were alternative organisational forms possible (with the semi-autonomous workgroup as a prototypical solution) that significantly reduced these effects. Such solutions achieved quite significant improvements in economic performance and quality of work in a way that used the potentials of new technology.

The Durham case is paradigmatic because it indicated that available technology does not one-directionally and inevitably force the design of organisations into a particular direction. As Leonardi[41] points out, the Durham case suggests that there is an indeterminate relationship between task and technologies, "(...) such that a technology's fixed materiality could support multiple task structures depending on people's desires and goals". Therefore, sociotechnical theorists positioned themselves in opposition to what they saw as the technological determinism of Taylor's scientific management philosophy[42], according to which – for example – the traditional conveyer belt structure would be seen as an inevitable result of prevailing technology. Challenging (implicit) technological determinism in strategies for implementing technology, making the connection between different aspects of an organisation's functioning (technological, economic, and social) in relation to environmental characteristics, adopting systems theory and developing new ways of designing organisations to facilitate human creativity and flexibility, arguably were the most important contributions of first-generation sociotechnical theory, which are still broadly supported in organisation science[43].

The emphasis on "organisational choice" remains relevant in the context of digital transformation. For example, Zuboff argues that the same information technology can often be used to both "automate and informate"[44]: it can either create tighter control over work at the operational levels or create possibilities for self-organisation and human judgement in the same positions[45]. Depending on "choice", very different organisations may result from "digital innovation". There is – as it were – a river that divides different philosophies of using the same technology. A major complication in exploiting the potential of new technology is constituted by the potential conflicts with the demands of existing social, economic, and political interests. Increasing abilities to self-organise at lower hierarchical positions might make hierarchical layers and managerial positions superfluous, which might explain why managerial action often "flows along the path of least resistance"[46], without a serious attempt at exploring the possibilities to "informate". This is an explanation for why innovation strategies often remain restricted to what is often colloquially referred to as "bolting new technology to an old organisation".

One particular example from the military domain in which "choice" has been key to the design of military organisations is the development of the concept *Mission Command*. Van Crefeld has pointed out, military history shows that improvements in communications have often been implemented in ways to reduce initiative and possibilities for self-organisation and initiative at lower levels[47]. He argues that such choices have significantly influenced the effectiveness of military organisations[48]:

> "The fact that, historically speaking, those armies have been most successful which did not turn their troops into automatons, did not attempt to control everything from the top, and allowed subordinate commanders considerable latitude has been abundantly demonstrated."

In its various forms and guises, *Mission Command* essentially is about enabling flexible response to dynamically complex environments by creating local conditions for self-management[49]. This idea about military command revolves around a choice to either "automate" or "informate" and is on par with the sociotechnical idea that environmental variety is best controlled closest to the locations in which it appears[50]. Subsequently, this core idea can be used to think about how to design organisations that best facilitate such self-management and could also be fundamental to the way digital transformation is approached.

1.5. The conceptualisation of sociotechnical relations

If the principle of organisational choice defines the discretionary space that can be used by organisations to develop into a desirable direction, sociotechnical relations are the unit of analysis on which the sociotechnical approach focuses. This leads to the question of what "viewing the social and technological systems of organizations in concert"[51] specifically entails. The early sociotechnical innovators indicated that in organisations the human system could not be understood without also understanding the technical system[52]. This also goes the other way around[53]:

"The problem was not of simply "adjusting" people to technology nor technology to people but of organizing the interface so that the best match could be obtained between them. Only the sociotechnical whole could be optimized.""

Against the background of this idea, the early sociotechnical innovators emphasised "joint optimization"[54]:

"Joint optimization means the best possible matching together of the people in any unit and the way their jobs are organized, with the physical equipment and the material resources in that unit".

Within the applied sociotechnical approach, Ulbo de Sitter formulated a critique on the idea of joint optimisation of a separate "social" and a "technical" system. According to De Sitter[55], such a distinction essentially is an abstract dichotomy, while in organisations social and technical aspects are always and inevitably interwoven. This critique on the traditional concept of "joint optimization" was for De Sitter the basis for developing integral sociotechnical theory, which emphasises the ways in which architectural characteristics affect different aspects of an organisation's functioning. Outside the realm of applied theory, Orlikowski emphasises the idea of "constitutive entanglement" according to which[56] "(....) the social and the material are constitutively entangled in everyday life". This emphasis on entangled sociotechnical relations indicates that the rejection of technological determinism does not imply the rejection of any influence of technology. Instead, the idea of entangled sociotechnical relations offers a middle ground between technological determinism and social determinism.

When sociotechnical relations are entangled, an altogether more complicated view of the process of sociotechnical change emerges. If technology co-shapes the context in which humans act and helps to shape the kind of choices human beings make, a dynamic is suggested that is difficult to predict with precision. As a result, a process of disruptive sociotechnical transformation cannot be tightly controlled. This is the essence of Collingridge's[57] dilemma of control, according to which we lack the evidence to govern technologies before "lock in" occurs because it is hard to understand the wider implications of innovations – and therefore hard to imagine relevant ways to "control" impacts – while after lock in the chance of establishing such control may be gone[58]. Such a difficult process constitutes a main challenge in sociotechnical transformation: implementation of new tools may have lasting secondary effects before we were able to understand them.

Such notions about the way sociotechnical relations are interwoven is relevant to an applied perspective on digital transformation. Specific digital tools are more than mere material artifacts that are used for a particular functional purpose.

Their use may have significant – perhaps even existential – transformative effects on organisations and even more in general, human beings and society. Zuboff[59] emphasised that technological innovation can be seen "(...) as an alteration of the material horizon of our world with transformative implications for both the contours and the interior texture of our lives". The consequence of this view is that "affordances" of digital technology are not a given but have to be discovered. The effects of digital innovations cannot just be deduced from functional specifications, but they appear when they are used. For an applied approach that means that not only an understanding of technology itself is important, but also the ability to envision how technological artefacts might connect to human practice and the ability to learn from experience.

As we have seen, digital tools may change the opportunities for communication and may pose new demands for interacting with technological artefacts, new ways for groups to cooperate; they may make long distance coordination possible, influence the way hierarchies operate, may change the design of a primary process and the strategic orientation of an organisation. Seen in this light, the characteristic flexibility of digital tools creates an abundance of opportunity, but such opportunities – affordances – can only be discovered in a process in which experimentation, imagination, envisioning and developing a deep understanding of how social practice and technological artefacts may become intertwined. As was argued in the previous section, a clear idea about "choice" can be essential to navigate such abundance of opportunity and the potential influence on choice and organisational "form and function".

1.6. Using creative, problem-solving potential

If sociotechnical transformation requires organisational choice, the question is what the aim of such choice should be. If there is choice, what should we choose, or what direction should our choices be aimed at? These are normative questions. For the sociotechnical approach, a key normative idea is that enabling the creative, problem-solving potential of human beings can simultaneously create organisations that generate more desirable outcomes (better quality of working life, more desirable from a societal perspective) as well organisations that are preferable for more functional reasons (more effective, efficient, flexible and innovative).

From the outset, the Tavistock institute strongly focused on democratisation and participation also given their backgrounds in (psychological) health care. Within organisational sciences this has evolved in a concern for "the quality of working life", which aims at improving the lives of people working within the organisation. The emphasis on self-organisation is important in this respect and may be contrasted with the anthropological assumptions of more classical perspectives on organisations which view human beings[60]: "(...) as a kind of trivial machines; programmable and re-programmable by operand conditioning, "motivated" by fear, performing operations contributing to contingent goals and possibly unknown to them, and examined as "cases" to increase their performance". Attention in the sociotechnical approach to ethics is not limited to intra-organisational concerns for the quality of working life. For example, Achterbergh and Vriens[61] bring forward the idea of "rich survival" of organisations which underlines the significance of the societal impacts of an organisation's pursuit. So moral concerns in the sociotechnical approach are not only directed inside (quality of working life) but also outside (societal impacts).

Taking the key insight of this lens into account, a specific perspective on contemporary discussions about military technology – enabled by the possibilities of digitalisation – emerges. One of the most important contemporary discussions in this realm is about "autonomous weapons", which is a development that is afforded by digital technology. A central concept in discussions about the ethical aspects of autonomous weapons is "meaningful human control", which should according to the Dutch *Advisory Council on International Affairs*[62] be understood in a comprehensive way in which the control of human operators on weapon systems is adequate and effective. Ekelhof[63] worked out a perspective on "meaningful human control" which takes the design of the sociotechnical system in which weapons are implemented into account. She brings forward that it is confusing to label a weapon system autonomous by narrowly focusing on characteristics of the technological system. Therefore, she focuses on the wider (organisational) targeting process to which an autonomous weapon system is connected[64]: "(...) asking whether the operator exercises meaningful human control over the weapon may be the wrong question here. Instead, we should ask how the military organization as such can or cannot ensure meaningful human control over important targeting decisions." Seen in this perspective, an autonomous weapon is a weapon system with potential autonomy in crucial functions in the targeting process. Conversely, the question of meaningful human control is related to the role of human operators in the targeting process, and this role is developed based on "choice".

This indicates that Ekelhof connects the issue of meaningful human control to the question of how technology functions in the context of a sociotechnical system. The sociotechnical approach has shown that the potential for such control is in essential elements related to the design of the sociotechnical network. She takes this issue up by referring to a targeting cycle, which essentially is a control loop. Developing meaningful human control starts according to her in a design phase, in which parts of this targeting control loop are connected to human actors. Meaningful human control implies control over a part of the control loop. However, as Kramer (2020) indicated, seen from a sociotechnical perspective meaningful human control is not about controlling one aspect of a control loop, but about controlling an integrated set of steps[65]. Moreover, controlling such an integrated set of steps in a control loop for making control meaningful. Enabling human judgement requires more than the mere presence and programmed activity of human beings, but instead requires control of integrated regulatory steps.

Therefore, enabling self-organisation to "ensure meaningful human control over important targeting decisions" requires an organisation design strategy that is typically not used in traditional bureaucracy and seems not to be part of discussions about "meaningful human control". A sociotechnical design perspective therefore relates to important dimensions of ethical decision making[66], which might be overlooked from a narrow technological perspective, but also from a narrow social perspective (as long as it is a human being that presses the button, it is OK).

1.7. Multilevel perspective

The prospects for a design philosophy that aims at creating the conditions for self-organisation would be gloomy if it could only be defended from the perspective of moral concern. A core idea of the sociotechnical perspective is that it is possible to find ways to simultaneously design more "humane" organisations and organisations that are more effective. This intention poses a design challenge in larger organisations. Early sociotechnical experiments indicated that semi-autonomous workgroups often remained an anomalous entity in what was in every other respect an organisation built up of Taylorist principles[67]. This observation led to the development of integral design theory, which is focused on establishing the structural conditions for self-organisation and therefore approaches organisations as a multilevel phenomenon. Structural design of organisations plays an essential role in achieving sociotechnical ambitions at the level of the semi-autonomous workgroup. Sociotechnical design is based on the – to some counterintuitive – idea that design characteristics of the organisational context significantly influence the ability of groups and individuals to respond creatively and effectively to environmental demands.

What is new in the present era is that digital tools allow for opportunities for self-organisation that were not possible previously. This potentially has implications for how to approach organisation design. As Winby and Mohrman[68] emphasise: "Exponential technology advances pose great challenges to traditional forms of organizing that have not been designed to take advantage of this fourth-generation technology. The technical in the sociotechnical equation has changed fundamentally in scope and impact on social organization, driving new ways of organizing, working together, and meeting human needs, *or not*" (italics in original). The former indicates the potential integral effects of new sociotechnical entanglements. Particularly the opportunities for communication, coordination and knowledge generation can have such pervasive effects that strategy, business models, designs and work systems can all be affected by the affordances of digital technology.

It can be propositioned that the sociotechnical challenge increases in complexity if the disruptive potential of new technology is greater. While hardly controversial, this proposition may have implications for organisation design. These can be illustrated by translating Henderson and Clark's[69] distinction of four different types of innovation into four types of sociotechnical innovation in an individual organisation (see table 1.2 below).

Table 1.2. Henderson and Clark's types of innovation vs sociotechnical innovation

Henderson and Clark's innovation type	Sociotechnical change	
	System level	*Type of transformation*
Incremental innovation	Individual level	Limited change
Modular innovation	Group level	Change in work process
Architectural innovation	Macro level	Change in organisation design
Radical innovation	Strategic level	Transformation in output

More specifically,

- *Incremental sociotechnical innovation* refers to different improvements that have no direct implications for the design of an organisation. For example, while the design of a new pen might be a great technological achievement, it is not likely to change much in the inner workings of an organisation. So, it might be radical from a technological perspective, but incremental from a sociotechnical perspective.
- *Modular sociotechnical innovation* refers to changes within the boundaries of a subsystem (e.g., a department), while the organisational design as such remains constant. In the military organisation, for example, this might involve a redesigned maintenance system with changes in tasks and responsibilities within a designated maintenance group. De Sitter[70] uses the term "local process innovation" to refer to this kind of sociotechnical innovation.
- *Architectural sociotechnical innovation* refers to the reconfiguration of an existing organisation. It may involve the design of a different transformation process that requires a different organisational design. For example, in the military organisation new digital ways of communicating might translate into the development of a new Operational Framework for the infantry. De Sitter[71] uses the term "interlocal process innovation" to refer to this kind of sociotechnical

innovation. The more innovation affects the configuration of organisations, the more it has an integral effect on different aspects of the functioning of an organisation (logistics, HR, accounting, etc).

- *Radical sociotechnical innovation* is the most extensive form of innovation because it has strategic implications. This occurs when a technological change enables a different kind of transaction with the environment. This might be the result of a change in perspective on the environment (new kind of threats, new perspective on existing threats), but also when radically different possibilities arise because of technological developments. In the military domain the creation of cyber units in the last decade can be considered a radical sociotechnical innovation. De Sitter[72] refers to "product development" in this context.

As the previous makes clear, technological innovation might affect different levels of a sociotechnical system. Trying to find out what innovations potentially involve for different layers in organisations, different aspects of an organisation's functioning for the layering of systems itself, making a conscious choice about the use of those potential effects in a way that supports rather than undermines self-organisation is a key part of a sociotechnical design strategy.

What is significant is that if radical sociotechnical organisation transformation changes relationships with the environment, a design strategy may extend beyond the walls of an individual organization[73]: "In the past, we designed single organizations; in the future, we will design the rough outlines of an ecosystem of partners and contributors who work together in an interconnected fashion using technological platforms to achieve a shared purpose." Discussions about the design of ecosystems resemble the discussion about "netcentric" forms of organising as they have been studied in the military organisation for quite some time[74]. Such netcentric ways of organising have a distinct technological side and are increasingly finding their way in military technology. However, the insights on how to organise such digitally enabled networks seem to struggle to keep up with technological affordances[75].

Clearly, the sociotechnical challenge increases from incremental to radical, not just in a narrow sense, but also because a more complicated cooperative network of stakeholders is involved. The previously mentioned approach of "bolting new technology to an old organisation" remains stuck at the incremental level, perhaps with limited implications at the modular level and flows along the path of least resistance. As the implications of technological innovations affect more levels, the network of stakeholders becomes more complicated, which also means that the sociotechnical design process increasingly requires the ability to manage the socio-political implications of transformations. Managing sociotechnical implications of technological developments across multiple levels is a significant challenge, particularly when organisations are not just confronted with a single innovation, but with a wave of different technological developments at the same time with potential interactions. Particularly in such cases a sociotechnical compass to navigate multiple implications is needed.

1.8. Action research

The last sociotechnical lens that we bring forward here, focuses on the challenge of implementation and the approach to organisational change. It binds the previous theoretical perspectives together in a view on how to work on digital transformation and organisation design in concrete settings. At the core of this strategy is action research methodology. This methodology has been at the core of the sociotechnical approach from the outset[76]. Van Beinum et.al. state that[77]: "(...) action research represents a certain way of understanding and managing the relation between theory and practice, between the epistemic subject (the researcher), and the empirical object (the researched)". Essentially, an action research strategy involves managing a process of co-generative learning[78]. More specifically, Greenwood and Levin define action research as follows[79]: "*Action research* is social research carried out by a team that encompasses a professional action researcher and the members of an organization, community, or network ("stakeholders") who are seeking to improve the participants' situation" (emphasis in original). They furthermore emphasise three elements in an action research perspective: a focus on improving a situation ("action"), with the use of formal knowledge of organisations ("research"), with an ambition to foster participation in problem analysis and solving ("participation"). The focus on action research indicates that the sociotechnical approach was never meant as a narrow expert-approach.

Relating this to the previously discussed cornerstones of sociotechnical innovation, action research is relevant as a basis for a sociotechnical innovation strategy because the available space for organisational choice is difficult to define and requires a sophisticated search process (action coupled with research), which requires experimentation and a process of co-generative learning. Action research seeks to improve the organisation's ability to understand itself through participation, while at the same time generating scientific insights[80]. This is an essential step in learning to self-manage ("learning by design"[81]). Because the potential wider systemic effects of innovations require understanding the ways different layers of systems are affected and require the involvement of a broad group of stakeholders.

A particular example can clarify how the topic of digital organisation design can be connected to action research. This is the so-called STARlab initiative (SocioTechnical Action Research lab) developed by a group of sociotechnical practitioners and scientists in the US. Pasmore et al. bring the STARlab concept forward as a strategy for digital sociotechnical design[82]: "(...) the STARLab is a rapid organization design approach that involves a multi-stakeholder group working iteratively to create design solutions to address the changing realities of the business context and to generate socio-technically optimized organization prototypes." They argue that compared to more traditional sociotechnical design that focused on the work-system of a single organisation with fixed technology and with the focus of controlling variance, the affordances of digital technology create a shift in focus of sociotechnical design. Work systems have become complex, technologically enabled networked ecosystems that extend beyond an organisation and its employees and are geographically dispersed with a technology that continuously evolves. This requires a design strategy that incorporates external stakeholders (an "outside in" design approach), while at the same time a capacity for continuous evolution.

1.9. Discussion: between madness and foolishness

This chapter aimed to sketch an applied sociotechnical perspective on digital transformation, with the intention to be useful for discovering the affordances of digital technology in relation to structural, managerial, legal, and economic implications for organisations. The sociotechnical approach offers a particular kind of perspective that can serve as a compass to navigate these different interconnecting issues. While the compass that was worked out in this chapter remained at a more abstract level, it can be valuable as a basis to formulate an (action) research program that is oriented towards the specific challenges of the military organisation.

The sociotechnical approach as formulated in this chapter clearly relates to other well-known innovation strategies. Creating experimental settings, prototyping, establishing "agile" ways of working, concern for problems of scaling and – indeed – concern – for ethical implications certainly are not new or unique to a sociotechnical approach. What is typical for a sociotechnical approach is that these are part of an interconnected applied perspective that can serve as a compass to deal with a complex entangled issue like digital transformation. So, creating isolated design labs without structural connection to external stakeholders and without connection to the parts of the organisations in which the outcomes are to be implemented is likely to lead to dysfunctional outcomes. Avoiding taking strategic effects of technological innovations into account makes "bolting new technology to an existing organisation" a more likely outcome; attention for "social innovation" as a concern separate from other aspects of an organisation's everyday functioning (technical, economical) is likely to lead to development of cosmetic ornaments that initially might make people feel better but leave less visible structural determinants that affect people's well-being untouched.

At the same time, a sociotechnical innovation strategy is demanding not just cognitively – because of the range of interconnecting issues on multiple levels that have to be managed – but also politically – because of vested interests that might be affected by organisational transformation. What makes this particularly difficult is that because of the inherent uncertain affordances of "truly" radical technology, the outcomes of sociotechnical transformation are uncertain as well. The potential disruptive effects of digital technology may lead to "madness" as it *can* cause an endless regression of questions about the validity of every aspect of the existing worldview and existing societal order – without a guaranteed result. At the same time, the anxiety that is potentially generated by such madness may lead to the "foolishness" of avoiding difficult questions about the viability of the existing organisation and to choosing the path of least resistance and hoping for the best[83]. A sociotechnical innovation strategy therefore aims to offer a pragmatic way of balancing such madness and foolishness.

Notes

[1]   André Hanelt, René Bohnsack, David Marz, and Claudiá Antunes Marante, "Systematic review of the literature on digital transformation: insights and implications for strategy and organizational change." *Journal of Management Studies*, 58, 5 (2020): 1159–1197.

[2]   Erik Brynjolfsson, and Andrew McAfee, *The Second Machine Age: Work, Progress and Prosperity in a Time of Brilliant Technologies*. (New York: W.W. Norton & Company, 2014); Richard Susskind, and Daniel Susskind. The Future of the Professions. *How Technology Will Transform the Work of Human Experts*. (Oxford: Oxford University Press, 2015); Peter Oeij, Steven Dhondt, Diana Rus, Geert Van Hootegem, "The digital transformation requires workplace innovation: an introduction." *Int. J. Technology Transfer and Commercialisation*, 16, 3, 199. (2019): 199-207; David Teece and Greg Linden. Business models, value capture, and the digital enterprise. *Journal of Organization Design* 6:8 (2017): 1-14.; Stephen Barley, Beth Bechky, B. Frances Milliken, "The changing nature of work: careers, identities, and work lives in the 21st century". In: *Academy of Management Discoveries*, Vol. 3, No. 2, (2017): 11–115.; Gianvito Lanzolla, Annika Lorenz, Ella Miron-Spektor, Melissa Schilling, Giulia Solinas, Christopher L. Tucci, "Digital transformation: what is new if anything? Emerging Patterns and Management Research." *Academy of Management Discoveries* Vol. 6, No. 3 (2020), 341–350; Sascha Kraus, Paul Jones, Norbert Kailer, Alexandra Weinmann, Nuria Chaparro-Banegas, and Norat Roig-Tierno, "Digital transformation: an overview of the current state of the art of research", *SAGE open*, vol 1 (1) (2021): 1-15.

[3]   WRR. *Opgave AI. De nieuwe systeemtechnologie*, WRR-Rapport 105, Den Haag: WRR, 2021.

[4]   Carl Frey and Michael Osborne, "The future of employment: how susceptible are jobs to computerization?" In: *Technological Forecasting and Social Change* 114 (2017): 254–280.

[5]   See e.g Claudia Loebbecke and Arnold Picot, "Reflections on societal and business model transformation arising from digitization and big data analytics: A research agenda". *Journal of Strategic*

Information Systems, 24, 3, (2015): 149–157; Richard Susskind, and Daniel Susskind. The Future of the Professions. *How Technology Will Transform the Work of Human Experts.* (Oxford: Oxford University Press, 2015).

[6]   David Graeber, *Bullshit Jobs: A Theory*. New York: Simon and Schuster, 2018; Mary Gray and Siddarth Suri. *Ghost Work: How to Stop Silicon Valley from Building a New Global Underclass.* Boston: Eamon Dolan Books, 2019.

[7]   Soshana Zuboff, *The Age of Surveillance Capitalism. The Fight for a Human Future at the New Frontier of Power.* London: Profile Books, 2019; Katherine Kellogg, Melissa Valentine, and Angèle Christin, "Algorithms at work: the new contested terrain of control." In: *Academy of Management Annals*, 14, 1, (2020): 366-410.

[8]   Jannis Kallinikos, Paul Leonardi., Bonnie Nardi. The Challenge of Materiality: Origins, Scope, and Prospects. In: Paul Leonardi, Bonnie Nardi, Jannis Kallinikos (eds), *Materiality and Organizing. Social Interaction in a Technological World.* London: Oxford University Press, 2012, 3-22.

[9]   Leslie Willcocks, "Robo-apocalypse cancelled? reframing the automation and future of work debate". *Journal of Information Technology*, 35, 4, (2020): 286–302.

[10] Raymond Zammuto, Terri Griffith, Ann Majchrzak, Deborah Dougherty Samer Faraj, "Information technology and the changing fabric of organization". *Organization Science*, 18, 5, (2007): 749–762.

[11] Lauren Waardenburg, Marleen Huysman, "From coexistence to co-creation: blurring boundaries in the age of AI". *Information and Organization.* 32 (2022) 100432.

[12] Ibid, p.752.

[13] Marleen Huysman, "Information systems research on artificial intelligence and work: A commentary on "Robo-Apocalypse cancelled? Reframing the automation and future of work debate"". *Journal of Information Technology,* Vol. 35, 4, (2020): 307–309.

[14] Frans van Eijnatten, Rami Shani and Myleen Leary, "Socio-technical systems: designing and managing sustainable organizations." Thomas Cummings (Ed.), *Handbook of Organization Development* (pp. 277-310). Thousand Oaks, CA: Sage, 2008; William Pasmore, *Designing Effective Organizations. The Sociotechnical Perspective.* New York: John Wiley and Sons, 1988; Albert Cherns, "Principles of sociotechnical design revisited." *Human relations*, 40, 3, (1987), 153-161.

[15] Tobias Kretschmer, and Pooyan Khashabi, "Digital transformation and organization design: An integrated approach", *California Management Review*, Vol. 62, 4, (2020): 86–104; André Hanelt, René Bohnsack, David Marz, and Claudiá Antunes Marante. "Systematic review of the literature on digital transformation: insights and implications for strategy and organizational change." *Journal of Management Studies*, 58, 5 (2020): 1159–1197; Sascha Kraus, Paul Jones, Norbert Kailer, Alexandra Weinmann, Nuria Chaparro-Banegas, and Norat Roig-Tierno. "Digital transformation: an overview of the current state of the art of research", *SAGE open*, vol 1 (1) (2021): 1-15; Lauri Wessel, Abayomi Baiyere, Roxana Ologeanu-Taddei, Jonghyuk Cha, and Tina Blegind-Jensen, "Unpacking the difference between digital transformation and IT-enabled organizational transformation." *Journal of the Association for Information Systems*, Vol. 22, 1 (2021): 1-57.

[16] Wanda Orlikowski, "Sociomaterial practices: exploring technology at work". *Organization Studies*, 28, (2007): 1435-1448; Patrick Bessen, and Frantz Rowe, 'Strategizing information systems-enabled organizational transformation: A transdisciplinary review and new directions'. *Journal of Strategic Information Systems*, 21, 2, (2012): 103–124; Day-Yang Liu, Shou-Wei Chen, and Tzu-Chuan Chou, "Resource fit in digital transformation: lessons learned from the CBC Bank global e-banking project". *Management Decisions*, 49,10, (2011), 1728–1742; Lauri Wessel, Abayomi Baiyere, Roxana Ologeanu-Taddei, R., Jonghyuk Cha. and Tina Blegind-Jensen, "Unpacking the difference between digital transformation and IT-enabled organizational transformation." *Journal of the Association for Information Systems*, Vol. 22, 1 (2021): 1-57.

17 Gianvito Lanzolla, Danilo Pesce. and Christopher Tucci, "The digital transformation of search and recombination in the innovation function: tensions and an integrative framework", *Journal of Product Innovation Management*, Vol 38, 1, (2021): 90–113; Lauri Wessel, Abayomi Baiyere, Roxana Ologeanu-Taddei, R., Jonghyuk Cha. and Tina Blegind-Jensen. "Unpacking the difference between digital transformation and IT-enabled organizational transformation." *Journal of the Association for Information Systems*, Vol. 22, 1 (2021): 1-57.

18 André Hanelt, René Bohnsack, David Marz, and Claudiá Antunes Marante, "A systematic review of the literature on digital transformation: insights and implications for strategy and organizational change." *Journal of Management Studies*, 58, 5 (2020): 1159–1197.

19 Sascha Kraus, Paul Jones, Norbert Kailer, Alexandra Weinmann, Nuria Chaparro-Banegas, and Norat Roig-Tierno, "Digital transformation: an overview of the current state of the art of research", *SAGE open*, vol 1 (1) (2021): 1-15; Lauri Wessel, Abayomi Baiyere, Roxana Ologeanu-Taddei, R., Jonghyuk Cha. and Tina Blegind-Jensen, "Unpacking the difference between digital transformation and IT-enabled organizational transformation." *Journal of the Association for Information Systems*, Vol. 22, 1 (2021): 1-57.

20 André Hanelt, René Bohnsack, David Marz, and Claudiá Antunes Marante, "A systematic review of the literature on digital transformation: insights and implications for strategy and organizational change." *Journal of Management Studies*, 58, 5 (2020): 1159–1197.

21 Sascha Kraus, Paul Jones, Norbert Kailer, Alexandra Weinmann, Nuria Chaparro-Banegas, and Norat Roig-Tierno, "Digital transformation: an overview of the current state of the art of research", *SAGE open*, vol 1 (1) (2021): 1-15.

22 Jimmy Huang, Ola Henfridsson, Martin Liu and Sue Newell, "Growing on steroids: rapidly scaling the user base of digital ventures through digital innovation". *MIS Quarterly*, 41, (2017): 301–314; Kretschmer, Tobias and Pooyan Khashabi, "Digital transformation and organization design: An integrated approach", *California Management Review*, Vol. 62, 4, (2020): 86–104.

23 Patrick Bessen, and Frantz Rowe, 'Strategizing information systems-enabled organizational transformation: A transdisciplinary review and new directions. *Journal of Strategic Information Systems*, 21, 2, (2012): 103–124.

24 Tobias Kretschmer, and Pooyan Khashabi, "Digital transformation and organization design: an integrated approach", *California Management Review*, Vol. 62, 4, (2020): 86–104; Lauri Wessel, Abayomi Baiyere, Roxana Ologeanu-Taddei, R., Jonghyuk Cha. and Tina Blegind-Jensen. "Unpacking the difference between digital transformation and IT-enabled organizational transformation." *Journal of the Association for Information Systems*, Vol. 22, 1 (2021): 1-57.

25 Tobias Kretschmer and Pooyan Khashabi, "Digital transformation and organization design: an integrated approach", *California Management Review*, Vol. 62, 4, (2020): 86–104.

26 Ian Hutchby, "Technology, texts and affordances". *Sociology*, Vol. 35 (2) (2001): 441-456; Gianvito Lanzolla, Danilo Pesce and Christopher Tucci. "The digital transformation of search and recombination in the innovation function: Tensions and an integrative framework", *Journal of Product Innovation Management*, Vol 38, 1, (2021): 90–113; Lauri Wessel, Abayomi Baiyere, Roxana Ologeanu-Taddei, R., Jonghyuk Cha. and Tina Blegind-Jensen. "Unpacking the difference between digital transformation and IT-enabled organizational transformation." *Journal of the Association for Information Systems*, Vol. 22, 1 (2021): 1-57.

27 Sascha Kraus, Paul Jones, Norbert Kailer, Alexandra Weinmann, Nuria Chaparro-Banegas, and Norat Roig-Tierno, "Digital transformation: An overview of the current state of the art of research", *SAGE open*, vol 1 (1) (2021): 1-15.

28   Ian Hutchby, Technology, texts and affordances. *Sociology*, Vol. 35 (2) (2001): 441-456; Sascha Kraus, Paul Jones, Norbert Kailer, Alexandra Weinmann, Nuria Chaparro-Banegas, and Norat Roig-Tierno, "Digital transformation: An overview of the current state of the art of research", *SAGE open*, vol 1 (1) (2021): 1-15.

29   Ian Hutchby, Technology, texts and affordances. *Sociology*, Vol. 35 (2) (2001): 441-456; Gianvito Lanzolla, Danilo Pesce. and Christopher Tucci. "The digital transformation of search and recombination in the innovation function: tensions and an integrative framework", *Journal of Product Innovation Management*, Vol 38, 1, (2021): 90–113.

30   André Hanelt, René Bohnsack, David Marz, and Claudiá Antunes Marante, "Systematic review of the literature on digital transformation: insights and implications for strategy and organizational change." *Journal of Management Studies*, 58, 5 (2020): 1159–1197; Sascha Kraus, Paul Jones, Norbert Kailer, Alexandra Weinmann, Nuria Chaparro-Banegas, and Norat Roig-Tierno, "Digital transformation: an overview of the current state of the art of research", *SAGE open*, vol 1 (1) (2021): 1-15; Lauri Wessel, Abayomi Baiyere, Roxana Ologeanu-Taddei, R., Jonghyuk Cha. and Tina Blegind-Jensen, "Unpacking the difference between digital transformation and IT-enabled organizational transformation." *Journal of the Association for Information Systems*, Vol. 22, 1 (2021): 1-57.

31   Patrick Bessen and Frantz Rowe. 'Strategizing information systems-enabled organizational transformation: A transdisciplinary review and new directions'. *Journal of Strategic Information Systems*, 21, 2, (2012): 103–124; André Hanelt, René Bohnsack, David Marz, and Claudiá Antunes Marante. 'A systematic review of the literature on digital transformation: insights and implications for strategy and organizational change.' *Journal of Management Studies*, 58, 5 (2020): 1159–1197.

32   Wanda Orlikowski, "Sociomaterial practices: exploring technology at work". *Organization Studies*, 28, (2007): 1435-1448; Gianvito Lanzolla, Danilo Pesce. and Christopher Tucci. "The digital transformation of search and recombination in the innovation function: tensions and an integrative framework", *Journal of Product Innovation Management*, Vol 38, 1, (2021): 90–113.

33   André Hanelt, René Bohnsack, David Marz, and Claudiá Antunes Marante, 'A systematic review of the literature on digital transformation: insights and implications for strategy and organizational change." *Journal of Management Studies*, 58, 5 (2020): 1159–1197.

34   Eric Trist, Gurth Higgin, Hugh Murray and Alex Pollock, *Organizational Choice*. London: Tavistock, 1963.

35   See e.g. Carl Pava, "Redesigning sociotechnical systems design: concepts and methods for the 1990s." *The Journal of Applied Behavioral Science*, 22, 3, 1986: 201-221; Raymond Zammuto, Terri Griffith, Ann Majchrzak, Deborah Dougherty Samer Faraj. "Information technology and the changing fabric of organization". In: *Organization Science*, 18, 5, (2007): 749–762; William Pasmore, Stuart Winby, Sue Mohrman, Rick Vanasse. "Reflections: sociotechnical systems design and organization change". *Journal of Change Management*, 19, 2, (2019): 67-85.

36   Pieter van Strien, "Towards a methodology of psychological practice", *Theory and Psychology*. Vol. 7(5) (1997): 683-700.

37   Eric Trist, Gurth Higgin, Hugh Murray and Alex Pollock, *Organizational Choice*. London: Tavistock, 1963; Garreth Morgan,. "Organizational Choice and the New Technology". In: Eric Trist and Hugh Murray. *The Social Engagement of Science. A Tavistock Anthology*. Philadelphia: University of Pennsylvania Press, 1993, 354-368.

38   c.f. Ulbo De Sitter, "Synergetisch produceren: Human Resources Mobilisation in de produktie een inleiding in structuurbouw." (Assen: Van Gorcum, 2000).

39   Eric Trist, and Ken Bamforth, "Some social and psychological consequences of the Longwall-method of goalgetting", *Human Relations*, 4 (1), (1951): 6-24;

40   Eric Trist, and Hugh Murray, *The Social Engagement of Science. A Tavistock Anthology*. Philadelphia: University of Pennsylvania Press 1993:pp. 580-598.

41 Paul Leonardi. "Materiality, Sociomateriality, and Socio-Technical Systems: What do these mean? How are they different? Do We Need Them?" Paul Leonardi, Bonnie Nardi and Jannis Kallinikos, (eds). *Materiality and Organizing: Social Interaction in a Technological World.* London: Oxford University Press. pp. 3-22, 2012.

42 Eric Trist, A Socio-Technical Critique of Scientific Management. In: Eric Trist and Hugh Murray. *The Social Engagement of Science. A Tavistock Anthology.* Philadelphia: University of Pennsylvania Press, 1993, 580-598.

43 Jannis Kallinikos, Paul Leonardi., Bonnie Nardi. The Challenge of Materiality: Origins, Scope, and Prospects. In: Paul Leonardi, Bonnie Nardi, Jannis Kallinikos (eds). *Materiality and Organizing: Social Interaction in a Technological World.* London: Oxford University Press, 2012, 3-22.

44 Soshana Zuboff, *In the Age of the Smart Machine. The Future of Work and Power.* New York: Basic Books, 1988, 7, pp. 389-391.

45 Paul Adler and Brian Borys, "Two types of bureaucracy: enabling and coercive". *Administrative Science Quarterly* 41 (1996): 61-89.

46 Soshana Zuboff, *In the Age of the Smart Machine: The Future of Work and Power.* New York: Basic Books, 1988, 390.

47 Martin van Creveld, *Command in War.* Cambridge, MA: Harvard University Press, 1985.

48 Ibid, p. 270.

49 Bart van Bezooijen and Eric-Hans Kramer, "Mission command in the information age: A normal accidents perspective to networked military operations". *Journal of Strategic Studies,* (2014): pp. 445-466.

50 William Pasmore, *Designing Effective Organizations. The Sociotechnical Perspective.* New York: John Wiley and Sons, 1988.

51 Raymond Zammuto, Terri Griffith, Ann Majchrzak, Deborah Dougherty Samer Faraj, "Information technology and the changing fabric of organization". *Organization Science,* 18, 5, (2007): 749–762.

52 William Pasmore, *Designing Effective Organizations. The Sociotechnical Perspective.* New York: John Wiley and Sons, 1988, p.2.

53 Eric Trist, "A Socio-Technical Critique of Scientific Management". In: Eric Trist, and Hugh Murray. *The Social Engagement of Science. A Tavistock Anthology.* Philadelphia: University of Pennsylvania Press, 1993, 587.

54 Paul Hill and Fred Emery, "Toward a new philosophy of management". Eric Trist, and Hugh Murray. *The Social Engagement of Science. A Tavistock Anthology.* Philadelphia: University of Pennsylvania Press, 1993, 266.

55 Ulbo de Sitter, *Synergetisch produceren: Human Resources Mobilisation in de produktie een inleiding in structuurbouw.* Assen: Van Gorcum, 2000.

56 Wanda Orlikowski, "Sociomaterial practices: exploring technology at work". *Organization Studies,* 28, (2007): 1435-1448, p.1437).

57 David Collingridge, *The Social Control of Technology.* Open University Press: Milton Keynes, UK, 1980.

58 Jack Stilgoe, Richard Owen, Phil Macnaghten, "Developing a framework for responsible innovation", *Research Policy.* 42 (2013): 1568–1580.

59 Soshana Zuboff, *In the Age of the Smart Machine. The Future of Work and Power.* New York: Basic Books, 1988.

60 Jan Achterbergh, and Dirk Vriens, *Organizations.* Berlin: Springer Verlag, 2010, 365.

61 Ibid.

62 AIV, *Autonome wapensystemen: Het belang van reguleren en investeren.* AIV-advies 119, CAVV-advies 383, 2021.

[63] Merel Ekelhof, "Autonome Wapens. Een verkenning van het concept Meaningful Human Control." *Militaire Spectator.* 184, 5 (2015): 232-245.; Merel Ekelhof, "Moving beyond semantics on autonomous weapons: meaningful human control in operation." *Global Policy,* (2019): 1-6.

[64] Merel Ekelhof, "Moving beyond semantics on autonomous weapons: meaningful human control in operation." *Global Policy,* (2019): 347.

[65] Herman Kuipers Pierre van Amelsvoort, Eric-Hans Kramer, *New Ways of Organizing; Alternatives to bureaucracy.* Leuven: Acco, 2020.

[66] Eric-Hans Kramer, Herman Kuipers and Miriam de Graaff, "An Organizational Perspective on Military Ethics". Desiree Verweij, Peter Olsthoorn and Eva van Baarle (eds). *Ethics and Military Practice,* Leiden: Brill.

[67] Herman Kuipers Pierre van Amelsvoort, Eric-Hans Kramer, *New Ways of Organizing. Alternatives to Bureaucracy.* Leuven: Acco, 2020.

[68] Stuart Winby and Sue Mohrman, "Digital sociotechnical system design" *The Journal of Applied Behavioral Science,* (2018): 1–25.

[69] Rebecca Henderson and Kim Clark, "Architectural innovation: The reconfiguration of existing product technologies and the failure of established firms." *Administrative Science Quarterly, 35,* 1, (1990): 9-30.

[70] Ulbo de Sitter, *Synergetisch produceren: Human Resources Mobilisation in de produktie een inleiding in structuurbouw.* Assen: Van Gorcum, 2000.

[71] Ibid.

[72] Ibid.

[73] William Pasmore, Stuart Winby, Sue Mohrman, Rick Vanasse, "Reflections: sociotechnical systems design and organization change". *Journal of Change Management,* 19, 2, (2019): 67-85.

[74] Bart van Bezooijen, and Eric-Hans Kramer, "Mission command in the Information Age: A normal accidents perspective to networked military operations". *Journal of Strategic Studies,* (2014): pp. 445-466.

[75] Antoine Bousquet, *The Scientific Way of Warfare. Order and Chaos on the Battlefield of Modernity.* London: Hurst Publishers, 2009, 234.

[76] Max Elden and Rupert Chisholm, "Emerging varieties of action research." *Human Relations, 46,* 2 (1993): 121-142; Frans van Eijnatten, Rami Shani and Myleen Leary. "Socio-technical systems: Designing and managing sustainable organizations." Cummings, Thomas (Ed.), *Handbook of Organization Development.* Thousand Oaks, CA: Sage, 2008, 277-310.

[77] Hans van Beinum, Claude Faucheux and René van der Vlist, "Reflections on the Epigenetic Significance of Action Research." In: Steven Toulmin and Björn Gustavsen, *Beyond Theory: Changing Organizations through Participation.* Amsterdam: John Benjamins Publishing Co., 1996, 180.

[78] David Greenwood and Mortin Levin, *Introduction to Action Research. Social Research for Social Change.* London: Sage publications, 2007, 93.

[79] Ibid, p. 3.

[80] Gerald Susman and Roger Evered, "An assessment of the scientific merits of action research". *Administrative Science Quarterly, 23,* (1978): 582-603.

[81] Susan Mohrman and Thomas Cummings, *Self-designing Organizations: Learning How to Create High Performance.* Reading, MA: Addison-Wesley, 1989.

[82] William Pasmore, Stuart Winby, Sue Mohrman, Rick Vanasse, "Reflections: sociotechnical systems design and organization change". *Journal of Change Management,* 19, 2, (2019): 73.

[83] Adapted from Stijn Vanheule. *Waarom een psychose niet zo gek is.* Tielt: Lannoo Campus, 2021.

References

Achterbergh, Jan and Dirk Vriens. *Organizations*. Berlin: Springer Verlag, (2010).

AIV, *Autonome wapensystemen: Het belang van reguleren en investeren.* AIV-advies 119, CAVV-advies 383, 2021.

Adler, Paul. S. and Brian Borys. "Two types of bureaucracy: enabling and coercive". *Administrative Science Quarterly* 41 (1996): 61-89.

Barley, Stephen, Beth Bechky, Frances Milliken. "The changing nature of work: careers, identities, and work lives in the 21st century". *Academy of Management Discoveries*, Vol. 3, No. 2, (2017): 11–115.

Bessen, Patrick, and Frantz Rowe. "Strategizing information systems-enabled organizational transformation: A transdisciplinary review and new directions". *Journal of Strategic Information Systems*, 21, 2, (2012): 103–124.

Bezooijen, Bart Van, and Eric-Hans Kramer. "Mission command in the Information Age: a normal accidents perspective to networked military operations". *Journal of Strategic Studies*, (2014): pp. 445-466.

Beinum, Hans Van, Claude Faucheux and René van der Vlist. "Reflections on the Epigenetic Significance of Action Research." In: Steven Toulmin and Björn Gustavsen, *Beyond Theory. Changing Organizations through Participation.* Amsterdam: John Benjamins Publishing Co., 1996.

Bousquet, Antoine. *The Scientific Way of Warfare. Order and Chaos on the Battlefield of Modernity*. London: Hurst Publishers, 2009.

Brynjolfsson, Erik and Andrew McAfee. *The Second Machine Age: Work, Progress and Prosperity in a Time of Brilliant Technologies*. New York: W.W. Norton & Company, 2014.

Cherns, Albert. "Principles of sociotechnical design revisited." *Human Relations, 40* 3, (1987), 153-161.

Collingridge, David. *The Social Control of Technology*. Open University Press: Milton Keynes, UK, 1980.

Creveld, Martin van. *Command in War*. Cambridge, MA: Harvard University Press, 1985.

Eijnatten, Frans van, Rami Shani and Myleen Leary. "Socio-technical systems: Designing and managing sustainable organizations." In: Thomas Cummings (Ed.), *Handbook of Organization Development* (pp. 277-310). Thousand Oaks, CA: Sage, 2008.

Ekelhof, Merel. "Autonome Wapens. Een verkenning van het concept Meaningful Human Control." *Militaire Spectator.* 184, 5 (2015): 232-245.

Ekelhof, Merel. "Moving beyond semantics on autonomous weapons: meaningful human control in operation." *Global Policy*, (2019): 1-6.

Elden, Max and Rupert Chisholm. "Emerging varieties of action research." *Human Relations*, 46, 2 (1993): 121-142.

Frey, Carl and Michael Osborne. "The future of employment: How susceptible are jobs to computerization?" *Technological Forecasting and Social Change* 114 (2017): 254–280.

Gray, Mary and Siddarth Suri. *Ghost Work: How to Stop Silicon Valley from Building a New Global Underclass*. Boston: Eamon Dolan Books, 2019.

Graeber, David. *Bullshit Jobs: A Theory*. New York: Simons and Schuster, 2018.

Greenwood, David. and Mortin Levin. *Introduction to Action Research: Social Research for Social Change.* London: Sage publications, 2007.

Hanelt, André, René Bohnsack, David Marz, and Claudiá Antunes Marante. "A systematic review of the literature on digital transformation: insights and implications for strategy and organizational change." *Journal of Management Studies*, 58, 5 (2020): 1159–1197.

Henderson, Rebecca, and Kim Clark. "Architectural innovation: the reconfiguration of existing product technologies and the failure of established firms." *Administrative Science Quarterly*, *35*, 1, (1990): 9-30.

Hill, Paul, and Fred Emery. "Toward a new philosophy of management". In: Eric Trist and Hugh Murray. *The Social Engagement of Science. A Tavistock Anthology*. Philadelphia: University of Pennsylvania Press. pp. 259-282, 1993.

Huang, Jimmy, Ola Henfridsson, Martin Liu and Sue Newell.. "Growing on steroids: Rapidly scaling the user base of digital ventures through digital innovation". *MIS Quarterly*, 41, (2017): 301–314.

Hutchby, Ian. "Technology, texts and affordances". *Sociology*, Vol. 35 (2) (2001): 441-456.

Huysman, Marleen. "Information systems research on artificial intelligence and work: A commentary on "Robo-Apocalypse cancelled? Reframing the automation and future of work debate"". *Journal of Information Technology*, Vol. 35, 4, (2020): 307–309.

Kallinikos, Jannis, Paul Leonardi, Bonnie Nardi. "The Challenge of Materiality: Origins, Scope, and Prospects". In: Paul Leonardi, Bonnie Nardi, Jannis Kallinikos (eds). *Materiality and Organizing: Social Interaction in a Technological World*. London: Oxford University Press. pp. 3-22, 2012.

Kellogg, Katherine, Melissa Valentine, and Angèle Christin. "Algorithms at work: the new contested terrain of control." *Academy of Management Annals*, 14, 1, (2020): 366-410.

Kramer, Eric-Hans. *De Militaire Bedrijfswetenschappen in Tien Stellingen. Een Sociotechnisch Perspectief*. Breda: FMW; Inaugural lecture, 2020.

Kramer, Eric-Hans, Herman Kuipers, and Miriam de Graaff. "An Organizational Perspective on Military Ethics". In: Desiree Verweij, Peter Olsthoorn and Eva van Baarle (eds). *Ethics and Military Practice*, Leiden: Brill, 2022.

Kraus, Sascha, Paul Jones, Norbert Kailer, Alexandra Weinmann, Nuria Chaparro-Banegas, and Norat Roig-Tierno. "Digital transformation: an overview of the current state of the art of research", *SAGE open*, vol 1 (1) (2021): 1-15.

Kretschmer, Tobias and Pooyan Khashabi. "Digital transformation and organization design: An integrated approach", *California Management Review*, Vol. 62, 4, (2020): 86–104.

Kuipers, Herman, Pierre van Amelsvoort, and Eric-Hans Kramer. *New Ways of Organizing: Alternatives to Bureaucracy*. Leuven: Acco, 2020.

Lanzolla, Gianvito, Annika Lorenz, Ella Miron-Spektor, Melissa Schilling, Giulia Solinas, and Christopher L. Tucci. "Digital transformation: what Is new if anything? Emerging patterns and management research." *Academy of Management Discoveries* Vol. 6, No. 3 (2020), 341–350.

Lanzolla, Gianvito, Danilo Pesce and Christopher Tucci. "The digital transformation of search and recombination in the innovation function: tensions and an integrative framework", *Journal of Product Innovation Management*, Vol 38, 1, (2021): 90–113.

Leonardi, Paul. "Materiality, Sociomateriality, and Socio-Technical Systems: What do these mean? How are they different? Do We Need Them?" In: Paul Leonardi, Bonnie Nardi and Jannis Kallinikos,

J. (eds). *Materiality and Organizing: Social Interaction in a Technological World.* London: Oxford University Press. pp. 3-22, 2012.

Liu, Day-Yang, Shou-Wei Chen, and Tzu-Chuan Chou. "Resource fit in digital transformation: Lessons learned from the CBC Bank global e-banking project". *Management Decision*, 49,10, (2011), 1728–1742.

Loebbecke, Claudia, and Arnold Picot. "Reflections on societal and business model transformation arising from digitization and big data analytics: A research agenda". *Journal of Strategic Information Systems*, 24, 3, (2015): 149–157.

Mohrman, Susan, and Thomas Cummings. *Self-designing Organizations: Learning How to Create High Performance.* Reading, MA: Addison-Wesley, 1989.

Morgan, Garreth. "Organizational Choice and the New Technology". In: Eric Trist and Hugh Murray. *The Social Engagement of Science. A Tavistock Anthology.* Philadelphia: University of Pennsylvania Press. pp. 354-368, 1993.

Oeij, Peter, Steven Dhondt, Diana Rus, and Geert Van Hootegem. "The digital transformation requires workplace innovation: an introduction." *Int. J. Technology Transfer and Commercialisation*, 16, 3, 199. (2019): 199-207.

Orlikowski, Wanda. "Sociomaterial practices: exploring technology at work". *Organization Studies*, 28, (2007): 1435-1448.

Pasmore, William. *Designing Effective Organizations. The Sociotechnical Perspective.* New York: John Wiley and Sons, 1988.

Pasmore, William, Stuart Winby, Sue Mohrman, and Rick Vanasse. "Reflections: sociotechnical systems design and organization change". *Journal of Change Management*, 19, 2, (2019): 67-85.

Pava, Carl. "Redesigning sociotechnical systems design: concepts and methods for the 1990s." *The Journal of Applied Behavioral Science*, 22, 3, 1986: 201-221.

Teece, David and Greg Linden. "Business models, value capture, and the digital enterprise". *Journal of Organization Design* 6:8 (2017): 1-14.

Trist, Eric, Gurth Higgin, Hugh Murray and Alex Pollock. *Organizational Choice.* London: Tavistock, 1963.

Trist, Eric. "A Socio-Technical Critique of Scientific Management". In: Eric Trist and Hugh Murray. *The Social Engagement of Science. A Tavistock Anthology.* Philadelphia: University of Pennsylvania Press (1993): pp. 580-598.

Trist, Eric and Ken Bamforth. "Some social and psychological consequences of the Longwall-method of goalgetting", *Human Relations*, 4 (1), (1951): 6-24.

Sitter, Ulbo, De. *Synergetisch produceren: Human Resources Mobilisation in de produktie een inleiding in structuurbouw.* Assen: Van Gorcum, 2000.

Stilgoe, Jack, Richard Owen, and Phil Macnaghten. "Developing a framework for responsible innovation", *Research Policy.* 42 (2013): 1568–1580.

Strien, Pieter van. "Towards a methodology of psychological practice". *Theory and Psychology.* Vol. 7(5) (1997): 683-700.

Susman, Gerald and Roger Evered. "An assessment of the scientific merits of action research". *Administrative Science Quarterly*, 23, (1978): 582-603.

Susskind, Richard and Daniel Susskind. *The Future of the Professions. How Technology Will Transform the Work of Human Experts*. Oxford: Oxford University Press, 2015.

Vanheule, Stijn. *Waarom een psychose niet zo gek is*. Tielt: Lannoo Campus, 2021.

Waardenburg, Lauren and Marleen Huysman. "From coexistence to co-creation: Blurring boundaries in the age of AI". *Information and Organization*. 32 (2022) 100432.

Wessel, Lauri, Abayomi Baiyere, Roxana Ologeanu-Taddei, R., Jonghyuk Cha and Tina Blegind-Jensen. "Unpacking the difference between digital transformation and IT-enabled organizational transformation." *Journal of the Association for Information Systems*, Vol. 22, 1 (2021): 1-57.

Winner, Langdon. "Do artefacts have politics?" *Daedalus*, 109,1 (1986): 121-136.

Winby, Stuart and Sue Mohrman. "Digital sociotechnical system design" *The Journal of Applied Behavioral Science*, (2018): 1–25.

WRR *Opgave AI. De nieuwe systeemtechnologie*, WRR-Rapport 105, Den Haag: WRR, 2021.

Zammuto, Raymond, Terri Griffith, Ann Majchrzak, Deborah Dougherty and Samer Faraj. "Information technology and the changing fabric of organization". *Organization Science*, 18, 5, (2007): 749–762.

Zuboff, Soshana. *In the Age of the Smart Machine: The Future of Work and Power*. New York: Basic Books, 1988.

Zuboff, Soshana. *The Age of Surveillance Capitalism: The Fight for a Human Future at the New Frontier of Power*. London: Profile Books, 2019.

CHAPTER 2

Data analytics in human resource management

Benefits and challenges

Tessa op den Buijs

Abstract

In recent years, the use of big data analytics in Human Resource Management (HRM) has become enormously popular. It supports and directs decision-making in HRM. The changes in the external and also internal environment of the armed forces (technology, labour market, aging population, personnel recruitment) are major challenges for personnel management and a data analytic approach can provide many benefits to apply to complex personnel issues. In this way, the Netherlands armed forces are able to gain better insight into the recruitment, employability and retention (inflow and outflow) of employees in relation to the desired situation of the Defence Vision 2035 and the implementation of a new HRM system. This chapter provides insight into the challenges facing the Netherlands armed forces today with respect to data analytics in HRM and starts with an overview of the movement in HRM towards data analytics. It then describes data analytics in general, followed by the specific application of data analytics in HRM, and in particular the benefits and challenges of applying data analytics for HR and personnel issues in the armed forces. The chapter concludes with how the armed forces can meet the challenges of a data analytic approach in HRM in the future.

Keywords: HRM system, Decision-making, Data analytics, Employability, Inflow-outflow.

2.1. Introduction

Today, organisations are aware that (big) data analytics can be applied to provide insight into trends and events and thus can be effective for strategic decision-making. This insight has grown tremendously, and this involves various fields of analysis and big data.[1,2] Organisations continually explore trends and transform business processes to cope with organisational challenges and improve decision-making.[3] Big data, data analytics and technology are important tools for strategic decision-making. Advanced analytic techniques on big data predict and influence data-driven choices in organisations. Information technology spreads fast and new methods are constantly emerging to deal with large scale data for strategic decision-making.[4]

Because of the (future) ambitions of the Netherlands armed forces in terms of technological developments and information-driven operations, it is, like industry, engaged in developments in the field of data analytics, partly due to the increasing importance of cyberspace. Also, the use of sensor data, satellite images, social media, videos, photos has become increasingly important. These developments allow the armed forces to better guide decision-making processes related to (day-to-day) operations.[5,6]

Data analytics in the Human Resource Management (HRM) domain has also become an important new direction and companies are actively looking to implement it, as organisations are always looking for metrics and methods to maximise the effectiveness of the workforce and optimise business results.[7] It is very valuable for the HR departments and the organisation to be able to measure and predict the strategic impact of their activities in the organisation. The Netherlands armed forces are also exploring this, as they are currently introducing a new HRM system in relation to the new Defence Vision 2035. Data analytics could help to better anticipate and respond to decision-making on current personnel issues, for example recruitment and selection, given the many vacancies the armed forces are facing.[8] Recruitment has become one of the critical success factors and in an all-volunteer force, the challenge is not to miss potential qualified volunteers.[9] Therefore it is important to get insight in the trends of the labour market, because a military organisation that consists of volunteers must compete with other organisations.

2.2. Traditional HRM and the rise of analytics in HRM

Employees with their knowledge, skills and abilities are called the human resources or human capital of organisations.[10] Human Resource Management can be referred to as the function that is responsible for staff selection and recruitment, training, performance appraisal, career development, disciplinary procedures, pre-retirement advisory work, equal opportunities policy and pay negotiations. A commonly used definition by Storey states that it is *"an approach to employee management that seeks to achieve competitive advantage through the strategic deployment of a highly engaged and capable workforce, using a range of cultural, structural and people management techniques"*[11]. HRM influences the performance of organisations in a positive way,[12] because there will be competitive advantage through people[13]: the employees of an organisation determine the success of the organisation. Thus, employee management is an important factor in achieving organisational goals.[14] To attract, deploy, and retain employees efficiently and effectively to gain and predict competitive advantage over other organisations and perform well is important.[15,16]

Since the 1980s, many academic studies in the field of HRM have been conducted and many theories have been described to motivate employees to adopt the most appropriate organisational behaviour. One of the pioneers in this field was Huselid, who scientifically substantiated HRM practices and their impact on organisational performance.[17] He showed that high performance work (HPW) systems go hand in hand with lower personnel turnover, increased market value and higher profits for the organisation.[18] HPW systems include refined selection and training methods, formal performance evaluations, and sound compensation plans and employee participation programs. Although many studies have shown that HRM can effectively improve the performance of organisations, there are still debates about the exact impact of HRM and causality – about the influence of HRM on performance – or about the importance of the relationship between an employer and an employee – employees are not only a resource, but also the human part is important -, and finally about the role of the context of organisations.[19,20]

Regarding the role of the context, studies have demonstrated two perspectives. The so-called universalist or "best practices" perspective of Pfeffer[21], states that organisations perform well when they implement the best practices in HRM.[22] However, the other theoretical perspective is that of the "best fit." This perspective assumes that organisations must align their HRM policies and practices with their environment for a positive impact on organisational outcomes.[23,24,25] Similarly, within organisations, there may be different HR practices or investments tailored to exclusive types of (strategic) employee groups (i.e. exclusive approach) or to all categories of personnel (i.e. the inclusive approach).[26] So we can conclude there is some influence of the internal and external context of organisations on HRM.[27]

However, theory is not always considered important in decision-making in practice, and sometimes HR decisions are based on subjective decisions due to, for example, personal interests or unsystematic practical experiences. Accordingly, in recent years there has been an increasing interest in evidence-based approaches to decision-making in organisations: *"translating principles based on the best evidence into organisational practices."*[28] It is important for organisations and managers to base their decision-making on the best scientific evidence combined with critical thinking and available business information. The principles of human behaviour and organisational processes are translated into more effective management practices. Four sources of information are important in evidence-based decision-making (1) professional reflection and judgment, (2) stakeholder concerns, (3) scientific evidence, and (4) reliable and valid organisational metrics.[29] The first two sources of information are almost always present in HRM. Scientific evidence, however, has become more important over time and several studies have appeared from management and organisational sciences, social sciences, and behavioural psychology.[30]

In the fourth source of information (i.e. reliable and valid organisational metrics), we see data analytics emerging as it proves helpful for HR decisions in practice. HRM specialists and policymakers translate scientific evidence into policies, which are then implemented and monitored to see if the intervention has been effective.[31] Furthermore, within and across organisations, contextual influences differ, and this implies that once implemented in practice, the effects of HRM can thus vary significantly. A data analytic approach uses the available data and measurement methods in a more inductive way to discover the effects achieved, rather than randomly as in traditional data analysis. In traditional HRM research, a more top-down deductive process based on theory is applied.[32] Thus an analytic approach differs from a sound scientific approach in practice and follows a distinct statistical path because it has a different goal.[33]

This fourth information aspect of evidence-based HRM is especially the biggest challenge facing the HRM function today, because HRM specialists often lack the ability to measure effects in the organisation using data analytics.[34] The user should be an expert in using large scale data and in discovering new business facts and trends. At the same time, there is a shift towards the development of methods and effective technologies to analyse big data. Frameworks and conceptual models for data analysis are being developed, for example, to evaluate the impact of data analysis on decision-making or to develop a framework for management, and their advantages and disadvantages are discussed in various studies.[35,36,37,38]

2.3. Data analytics

Big data and data analytics have been in the interest of scientists and practitioners for several years, but scientific studies do not yet agree on the effectiveness of the analytical approach in HRM because of many challenges. Nevertheless, there is an increase of the data analytic approach in the HRM domain, and some large companies have already had good experiences with it, for example Google.com, or Booking.com, Walmart, some healthcare companies,[39] but also the Dutch company Post-NL.[40] They apply data analytics to influence the behaviour of their employees and to improve decision-making in the organisation.[41]

Defence organisations, including the Netherlands armed forces, have also introduced data analytics, and although data analytics is not yet applied in a very structured way within all units and divisions, it has certainly already gained a foothold in the Netherlands armed forces.[42,43]

But what is (big) data analytics? We all know examples of big data such as email, social data, XML data, audio files, photos, videos, GPS data, satellite images, sensor data, as well as spreadsheets, and web log data from the internet or mobile data. They all refer to large amounts of data and definitions also focus on the data in storage (volume) and other attributes like data variety, data velocity.[44,45] We actually speak of data analytics when traditional computing capabilities (e.g., storage, analysis, transfer networks and, for example, visualisation) cannot be applied because the amount, speed, complexity or quality of the data is beyond the user's control. It is clear that big data refer to volume. This is a characteristic of big data (e.g. expressed in terabytes or even zettabytes)[46] and the data consist of transactions (i.e. bank or store), tables and files or other types. The velocity or speed refers to the frequency and the speed of data delivery by each device (think of sensors on vehicles, robotic machines, microphones that record sound).[47] The third characteristic, data variety, addresses the types of sources that are tapped, for example social media, logs, blogs, call centre applications, text data, semi-structured data, geographical data, or sensor data.[48] Which data sets and from what sources they are collected could vary from organisation to organisation.[49] Some data concern the whole organisation (business analytics) and other data only the financial unit (financial analytics). There is no 'one size fits all'. In line with these characteristics, it is easy to understand that the benefits of data analytics are numerous. Data are voluminous, delivered quickly, and provided by a variety of sources.

By combining analytics of big data with more traditional data analysis, organisations are able to gain new insights and a deeper understanding of organisational dynamics. It opens up new opportunities and optimises decision-making, discovers trends and predicts future events. In telecommunications and transportation companies, for example, big data analytics is used to better predict customer usage. Similarly, supermarkets are able to better understand their customers and predict which products will sell.[50,51]

An analytical approach also allows for more complete and accurate responses and can improve data processing performance because more types of data are available and will be combined, such as text and semi-structured data. In traditional scientific studies, one would say that this mixed-methods approach increases the quality of a study.[52] Another benefit is that an organisation is able to act more quickly because data are collected in (near) real time as the speed of big data reflects. Data analytics could optimise organisations' decision-making process because more accurate data are available. The use of IT programs (e.g., IBM SPSS) that simultaneously support data analytics and the more traditional data analysis can provide benefits and is less expensive.[53,54] In addition to investing in software and hardware packages, organisations also invest in setting up different units that deal with a different focus on data analytics and environments and platforms to share information. In the Netherlands armed forces, different units are engaged in a different focus on data analytics, also called data science departments.[55,56]

2.4. Data analytics in HRM

The definition of data analytics in HRM, like the different definitions about data analytics, is not very unambiguous and can be labelled in different ways, such as HR analytics[57] people analytics,[58] workforce analytics,[59,60] talent analytics,[61] or human capital analytics.[62,63] The definitions are very similar, but there is a nuanced difference between them. HR analytics, human capital, and people analytics are often used to refer to more general analytical topics that cover the whole range of HRM.[64] An overview study by Marler and Boudreau led to the following definition for people analytics: *"HR practice enabled by information technology that uses descriptive, visual, and statistical analysis of data related to HR processes, human capital, organisational performance, and external economic benchmarks to determine business impact and enable data-driven decision-making."*[65] Workforce and talent analytics, however, have a narrower scope as these terms focus more on employees, such as strategic workforce planning, recruitment and selection, and also development of employees.[66] Huselid defines workforce analytics as follows: *"Workforce analytics refers to the processes involved with understanding, quantifying, managing, and improving the role of talent in the execution of strategy and the creation of value. It includes not only a focus on metrics (e.g., what do we need to measure about our workforce?), but also analytics (e.g., how do we manage and improve the metrics we deem to be critical for business success?"*[67] A Dutch study refers to data analytics in HRM as making decisions about people and the organisation based on numbers.[68] Other definitions of workforce analytics include for example applying analysis, statistics and models to data related to employees, or aim at evidence-based results to improve employee behaviour and the organisational performance by optimising decision-making from the human side of the business.[69,70]

If we take a critical look at data analytics in HRM, we see that the data analytic approach in human resource management is still quite recent and focuses primarily on screening job applicants.[71] Organisations use special software to review job applications faster than in a manual fashion, and mostly to save on administrative costs. From the HRM function this seems logical, but from a strategic perspective there is much more to be gained from HR data analytics. For example, an analytical approach proves especially valuable for discovering and strengthening the strategic relationship between human resources and the organisation's overall strategy.[72] Data analytics can also be used to answer questions about how HRM improves employee knowledge and skills to enhance organisational performance and competitive advantage. It addresses motivation and training programs to develop talents in the organisation, and it can predict what may happen in stead of providing descriptive reports on what happened.[73] Big data sets also make it possible to predict future behaviour of employees and explore labour market trends. Not only can big data be useful for faster and more cost-effective selection, but also for recruiting new potential. The U.S. Army Recruiting Command, for example, has significantly expanded virtual recruiting and this has produced more data. Virtual recruiting takes place through digital systems (e.g., email, internet, and social media applications). Virtual recruitment teams (VRTs) exclusively use virtual recruitment processes to attract personnel, not only for selection, but also for recruiting new employees.[74]

There are also scholars who claim the opposite. They argue that data analytics does not exist in HRM. The employee data used are not large in volume – there are not millions of people working for organisations – the data have no special variety because mostly internal data are used.[75,76] So why should there be any reason to use special software and methods associated with big data analytics? The challenge lies not in applying data analytics but in analysing the data and interpreting them. So, we need to ask *how* big data can support HRM. Some scholars argue that for HRM we should no longer talk about "big" data but about "smart" data, because the insights that the data can provide are more important than, for example, volume and speed.[77] HR managers need to focus on the "smart" HR data available in the organisation for decision-making, rather than on "big" data.[78]

2.5. Data analytics in HRM: benefits for the armed forces

In September 2019, the Netherlands armed forces introduced a new HR model and with so-called "pilots" new ways of recruiting personnel (more decentralised) have been implemented and more customised contracts have been offered so far. Cooperation in the field of recruitment and training has also been professionalised, with close contact being maintained with regional civilian partners. In this development, there is a focus on social anchoring and cooperation with educational institutions, local businesses and governments. The wishes of personnel will also be taken into account. This should enable the Netherlands armed forces to meet future personnel challenges as stated in the Defence Vision 2035[79] and to reach even more workforce diversity.[80]

2.5.1. Data analytics in HRM

Some years ago, the Netherlands armed forces introduced data analytics in HRM. There are some references to courses, some initial data, explanatory notes, and manuals and other documents available on platforms provided by different units. A platform has also been introduced by the armed forces Trends, Research and Statistics (TOS) department. This department has been responsible for the analysis of data in the field of HRM in the armed forces for a number of years.

An interview with representatives of the Trends, Research and Statistics (TOS) department that is associated with the Business Information Personnel & Organisation department (BI P&O) reveals that TOS conducts Social Scientific Research (SWO) and carries out, among other things, work perception and job satisfaction surveys on a regular basis. The Business Information Personnel & Organisation (BI P&O) extracts data from the defence information systems.[81] The services have been working together for a number of years, combining figures from management information systems with social scientific research, to give greater depth to findings on the exit intentions of military personnel, for example. This cooperation enhances the credibility of the armed forces' approach to personnel issues.[82]

Organisations often try to address retention issues through in-depth retrospective interviews (so-called exit interviews), but by applying an analytical approach, companies can be predictive and respond proactively rather than reactively. For example, by conducting predictive behavioural analysis, a company could discover that the expensive bonuses it had started handing out because of staff turnover was an ineffective measure.[83] The Netherlands military therefore also combines quantitative and qualitative data as part of exit analysis.[84]

Furthermore, TOS carries out and monitors social scientific questionnaire research on HR topics and regularly publishes reports and other documentation consisting of visuals and infographics. Accurate information and analyses and timely delivery are central to this. An example of a report is the HR Monitor, for which there was a great need within the armed forces.[85] The HR Monitor consists of a logically structured 'catchy' report. It has a 'flow' structure with many infographics. This way, people start making connections and asking deeper questions. While this monitor is a great example of data analytics, the HR analytic approach in the armed forces is still developing. In the future, the armed forces will combine more data on intentions to leave and reasons for doing so.

Within the Netherlands armed forces, the idea of HR analytics on sensitive health data of military personnel to make health predictions has also been a "hot topic" for several years. One advantage may be that the armed forces are able to better respond to the medical condition of its personnel and, for example, gain insight into the health situation of military employees. However, the privacy and security of these data seem to be challenges for the armed forces, especially when it concerns sensitive and personal (health) data.[86,87] Therefore, it is important to develop a sound ethical practice for data analytics with respect to personal (health) data. In the U.S., for example, ethics is broken down into discrimination or exclusion of specific target groups in data. On the other hand, there are concerns that data analytics can lead to a breach of privacy.[88] For example, if tracking systems or audio, video recordings are used and if these data identify certain employees, employees may start behaving differently or the system may not be accepted. There are many laws in Europe that are similar to the U.S. laws, such as the General Data Protection Regulation (GDPR).[89] This GDPR focuses on all data that are processed (e.g., input, transmission of data, output, stored data) even if no European citizen is involved. This law consists of a set of rules for the (automatic) collection, processing and storage of personal data. Organisations must be careful and responsible when storing and securing employees' personal and medical data, so this can be a challenge when working with (big) data. Data should be carefully processed by the employer and this has to be a transparent process. In social sciences, organisational sciences, (behavioural) psychology and medical sciences, for example, it is common to use informed consent forms that explain to participants how their data will be processed and stored.[90]

The military organisation is aware of GDPR in health studies and other studies that analyse personal data. GDPR coordinators have also been appointed to advise on compliance and various information documents on the application of the GDPR are available. With regard to scientific research in the armed forces, ethical protocols have also been developed for compliance with the ethical aspects of research. Such protocols are obviously also applicable to (big) data analytics.

2.5.2. Different analytic approaches in HRM

Analytics in HRM uses various sources of (big) data, in addition to the three characteristics of big data already described above. In HR analytics in the Netherlands armed forces, a distinction is made between HR data sources (e.g., internal recruiting data, retention data, demographic data, absence data, data on compensation and benefits, data on performance, on engagement, exit reasons and well-being), and diverse data analysis tools are used to discover valuable insights and trends in data sources. As previously described, platforms have been developed and visualisations created in the Netherlands armed forces to share information.[91] This information focuses on HR analytics, workforce analytics and people analytics.

Workforce analytics and people analytics still seem to lag behind in relation to the HR analytics information displayed on the intranet platform of the Netherlands armed forces. But this platform is still under development. However, there is an online platform to share information and that is important. The importance of a platform for sharing information is also evident in a Dutch study on data analytics.[92] An internal online platform is important for the transfer of know-how, it is crucial to learn from, and it is important for sharing ideas, experiences and insights.[93]

Workforce analytics is described on the platform and focuses on the workforce in relation to the organisation as a whole. Here, descriptive internal (e.g., data of employees) and external (e.g., labour market trends) data and predictive data are used primarily to support strategic workforce planning. So, workforce analytics aims at a wide range of activities and processes, which are constantly evolving. Workforce planning is about building strategies to ensure the organisation has the appropriate workforce to deliver the organisation strategy (e.g. demand and supply).[94] This can be an important approach for the Netherlands armed forces in connection with the strategic personnel planning in the new HRM system. The main focus of workforce analytics is on strategic issues relating to future bottlenecks in the workforce, all kinds of diversity issues, or questions relating to inflow through outflow. In view of the current vacancies in the armed forces, these are important issues that can be better predicted by using workforce analytics.

Workforce planning is related to workforce analytics and focuses on operational or short-term planning topics like the number of filled positions or the sickness absenteeism rate in certain units. Short-term planning provides commanders with information about their unit, for example when preparing for an operational situation abroad. Such analyses inform commanders about the current or changing workforce and influence their decisions. This analytical approach determines which metrics, or what data, are relevant and need to be documented, and how these data should be linked.

Long-term planning is also described on the defence platform, and this is related to the Defence Vision 2035. The central questions here are: how can workforce analytics contribute to optimising recruitment methods, what can be done differently with a view to the future, and how can valuable staff be retained? With workforce analytics, the translation of data into metrics and the analysis of changes and relationships in these metrics, it is possible to make predictions and discover trends that allow the organisation to actively respond to changes in the labour market.[95]

In both workforce analytics and planning in the armed forces, the use of internal and external data is important. External data or metrics are related to the context of the armed forces and are very easily accessible. Since the context of the armed forces is important, many trends for the recruitment policy of the Netherlands armed forces can easily be predicted based on the expected developments in the labour market, the skills needed by employees in the future and the development of scarcity in the labour market. Internal data, or HR metrics, consist of data (numbers) that describe certain processes or HRM outcomes in detail, for example, the success in recruiting new employees, the sickness absence rate, the employee turnover, or the costs of training programs.[96] Valuable combinations can be made with internal and external data. According to some researchers, traditional internal and external data should also be combined with "larger" data on personnel, for example location data from mobile phones, information from employees' electronic calendars or internet activities, or information about the people they communicate with via mobile phone and emails and the content of conversations.[97] However, internal data and "larger" data are more difficult to achieve. They consist of sensitive personal data, for example the sickness absence rate in certain units (this could be derived from the personal data platform) or the location data of cell phones. However, as described previously, the armed forces must comply with privacy and data protection legislation based on the European GDPR (General Data Protection Regulation).[98,99]

When it comes to people analytics, another analytic approach, organisations are increasingly aware of the potential and value of their workforce, and business value is increasingly associated with having a highly talented workforce that can bring radical innovations, so-called "talents." In the literature, the application of new methods and new ways of thinking in the field of talent management seem to be innovative topics. HR tools such as staff training and talent development programs are important instruments for HRM. The deployment of drones, new weapon systems and cyber developments have ensured the armed forces need a different type of qualified personnel (see also Defence Vision 2035). The armed forces need talents for all kinds of technological and technical positions. Therefore, the new HRM system of the Netherlands armed forces will focus on talent management both in the internal and external labour market.[100] We also see this development in other civilian companies and there is a high demand for qualified personnel. It is hard to find employees in the labour market who possess exactly the skills and qualifications required by the military. Once hired, it is vital to ensure that the necessary talent is retained, developed and motivated to work towards the organisation's strategic goals.[101] Talent management aligns individual capabilities with the needs of the armed forces and optimises human performance and engagement (the 'best' soldier at the right position at the right time). It integrates recruitment, development, employment, and retention strategies.[102] For example, the U.S. army's personnel system includes a more personalised management approach that is more likely to motivate and retain officers by preparing them for leadership roles.[103] Talent analytics may support issues on the relationship between training, development, productivity and reducing employee turnover. Further, it contributes to employee wellness programs, or addresses questions on specific degrees of employees and productivity.[104]

Talent analytics can be more important for certain categories of employees than for others. Typically, the focus is on employees in strategic positions who have relatively important accomplishments for the organisation. Talent analytics could identify these talents and contribute to optimising strategic decision-making and increasing organisational success.[105] Many organisations already distinguish high potentials or talents among their employees and differentiate their HRM investments (exclusive approach).[106] The Netherlands armed forces deploy talent management as one of the instruments in the HR transition. This approach is important for the inflow and outflow of personnel, where the focus is on all personnel and not on special categories (inclusive approach).[107] The idea is that the employees, based on their knowledge, competences and needs, are central and therefore of added value to the organisation.[108] These changes are only possible in a culture and leadership climate in which coaching, education and training are important. This means that the mind-set and management of the armed forces must change. This will enable the armed forces to attract and retain more talented, committed and deployable personnel.

All in all, the development of analytic approaches in HRM seems to fit the timeline of the introduction of the Defence Vision 2035 and the HR transition in the Netherlands armed forces.[109,110] This process is expected to be achieved by the end of 2024. In relation to this development, a number of "pilots" with data analytics in the field of workforce and talent management have already been carried out at specific units within the branches of the armed forces, which, as far as is known, were positive. Other units, however, for example the *Royal Netherlands Marechaussee*, are still in the process of developing roadmaps for the practical implementation of talent management.[111]

2.6. Challenges in data analytics in HRM

Traditional HR in practice has shifted to the growing application of data analytics in HRM and today it is even a "hot topic". Although it seems to be a promising development in the HR domain, there are still certain challenges for organisations and for the armed forces that need to be addressed in order to implement data analytics in HRM in an appropriate and effective way. Some key challenges for the Netherlands armed forces are described.

The first challenge is the fact that data analytics allows more and "larger" data to be collected and combined, but more data does not always mean more useful information. Sometimes very complex, unstructured data can produce noise, anomalies and ambiguity. It is then difficult to discover the patterns and trends that are relevant to the organisation's performance.[112]

In addition, the complexity of data requires other sophisticated IT programs and HR specialists need more analytical, technical and innovative knowledge and skills. They need to be educated and trained in HR intelligence, key figures, metrics and all kinds of financial metrics in order to manage HR strategy and make cause-and-effect relationships and future predictions about personnel investments and their effects.[113] One way to do this is to share knowledge by means of job rotation, also possible in the armed forces. In this way, practitioners of data analytics are introduced to different new activities in analytics practice.[114]

Another challenge is the fact that many data analytics models have been developed in recent years but, because organisations differ in structure and in contextual influences, there is no single uniform model for data analytics in HRM (just as there is no uniform definition of HR analytics). It would be preferable to develop a model that is able to measure and provide evidence of the relationship between the organisation and employees as human capital. Such a model could influence the assessment of big data implementation and decision-making.[115] Unfortunately, despite several promising studies, the evidence for a link to organisational performance is not yet very strong.[116,117,118,119]

A model with mechanisms for the governance of HR analytics would also be preferable. After all, in addition to an appropriate IT environment, the analytics approach requires a different culture, mind-set, leadership climate and governance structure.[120] This also applies to the armed forces. Although some preliminary results point towards an online platform (already being further developed in the armed forces) and the sharing of information through such a platform, further research is needed on the mechanisms for governance of data analytics.[121]

Moreover, many organisations, including military organisations, probably forget that people's social world can play a role. It is part of their daily lives and therefore these social data can be considered as big data. However, the question is: how can the social world of a network be analysed? This could be a future challenge for data analysis in HRM. The social world can be used to predict a user's behaviour and a social network could therefore be an important cohesive variable or predictor of military behaviour.[122] To identify the behaviour of certain units in the military, because there are different subcultures in the military, may not be trivial. The social network can be an important predictor of organisational behaviour in both data analytics and in traditional data analysis.[123]

Finally, security and privacy issues in relation to data analytic approaches can present significant challenges. They relate to a safe storage of data and the protection of the transmission of data. How can we prevent anyone from learning about privacy-sensitive information, or information about an employee's network as mentioned above? This is where the GDPR plays an important role, including in the armed forces. We can argue that the issues of data security and privacy in data analytics are fundamentally the same challenge as in traditional data analysis of personnel data. So, the question is how to deal with these issues in data analytics in HRM?[124,125]

2.7. Conclusion

The collection, storage, analysis and management of a huge amount of data in HRM is inevitable and will only increase in the future. Data analytic approaches in HRM focus on how employee actions directly affect organisational performance and should be transparent. They influence organisational performance in a positive way and can also positively change the employer-employee relationship. Researchers and HR professionals therefore look for opportunities to "operationalise" their research through data analytics and optimise human resources for better employee and organisational performance. Data analytics could support decision-making in various personnel-related and strategic questions, also in the Netherlands armed forces. In terms of a HR analytic approach, the Netherlands armed forces are already on the right track, but before big data analytics can actually deliver on promises in HRM, there is still work to do for the military organisation. After all, there are benefits and challenges at the same time.

Notes

[1]   George Gerard, Martine Haas and Alex Pentland. "Big data and management," *Academy of Management Journal* 57, no.2 (2014): 321-326.

[2]   Smaranika Mohapatra, Jharana Paikaray, Neelamani Samal, "Future trends in cloud computing and big data," *Journal of Computer Sciences and Applications* 3, no. 6 (2015):137-142. https://doi: 10.12691/jcsa-3-6-.

[3]   Samuel Fosso Wamba, Shahriar Akter, Andrew Edwards, Geoffrey Chopin and Denis Gnanzou, "How 'big data' can make a big impact: Findings from a systematic review and a longitudinal case study," *International Journal of Production Economics*, (2014): 1-34. 10.1016/j.ijpe.2014.12.031.

[4]   Chun-Wei Tsai, Chin-Feng Lai, and Han-Chieh Chao, "Big data analytics: a survey," *Journal of Big Data* 2, no. 21, (2015): 1-32.

[5]   Paul C. van Fenema, Jan.P.H. Kalden, Bas,.J.H. Rietjens, and Remco Schimmel, "Big data analytics en Defensie visie en aanpak," *Militaire spectator* 184, no. 9 (2015): 374- 387.

[6]   Recently, a data science knowledge and expertise centre has opened in the Netherlands armed forces (Center of Excellence; CoE) where all data science knowledge and artificial intelligence come together. For example, insights are gained from sensor data of vehicles, geographical data of satellites, and computer science and statistics combined with information from public sources.

[7]   Dag Øivind Madsen and Kare Slåtten, "The rise of HR analytics: A preliminary exploration" *Global Conference on Business and Finance Proceedings* 12, no. 1 (2017):148-159. Retrieved from: SSRN: https://ssrn.com/abstract=2896602

[8]   There were 8,000 vacancies in the Netherlands armed forces in 2019 and as many as 9,000 in 2020 out of a total population of 54,000 working in the armed forces. In: Tessa op den Buijs and Erik van Doorn. "What women want! A study on the motivation of young women to join the Netherlands armed forces. In press.

9   Tessa op den Buijs and Peter Olsthoorn. "Challenges in HRM for Military Organizations: trends in the military professionalization." In: *Handbook of Military Sciences*. In press.

10 Paul Boselie, *Strategic Human Resource Management, A Balanced Approach* London: McGraw-Hill: 2014: p. 212.

11 John Storey, "What is human resource management?", in John Storey, *Human Resource Management: A Critical Text*. London: Thomson, 2007: p. 7.

12 Boselie, *Strategic ...* (2014).

13 Jeffrey Pfeffer, *Competitive Advantage through People Unleashing the Power of the Workforce*. Boston: Harvard Business School Press, 1994.

14 Peter Boxall and John Purcell, *Strategy and Human Resource Management*. New York: Palgrave MacMillan, 2011.

15 Boselie, *Strategic ...* (2014).

16 Mark A. Huselid and Brian E. Becker, "Bridging micro and macro domains: Workforce differentiation and strategic human resource management [Editorial]", *Journal of Management* 37, no. 2 (2011): pp. 421–428. https://doi.org/10.1177/0149206310373400

17 Boselie, *Strategic ...* (2014).

18 Mark A. Huselid, "The impact of human resource management practices on turnover, productivity, and corporate financial performance", *Academy of Management Journal* 3, no. 38 (1995): 635-72.

19 Paul van der Laken, *Data-driven human resource management: The rise of people analytics and its application to expatriate management*, Doctoral thesis, (Alblasserdam: Ridderprint, 2018).

20 Kaifeng Jiang, David, P. Lepak, Jia Hu and Judith C. Baer, "How does human resource management influence organizational outcomes? A meta-analytic investigation of mediating mechanisms." *Academy of Management Journal* 55, no. 6 (2012): 1264–1294. https://doi.org/10.5465/amj.2011.0088

21 Jeffrey Pfeffer, "Seven practices of successful organizations," *California Management Review*, 40 (1998*)*: 96-124.10.2307/41165935.

22 Huselid, "The Impact ..." (1995); Pfeffer, "Seven ..." (1998).

23 Boxall and Purcell, *Strategy ...* (2011).

24 Jaap Paauwe and Elaine Farndale, *Strategy, HRM and Performance: A Contextual Approach* (second ed.) Oxford: University Press, 2017.

25 Dave Ulrich and James H. Dulebohn, "Are we there yet? What's next for HR?," *Human Resource Management Review* 25, no. 2 (2017): pp. 188–204. https://doi.org/10.1016/j.hrmr.2015.01.004

26 Boselie, *Strategic ...* (2014): p. 212.

27 Huselid and Becker, *Bridging micro...*(2011).

28 Denise M. Rousseau, "Is there such a thing as "evidence-based management?" *Academy of Management Review* 31, no. 2 (2006): p. 256.

29 Denise M. Rousseau and Eric G. R. Barends, "Becoming an evidence-based HR practitioner," *Human Resource Management Journal* 21, no. 3 (2011): 221-235. https://doi.org/10.1111/j.1748-8583.2011.00173.x

30 Boselie, *Strategic..* (2014).

31 Erik P. Piening, Alina M. Baluch, and Hans-Gerd Ridder, "Mind the intended-implemented gap: understanding employees' perceptions of HRM," *Human Resource Management* 53, no. 4 (2014): 545-567.

32 Van der Laken, *Data-driven...* (2018).

33 Van der Laken, *Data-driven...* (2018).

34 Paauwe and Farndale, *Strategy...* (2017).

35 Tsai, et al., "Big data ..." (2015).

36 Philippe Russom, *Big Data Analytics: TDWI best practices report*. The Data Warehousing Institute (TDWI) Research, 2011.

37   Celia Adrian, Rusli Abdullah, Rodziah Atan, and Yusmadi Yah Jusoh, "Conceptual model develop-
     ment of big data analytics implementation assessment effect on decision-making," *International
     Journal of Interactive Multimedia and Artificial Intelligence* 5, no. 1 (2018): 101-106.
38   Jeroen Baijens, Remko W. Helms and Tjeerd Velstra, "Towards a Framework for Data Analytics
     Governance Mechanisms,". *In Proceedings of the 28th European Conference on Information Systems
     (ECIS) (*An Online AIS Conference, June 15-17, 2020). https://aisel.aisnet.org/ecis2020_rp/81
39   IBM Corporation, *Business Analytics for big data. Unlock value to fuel performance* (United States:
     IBM Corporation, 2013), 1-10.
40   Toine Al and Irma Doze, *HR Analytics, waarde creëren met datagedreven HR-beleid.* Amsterdam:
     Mindcampus, 2018.
41   Wullianallur Raghupathi and Viju Raghupathi, "Big data analytics in healthcare: promise and
     potential," *Health Information Science and Systems* 2, no. 3 (2014): 1-10. https://doi.org/10.1186/2047-
     2501-2-3
42   Van Fenema, et al., "Big data analytics ..." (2015).
43   Ignazio Montiel-Sánchez, "Disruptive defence innovations: big data analytics for defence," *Article
     EDA.* 2021: 1. Retrieved from: https://eda.europa.eu/webzine/issue14/cover-story/big-data-analyt-
     ics-for-defence
44   Russom, *Big data..* (2011).
45   Tsai, et al., "Big data.." (2015).
46   IBM Corporation, *Business Analytics...* (2013).
47   Russom, *Big data...* (2011).
48   Van Fenema, et al., "Big data analytics..." (2015).
49   Russom, *Big data...* (2011).
50   Van Fenema, et al., "Big data analytics..." (2015).
51   IBM Corporation,. *Business Analytics...* (2013).
52   Alan Bryman, Tom Clark, Liam, Foster, Luke Sloan, *Bryman's Social Research Methods.* Oxford:
     University press, 2021.
53   IBM Corporation, *Business Analytics ...*(2013).
54   Baijens, et al., "Towards a Framework ..." (2020).
55   Baijens, et al., "Towards a Framework ..." (2020).
56   Beatrice Snel and Jeffrey Louvenberg, "Eén plus één is drie: het succes van twee samengevoegde
     afdelingen binnen Defensie. Hoe twee werelden samen een verhaal vertellen," *Best Practices
     HR-analytics bij het Rijk Interviewreeks* (2016/2017): 42-45.
57   Paauwe and Farndale, *Strategy ...* (2017)
58   David Green, "The best practices to excel at people analytics," *Journal of Organizational Effectiveness:
     People and Performance* 4, no. 2 (2017): 10.1108/JOEPP-03-2017-0027.
59   Kevin D. Carlson and Michael J. Kavanagh, "HR metrics and workforce analytics," *Business (2017).*
60   Thomas Rasmussen and Dave Ulrich, "Learning from practice: How HR analytics avoids being a
     management
     fad," *Organizational Dynamics* 44, no. 3 (2015): 1-7. 10.1016/j.orgdyn.2015.05.008.
61   Manuela Nocker and Vania Sena, "Big data and human resources management: The rise of talent
     analytics," *Social Sciences* 8, no.10 (2019): 1-19. https://doi.org/10.3390/socsci8100273
62   Morten, K. Andersen, "Human capital analytics: the winding road," *Journal of Organizational
     Effectiveness: People and Performance.* 4, no.2 (2017):133-136. 10.1108/JOEPP-03-2017-0024.
63   R.H. Hamilton and William A. Sodeman, "The questions we ask: Opportunities and challenges for
     using big data analytics to strategically manage human capital resources," *Business Horizons* 63,
     no. 1 (2020): 85-95. https://doi.org/10.1016/j.bushor.2019.10.001.

[64] Van der Laken, *Data-driven ...(2018)*.

[65] Janet Marler and John Boudreau, An evidence-based review of HR analytics. *The International Journal of Human Resource Management* 28 (2016): p. 15. 10.1080/09585192.2016.1244699.

[66] Van der Laken, *Data-driven ... (2018)*.

[67] Mark Huselid and Dana Minbaeva, "Big data and HRM." In: A, Wilkinson, N. Bacon, L. Lepak, & S. Snell, (Eds.) *Sage Handbook of Human Resource Management* (2nd edition), 2018): p. 7.

[68] Al and Doze, "HR Analytics ..." (2018).

[69] Fenna Piersma and Alicia Streppel, *HR Analytics. Een praktische inleiding* (Groningen: Noordhoff Uitgevers, 2019).

[70] Van der Laken,. *Data-driven...(2018)*.

[71] Hamilton and Sodeman, "The questions ..." (2020).

[72] Hamilton and Sodeman, "The questions ..." (2020).

[73] Hamilton and Sodeman, "The questions ..." (2020).

[74] Nelson Lim, Bruce R. Orvis, and Kimberly Curry Hall, *Leveraging Big Data Analytics to Improve Military Recruiting.* Santa Monica, CA: RAND Corporation, 2019. retrieved from: https://www.rand. org/pubs/research_reports/RR2621.html.

[75] Peter Cappelli. "There's no such thing as big data in HR." *Harvard Business Review.* (2017). Retrieved from: There's No Such Thing as Big Data in HR (hbr.org)

[76] Hamilton and Sodeman, "The questions ..." (2020).

[77] Huselid and Minbaeva, "Big data ... " (2018).

[78] Huselid and Minbaeva, "Big data ..." (2018).

[79] Ministry of Defence. *Defence Vision 2035 "Fighting for a secure future"* (the Hague: Ministry of Defence, 2020).

[80] The Defence Vision 2035 'Fighting for a secure future' states that by 2035, the Dutch defence organisation must be a smart and technologically advanced organisation; one that has a strong ability to adapt to situations and that acts on the basis of the best information. In addition, the defence organisation is a reliable partner for its national and international partners to jointly face threats. Retrieved from: https://english.defensie.nl/downloads/publications/2020/10/15/defence-vision-2035

[81] Snel and Louvenberg, "Eén plus één ..." (2016/2017).

[82] Hamilton and Sodeman,, "The questions ..." (2020).

[83] Bruce Fecheyr-Lippens, Bill Schaninger, and Karen Tanner, "Power to the new people analytics: Techniques used to mine consumer and industry data may also let HR tackle employee retention and dissatisfaction," *McKinsey Quarterly. McKinsey & Company* (2015): 1-2.

[84] Snel and Louvenberg. "Eén plus één ..." (2016/2017).

[85] Ministry of Defence. (2018, 2019 and 2021), *HR Monitor. DOSCO/Ministerie van Defensie* (The Hague: TOS. 2018, 2019, 2021).

[86] Wallinallur Raghupathi and Viju Raghupathi, "Big data analytics in healthcare: promise and potential," *Health Information Science and Systems* 2, no. 3 (2014): 1-10. https://doi.org/10.1186/2047-2501-2-3

[87] Hamilton and Sodeman, "The questions ..." (2020).

[88] Hamilton and Sodeman, "The questions ..." (2020).

[89] European Commission, *The GDPR: New opportunities, new obligations.* (2018). Retrieved from: Https://ec.europa.eu/commission/sites/beta-political/files/data-protection-factsheet-sme-obligations.

[90] Bryman. et al., *Bryman's ...* (2016).

[91] Ministry of Defence. Internal document. Retrieved from the intranet website. (2022).

[92] Baijens, et al., "Towards a Framework ..." (2020).

[93] Baijens, et al., "Towards a Framework ..." (2020).

94   Van der Laken, *Data-driven ...(2018)*.

95   Van der Laken, *Data-driven ... (2018)*.

96   Van der Laken, *Data-driven ... (2018)*.

97   Mark Stuart, David Angrave, Andy Charlwood, Ian Kirkpatrick and Mark Lawrence,. "HR and analytics: why HR is set to fail the big data challenge,." *Human Resource Management Journal.* 26, no. 1 (2016):1-11.

98   European Commission, *The GDPR ...* (2018).

99   Tsai, et al., "Big data ..." (2015).

100   Ministry of Defence, *Internal document ...* (2021).

101   Boselie, *Strategic ...* (2014).

102   Michael J. Arnold, *Talent management in the Army: Review, comment, and recommendation on talent management models.* White paper. Human dimension capabilities development task force (HDCDTF) (2015). Retrieved from: https://usacac.army.mil/sites/default/files/publications/HDCDTF_WhitePaper_Talent%20Management%20Models_Final_2015_04_17_2015.pdf

103   Arnold, *Talent management in ...* (2015).

104   Manuela Nocker and Vania Sena, "Big data and human resources management: The rise of talent analytics," *Social Sciences* 8, no.10 (2019): 1-19. https://doi.org/10.3390/socsci8100273

105   Nocker and Sena, "Big Data ..." (2019).

106   Boselie, *Strategic ...* (2014).

107   Boselie, *Strategic ...* (2014).

108   Ministry of Defence, *Internal document ...(2021)*.

109   Ministry of Defence. *Defence Vision 2035 ...* (2020).

110   Jessica Bode, "Werving en selectie op de schop." *Defensiekrant*, 01 (2020). Retrieved from: https://magazines.defensie.nl/defensiekrant/2020/36/01_werving_36?fbclid=IwAR3V2IsXJZeCAnE3Nvuvy-QysXqN7w45N72jt2nU4bS4e-O8iROMGLMWGgI.

111   Luuk van Lieshout, *Talent! en nu? Kwalitatief onderzoek naar de voorgenomen praktijk van talentmanagement binnen de koninklijke marechaussee*, Bachelor thesis in press. Breda: NLDA, 2022.

112   IBM Corporation, *Business Analytics ...(2013)*.

113   Baijens, et al., "Towards a Framework ..." (2020).

114   Baijens, et al., "Towards a Framework ..." (2020).

115   Adrian, et al., "Conceptual Model ... " (2018).

116   Janet Marler and John Boudreau, "An evidence-based review of HR analytics," *The International Journal of Human Resource Management.* 28, no.1 (2016): 15. 10.1080/09585192.2016.1244699.

117   Marler and Boudreau, "An evidence-based..." (2016).

118   Van der Laken, *Data-driven ...* (2018).

119   Tsai, et al., "Big data ..." (2015).

120   Baijens, et al., "Towards a Framework ..." (2020).

121   Baijens, et al., "Towards a Framework ..." (2020).

122   Tsai, et al., "Big data ..." (2015).

123   Tsai, et al., "Big data ..." (2015).

124   Tsai, et al., "Big data ..." (2015).

125   Hamilton and Sodeman, "The questions ..." (2020).

References

Adrian, Cecilia, Rusli Abdullah, Rodziah Atan, Yusmadi Yah Jusoh. "Conceptual model development of Big Data Analytics implementation assessment effect on decision-making." *International Journal of Interactive Multimedia and Artificial Intelligence*, 5, no. 1 (2018): 101-106.

Andersen, Morten, K. "Human capital analytics: the winding road." *Journal of Organizational Effectiveness: People and Performance*. 4, no. 2 (2017): 133-136. 10.1108/JOEPP-03-2017-0024.

Arnold, Michael, J. *Talent management in the Army. Review, comment, and recommendation on talent management models*. White paper. Human dimension capabilities development task force (HDCDTF), 2015. Retrieved from: https://usacac.army.mil/sites/default/files/publications/HDCDTF_WhitePaper_Talent%20Management%20Models_Final_2015_04_17_2015.pdf

Baijens, Jeroen, Remko W. Helms and Tjeerd Velstra. "Towards a Framework for Data Analytics Governance Mechanisms" (2020). *In Proceedings of the 28th European Conference on Information Systems (ECIS)*, An Online AIS Conference, June 15-17, 2020. Retrieved from: https://aisel.aisnet.org/ecis2020_rp/81

Bode, Jessica. "Werving en selectie op de schop." *Defensiekrant*, 01 (2020). Retrieved from: https://magazines.defensie.nl/defensiekrant/2020/36/01_werving_36?

Boselie, Paul. *Strategic Human Resource Management: A Balanced Approach*. London: McGraw-Hill, 2014.

Boxall, Peter and John Purcell. *Strategy and Human Resource Management*. New York: Palgrave MacMillan, 2011.

Bryman, Alan, Tom Clark, Liam, Foster, Luke Sloan. *Bryman's Social Research Methods*, 6th Edition. Oxford: University Press, 2021.

Cappelli, Peter. "There's no such thing as big data in HR." *Harvard Business Review*. (2017). Retrieved from: There's No Such Thing as Big Data in HR (hbr.org).

European Commission. *The GDPR: New opportunities, new obligations*. 2018. Retrieved from: Https://ec.europa.eu/commission/sites/beta-political/files/data-protection-factsheet-sme-obligations.

Fecheyr-Lippens, Bruce, Bill Schaninger, and Karen Tanner. "Power to the new people analytics: Techniques used to mine consumer and industry data may also let HR tackle employee retention and dissatisfaction." *McKinsey Quarterly. McKinsey & Company* (2015): 1-2.

Fosso Wamba, Samuel, Shahriar Akter, Andrew Edwards, Geoffrey Chopin and Denis Gnanzou. "How 'big data' can make big impact: findings from a systematic review and a longitudinal case study." *International Journal of Production Economics*. (2014): 1-34. 10.1016/j.ijpe.2014.12.031.

Gerard, George, Martine Haas, and Alex Pentland. "Big data and management." *Academy of Management Journal* 57, no. 2 (2014): 321-326.

Green, David. "The best practices to excel at people analytics." *Journal of Organizational Effectiveness: People and Performance* 4, no. 2 (2017): 10.1108/JOEPP-03-2017-0027.

Hamilton, R.H. and William A. Sodeman. "The questions we ask: Opportunities and challenges for using big data analytics to strategically manage human capital resources," *Business Horizons* 63, no. 1 (2020): 85-95. ttps://doi.org/10.1016/j.bushor.2019.10.001.

Huselid, Mark, A. "The impact of human resource management practices on turnover, productivity, and corporate financial performance." *Academy of Management Journal* 3, no. 38 (1995): 635-72.

Huselid, Mark, A. and Brian E. Becker. "Bridging micro and macro domains: workforce differentiation and strategic human resource management" [Editorial]. *Journal of Management* 37, no. 2 (2011): 421–428. https://doi.org/10.1177/0149206310373400

Huselid, Mark and Dana Minbaeva. "Big data and HRM." In Press, *Sage Handbook of Human Resource Management* (2nd edition), Wilkinson, A., Bacon, N, Lepak, L., & Snell, S. (Eds.), 2018.

IBM Corporation. *Business Analytics for big data: Unlock value to fuel performance.* United States: IBM Corporation, 2013.

Jiang, Kaifeng, David P. Lepak, Jia Hu and Judith C. Baer. "How does human resource management influence organizational outcomes? A meta-analytic investigation of mediating mechanisms." *Academy of Management Journal* 55, no. 6 (2012): 1264–1294. https://doi.org/10.5465/amj.2011.0088

Lim, Nelson, Bruce R. Orvis, and Kimberly Curry Hall. "Leveraging big data analytics to improve military recruiting." Santa Monica, CA: RAND Corporation, 2019. retrieved from: https://www.rand.org/pubs/research_reports/RR2621.html.

Madsen, Dag Øivind and Kare Slåtten. "The rise of HR analytics: A preliminary exploration" *Global Conference on Business and Finance Proceedings* 12, no. 1 (2017):148-159. Retrieved from SSRN: https://ssrn.com/abstract=2896602

Marler, Janet and John Boudreau. "An evidence-based review of HR analytics." *The International Journal of Human Resource Management* 28, no. 1 (2016): 3-26. 10.1080/09585192.2016.1244699.

Ministry of Defence. *HR Monitor. DOSCO/Ministerie van Defensie.* The Hague: TOS, 2018, 2019, 2021.

Ministry of Defence. Defence Vision 2035 "Fighting for a secure future." The Hague: Ministry of Defence, 2020.

Ministry of Defence. Internal document. Retrieved from the Defence intranet website, 2022.

Mohapatra, Smaranika, Jharana Paikaray, and Samal Neelamani. "Future trends in cloud computing and big data." *Journal of Computer Sciences and Applications* 3, no 6 (2015):137-142. http://doi: 10.12691/jcsa-3-6-6.

Montiel-Sánchez, Ignazio. "Disruptive defence innovations. big data analytics for defence" Article EDA (2021). Retrieved from: https://eda.europa.eu/webzine/issue14/cover-story/big-data-analytics-for-defence.

Nocker, Manuela and Vania Sena. "Big data and human resources management: The rise of talent analytics." *Social Sciences* 8, no. 10 (2019): 1-19. https://doi.org/10.3390/socsci8100273

Op den Buijs, Tessa and Erik van Doorn. "What women want! A study on the motives of young women to join the Netherlands armed forces." In press.

Op den Buijs, Tessa and Peter Olsthoorn. "Challenges in HRM for Military Organizations: trends in the military professionalization." In: *Handbook of Military Sciences.* In press.

Paauwe, Jaap and Elaine Farndale. *Strategy, HRM and Performance: A Contextual Approach.* (second ed.). Oxford: University Press, 2017.

Pfeffer, Jeffrey. *Competitive Advantage through People Unleashing the Power of the Workforce.* Boston: Harvard Business School Press, 1994.

Pfeffer, Jeffrey. "Seven practices of successful organizations." *California Management Review* 40, no. 2 (1998):96-124. 10.2307/41165935.

Piening, Erik, P., Alina M. Baluch, and Hans-Gerd Ridder. "Mind the intended-implemented gap: understanding employees' perceptions of HRM." *Human Resource Management* 53, no. 4 (2014): 545-567.

Piersma, Fenna and Alicia Streppel. *HR Analytics. Een praktische inleiding.* Groningen: Noordhoff Uitgevers, 2019.

Raghupathi, Wallinallur and Viju Raghupathi. "Big data analytics in healthcare: promise and potential." *Health Information Science and Systems* 2, no. 3 (2014): 1-10. https://doi.org/10.1186/2047-2501-2-3

Rasmussen, Thomas and Dave Ulrich. "Learning from practice: how HR analytics avoids being a management fad." *Organizational Dynamics* 44, no. 3 (2015): 1-7. 10.1016/j.orgdyn.2015.05.008

Rousseau, Denise, M. "Is there such a thing as "evidence-based management?" *Academy of Management Review* 31, no. 2 (2006): 256-269.

Rousseau, Denise M. and Eric G. R. Barends. "Becoming an evidence-based HR practitioner" *Human Resource Management Journal* 21, no. 3 (2011): 221-235. https://doi.org/10.1111/j.1748-8583.2011.00173.x

Russom, Philippe. *Big Data Analytics, TDWI best practices report.* The Data Warehousing Institute (TDWI) Research, 2011.

Snel, Beatrice and Jeffrey Louvenberg. "Eén plus één is drie: het succes van twee samengevoegde afdelingen binnen Defensie. Hoe twee werelden samen een verhaal vertellen." *Best Practices HR-analytics bij het Rijk Interviewreeks.* (2016/2017): 42-45.

Storey, John. "What is human resource management?" In Storey, J. (ed.). *Human Resource Management: A Critical Text.* London: Thomson, 2007.

Stuart, Mark, David Angrave, Andy Charlwood, Ian Kirkpatrick, and Mark Lawrence. "HR and analytics: why HR is set to fail the big data challenge." *Human Resource Management Journal* 26, no. 1 (2016):1-11.

Tsai, Chun-Wei, Chin-Feng Lai, and Han-Chieh Chao. "Big data analytics: a survey." *Journal of Big Data* 2, no. 21 (2015): 1-32.

Ulrich, Dave and James, H. Dulebohn. "Are we there yet? What's next for HR? *Human Resource Management Review* 25, no. 2 (2015): 188–204. https://doi.org/10.1016/j.hrmr.2015.01.004.

Van der Laken, Paul. *Data-driven human resource management: The rise of people analytics and its application to expatriate management.* Doctoral thesis. Alblasserdam: Ridderprint, 2018.

Van Fenema, Paul C., John, P.H. Kalden; Bas, J.H. Rietjens, and Remco Schimmel. "Big data analytics en Defensie visie en aanpak." *Militaire spectator* 184, no. 9 (2015): 374- 387.

Van Lieshout, Luuk. *Kwalitatief onderzoek naar de voorgenomen praktijk van talentmanagement binnen de koninklijke marechaussee,* Bachelor thesis in press. Breda: NLDA, 2022.

CHAPTER 3

Data-driven maintenance of military systems

Potential and challenges

Tiedo Tinga, Axel Homborg & Chris Rijsdijk

Abstract

The success of military missions is largely dependent on the reliability and availability of the systems that are used. In modern warfare, data is considered as an important weapon, both in offence and defence. However, collection and analysis of the proper data can also play a crucial role in reducing the number of system failures, and thus increase the system availability and military performance considerably. In this chapter, the concept of data-driven maintenance will be introduced. First, the various maturity levels, ranging from detection of failures and automated diagnostics to advanced condition monitoring and predictive maintenance are introduced. Then, the different types of data and associated decisions are discussed. And finally, six practical cases from the Dutch MoD will be used to demonstrate the benefits of this concept and discuss the challenges that are encountered in applying this in military practice.

Keywords: Data collection, Diagnosis, Prognosis, Data analytics, Condition monitoring

3.1. Introduction

Military systems like naval ships, aircraft, helicopters and armed vehicles are operated all around the world in harsh environments. Their quality is essential for the success of a military mission, and failure of the systems may have severe consequences. Therefore, maintenance is important to military organisations, and it has also attracted much research attention in the past decades. In recent years, the collection and storage of all kinds of data has become common practice. For the military domain, this yields many possibilities to use the data for offensive and defensive purposes. But data also offers a lot of opportunities to improve the maintenance and logistic processes, which ultimately leads to more reliable and better available military systems.

In this work, case studies from the Dutch Ministry of Defence (MoD) will be used to illustrate the potential of data-driven maintenance, but also to discuss the challenges encountered on implementing a data-driven maintenance policy. In section 3.2, the basic concepts of data-driven maintenance are introduced, while section 3.3 describes the different data types and types of applications. Section 3.4 to 3.6 then discuss a number of case studies that demonstrate the potential and challenges. Section 3.4 covers manually collected data, section 3.5 concerns autonomously collected control system data and section 3.6 discusses autonomously collected condition monitoring data. Finally, some conclusions are forwarded in section 3.7.

3.2. Concepts in data-driven maintenance

This section will first introduce the basic motivation for data-driven maintenance and will then discuss the different maturity or ambition levels.

3.2.1. Primer on data-driven maintenance

Society heavily relies on the quality of technology. A retained quality cannot be taken for granted, it typically requires deliberate effort that originates from decisions. This effort is known as maintenance. Maintenance can be defined as:

> *The combination of all technical and administrative actions, including supervision actions, to retain or to restore an item's quality.*

This definition paraphrases the CEN 2001[1] and IEC 1990[2] definitions of maintenance by introducing the term *quality*, which is defined as:

> *The degree to which a set of inherent characteristics of an object fulfils requirements (ISO norm[3]).*

In this definition, *inherent* means existing in and not assigned to an item or system. Therefore, inherent characteristics are measurable by sensors as the decision maker's mind state is not involved. However, a decision maker's mind state determines the *requirements* that express the needs or expectations for the system (performance). Requirements are not immediately observable which impedes data driven maintenance. So, quality is not immediately observable, but it excellently represents a decision maker's observable concern.

Decisions involving quality are typically group decisions. Such a group exists because its members have chosen to collaborate, which requires decision makers to agree on (*i*) their goals and on (*ii*) their actions. Decision makers can only align with the organisation's goals and with the organisation's decision rules once these goals and rules are made explicit. A notion of "quality" may convey the group's common sense. This common sense is essential to make quality observable. So, data driven maintenance relies on common sense about quality. Once quality has been defined properly for a specific system, data-driven maintenance can assist in retaining or restoring that system's quality efficiently and effectively. The various ways to do this, i.e. the different maturity levels of data-driven maintenance, will be discussed next.

3.2.2. Maturity levels of data-driven maintenance

Data can support (predictive) maintenance decision making in many different ways. To structure these various approaches, different maturity levels can be identified. In the world of industrial digitalisation (Industry 4.0), the concept of prescriptive or autonomous maintenance is considered as the ultimate maturity in data driven maintenance. In that case, the system automatically takes a diagnosis, predicts the remaining useful life of a specific component, and fully prescribes what maintenance task needs to be executed at what moment in time. And in case of autonomous maintenance, this task is also executed without human interaction. This situation can be visualised by a control system as in Figure 3.1. The system's operation is fully described by some control loop, where the control model (C) adequately responds to any disturbance faced by the process (P).

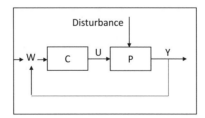

Fig. 3.1. Example of an autonomous control loop.

In practice, the control model (*C*) can only handle the a-priori known disturbances, implying that the control model (*C*) occasionally fails to provide the inputs (*U*) that steer the outputs (*Y*) towards the required value. In that case, a *fault* has occurred, which is defined as:

an anomaly that leads to a quality non-conformity.

Faults can therefore be considered as symptoms of impending quality nonconformities. Such a symptom may be represented by some usage, load or health
characteristic, which in some cases can also be observed or even measured.
Figure 3.2, that originates from classical Fault Detection and Diagnostics (FDD)
theory[4], shows that controlling these faults includes five steps: detection, isolation,
identification, prediction and recovery. Conventional FDD includes:

–   Fault detection: the determination of faults, present in a system and the time of
    detection;
–   Fault isolation: the determination of the kind, location, and time of detection of
    a fault;
–   Fault identification: the determination of the size and the time variant behav-
    iour of a fault;
–   Recovery: the actions, following from the identification, that restores the sys-
    tem quality;

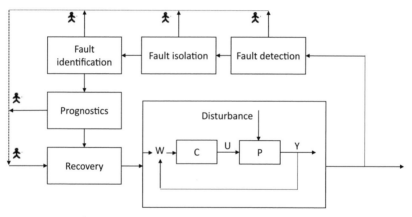

Fig. 3.2. Example of an autonomous control loop extended with fault control.

In the context of predictive maintenance, a new step can be inserted before the
recovery step, i.e. *Prediction*. In this step a prediction of the time to failure is made,
which not only allows for assessing the present health state of the system, but also
the future states. The latter allows a more accurate decision on the recovery actions.

The resulting first four steps of this modified FDD process then coincide with the
maturity levels typically considered in predictive maintenance[5]:

–   Detection: is something wrong?;
–   Diagnosis: what is wrong?;
–   Health assessment: to what extent is it wrong?;
–   Prognostics: how long does it take before it gets really wrong, i.e. fails?

Note further that rudimentary fault control is entirely human driven, which means that each of these steps requires a human analysis and decision to be taken (as represented by the human symbols in Figure 3.2).

Now, data-driven maintenance intends to reduce the human involvement in fault control by untangling the implicit human expertise and implementing this in algorithms. Data driven maintenance will thus make fault control more consistent and autonomous. Ultimately, all the steps in the fault control process may become fully autonomous, which yields the ambitioned prescriptive (autonomous) maintenance. Note that in modern (and critical) systems, reconfigurations, switch-overs and emergency precautions are already increasingly triggered autonomously.

Figure 3.3 shows a model of the steps an organisation that intends to mature in data driven maintenance needs to follow. This also shows that the minimal requirement is to ensure access to the data that allows for autonomous fault *detection* (being the lowest maturity step). As the locations of the data collection and the Fault Detection and Diagnostics (FDD) process may differ, data transport becomes essential. The Dutch MoD experienced that transporting data from the installed control systems of some weapon systems to an analysis environment is not trivial at all, as it triggered quite some issues related to data security and data ownership.

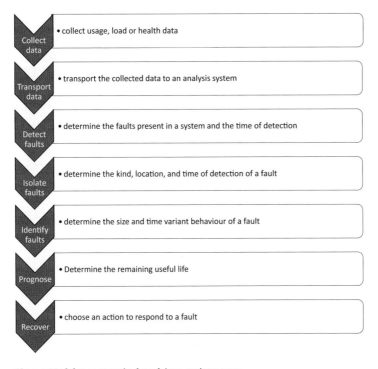

Fig. 3.3. Model to mature in data driven maintenance.

Maturity in data-driven maintenance heavily depends on the degree to which the first four tasks are performed autonomously. Figure 3.3 suggests that maturity growth in data-driven maintenance follows the flow of the fault control loop in Figure 3.2. So, maturity growth in data-driven maintenance should start with autonomous fault detection. Only if this level is accomplished, can the next step in maturity be made. This is confirmed by the framework proposed by Tiddens et al.[6,7], indicating that maintenance ambition level and type and amount of data should match to enable a successful increase in maintenance maturity level.

3.2.3. Data-driven maintenance model selection

As maintenance involves decisions that can only influence the yet-to-be-observed-future, a model is required that can go beyond a sample of historic data. The selection of such a model can be supported by several processes[8,9,10]. Figure 3.4 classifies the various model types. Knowledge-based models rely on knowledge of the physics that prescribes the variables, the parameters, and the structure of the model. Knowledge-based models are superior in making claims about unprecedented circumstances. A history-based model is a resort when knowledge of the model properties is lacking. It just employs data to estimate unknown model properties or to weigh some arbitrary set of candidate models. Because history-based models rely on data from the past, claims about unprecedented circumstances are risky. Figure 3.4 shows that history-based model selection may take place from labelled data or from unlabelled data. Here, the label indicates the item's quality (e.g. failed or healthy). Labelled data allows for supervised machine learning whereas unlabelled data only allows for unsupervised machine learning. Finally, advanced cases of data driven maintenance typically comprise a hybrid model selection as both historical data and expert knowledge are imperfect, but they may be complementary (see section 3.6.4).

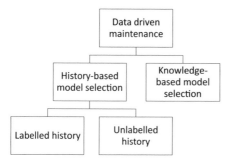

Fig. 3.4. Classification of data driven maintenance.

3.3. Data types and maintenance decisions

During the life cycle of a military system, comprising the design and build, the operational phase and the phase-out of the system, many decisions have to be taken. These decisions range from short term to long term, and from individual systems to complete fleets. Also, the range of data types that can be used to support these decisions is very wide and can originate from many different sources. Therefore, this section will first provide an overview of the various data types and sources, and then the various ways to apply data-driven decisions in the life cycle management (LCM) process will be discussed.

3.3.1. Different types of data

The previous subsections described the various maturity levels in data-driven maintenance. These all require at least some data, but large differences in the specific requirements to data exist. The various types of data and associated sources are clustered in the following three categories:

A. Manually collected system data
Data that is collected from registrations or inspections performed by persons. The data is typically entered into a computer system or database manually. This concerns the following subcategories:
– *A1. Maintenance data:* information on the executed (or planned) maintenance tasks (repair, replacement, lubrication), typically obtained from enterprise resource planning (ERP) or computerised maintenance management systems (CMMS);
– *A2. Inspection data:* obtained during periodic inspections, using sensors or measurement devices that are not installed in the system. This provides insight into the health of the system at the moment of inspection. This can be binary data (system up or down), as well as continuous data (e.g. remaining profile depth in tyres);
– *A3. Operational (or usage) data:* information on how the system has been used (machine settings, type of product, rates, ...)
– *A4. Configuration data:* provides information on the specific configuration of a system, like type, version, etc.;

B. Autonomously collected system data
Data that is collected by the system itself, and also directly stored into a computer system or database, without any human intervention or action. This concerns the following subcategories:

–   *B1. Control system (sensor) data*: obtained from sensors installed in the system, providing values for e.g. temperature, pressure, vibration level, flow, rpm, etc. This type of data is obtained from sensors present in the system for control purposes, e.g. in a supervisory control and data acquisition (SCADA) system. Also, event data is part of this category, providing registrations of events that occurred, like commands or notifications (of faults or quality nonconformities);
–   *B2. Condition monitoring (sensor) data:* obtained from sensors installed in the system, with the specific purpose of monitoring the condition of the system. Examples are vibration sensors, oil quality sensors, corrosion sensors, etc.;

C. Background data
This data provides insight into the context / surrounding / environment that the system is operated in, like:

–   *C1. Environmental data*: information on the environmental conditions in which the system is operated, like temperature, humidity, wind, climate. Can be obtained from (on-board) sensors, or from (weather) databases.
–   *C2. Location information*: provides insight into the (geographical) location of the system, e.g. specified by a GPS position or the type of surface;

For the development of data-driven maintenance, the autonomously collected (sensor) data (*B*) is the most important, as sensors typically provide data with a high sampling frequency, and are less affected by human errors. The latter largely reduces the quality of the manually registered category *A* data. It is observed that in recent years the volume of category *B* data collected within the MoD has grown significantly, e.g. through Flight Data Recorders in helicopters and data recorders in vehicles and on ships. This enables reaching the 1st maturity level (autonomous data collection) in Figure 3.3.

3.3.2. Application of data-driven maintenance in the LCM process

As was mentioned earlier, many different decisions have to be taken during the life cycle of a military system. Data can be used to support these decisions, but this can be applied in various ways. The control loop and associated decisions in Figure 3.2 are related to a single system: these will be referred to as maintenance decisions. But the collected data and derived insights can also be used to optimise long-term and fleet-wide maintenance programs, and the associated service supply network (spare parts, capacity planning). The latter type of decisions will be referred to as LCM decisions. Therefore, the following applications of data-driven maintenance can be distinguished:

I. Situational awareness & health assessment: for a military system it is very important that its present state and capabilities are known to the operator. Based on that information, the commander or operator can decide what type of missions or actions can (still) be performed. This type of decision is typically short term (what is the state *now*) and on the level of an individual system. Data can assist in doing such an assessment using the FDD process (Figure 3.2). This process can be fully manual but can also be made more autonomous using artificial intelligence to increase the efficiency and ensure a quicker response.

II. Short term maintenance planning: to prevent system downtime, maintenance must be performed, preferably before a failure occurs. To properly plan these types of maintenance tasks, the operator needs to know how long the system can be used before it fails, but also how the operation can be adapted to ensure a certain period of failure-free operation. Also, these types of decisions are rather short term (~ weeks) and are on the level of the individual system. The main difference with the previous decision type is the addition of a prediction (of the remaining useful life) to the health assessment. In this case, data can assist in prognostics (Figure 3.2 and sections 3.4.2 and 3.6.3).

III. Long-term maintenance planning & scheduling: the operator of a system has to plan the short-term maintenance tasks (see previous decision), but for a fleet of systems it might not be optimal to plan this on an individual basis, as this may lead to peaks in the maintenance workload. Therefore, long term planning and scheduling for the complete fleet is more efficient. This is therefore an LCM decision rather than a maintenance decision. However, the way the individual systems are used (severe missions or easy training program) determines their maintenance demand, which should be incorporated into the maintenance optimisation. In this case, data can assist in:
- Specifying the usage profiles of the systems (see section 3.5.1 and 3.5.2)
- Predictions of failures based on these profiles (see section 3.4.1)
- Optimising the maintenance intervals (planning and scheduling) to achieve the highest fleet-wide availability (see section 3.4.2)

IV. Optimising supply chain processes: once the maintenance process is properly optimised, the associated demand for work force, facilities and spare parts is also

known. This means that these LCM processes can also be optimised. Data can in this case assist in:
- Calculating the optimal inventory levels for spare parts
- Determining the best moments for purchasing parts
- Planning and scheduling of facilities and work force (see section 3.4.2)

In the Autonomic Logistics Information System[11] (ALIS) of the F-35 fighter aircraft several of these functionalities have already been implemented. This means that many logistic activities are (automatically) triggered by degradation and upcoming failures in the operational aircraft.

In the remainder of this work, a number of cases from the Dutch MoD[12] will be discussed in terms of potential and challenges. These cases (see Table 3.1) will be organised along the types of data introduced in 3.3.1: in section 3.4 the challenges of human entered data (A) will be discussed. Section 3.5 then addresses the potential and challenges of autonomously registered control data (B1), while section 3.6 focuses on condition monitoring sensor data (B2). Moreover, for each of the cases it will be specified which maturity level (1 – 4) is covered (detection, isolation, identification, prediction), and what type of maintenance decision (I – IV) is supported.

Table 3.1. Overview of case studies and associated data types, maturity levels and decision types (ST = short term, LT = long term)

Case study (section)	Data type	Maturity level	Decision
Vehicle inspections (4.1)	A	0 – No FDD	I – Situational awareness
Ship corrosion optimisation (4.2)	A	4 – Prediction	III / IV – LT Optimis./ planning
Vehicle speed registration (5.1)	B1	0 – Usage monitoring	II / III – ST + LT optimisation
Vessel water temp. registr. (5.2)	B1	2 – Fault isolation	I / II – Sit. awaren. + ST optimisation
Corrosion monitoring (6.1/6.2)	B2	3 – Fault identification	I / II / III ST + LT optimisation
Corrosion prediction (6.3)	B2	4 – Prediction	III – LT optimisation

3.4. Decisions based on manually collected system data

Traditionally, most of the registrations of either executed maintenance or the results of (visual) inspections within the MoD are entered manually in a Computerised Maintenance Management System (CMMS). Although some basic analyses can be done on this type of data, the two cases in this section will demonstrate that this data has some clear drawbacks.

3.4.1. Vehicle inspections using a hand-held device

The conventionally collected notifications of vehicle maintenance actions in a CMMS proved to be prone to human errors and they only incidentally entailed a narrative about the nonconformity in quality, let alone about underlying faults. The use of hand-held electronic devices appeared to be prosperous to increase the level of detail and to include the time stamp of visual inspections[13]. A pilot project with a hand-held application at the MoD illustrated that the arrival rate of quality nonconformities resulting from visual inspections significantly depended on the inspector and on the operating regime.

To illustrate this, in Figure 3.5 the number of detected quality nonconformities registered with the hand-held device is plotted versus the usage intensity (kilometres divided by time). This reveals that the relations between quality nonconformities and the usage intensity are very different for the two operating scenarios considered. This indicates a clear dependence of system degradation on usage intensity, which could be further explored and used for long term maintenance optimisation (LCM decision type III). Without the handheld application, the number of quality non-conformities was hard to retrieve efficiently, so this analysis would not be possible.

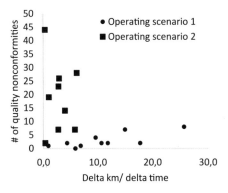

Fig. 3.5. Scatter plot of quality nonconformities versus usage intensity given two operating regimes.

It should be realised that this handheld application predominantly collected quality nonconformities that already occurred. Quantitative data that enables FDD, let alone predicts a specific quality nonconformity (*prognosis*), is not provided by the CMMS and neither by this hand-held application. To conclude, CMMS data alone appeared to be insufficient to mature in data-driven maintenance, while the structured data collection enabled by the hand-held device allowed a one step increase in maturity.

3.4.2. Naval ship corrosion maintenance optimisation

Navy ships require maintenance on the corrosion protection coating systems on a regular basis. If repairs or replacement of the coatings are performed too late, the underlying steel structures of the ship hull may be degrading. On the other hand, if maintenance is performed too early, unnecessary costs are made, and the often-required docking of the ship might lead to a long period of non-availability. To optimise the maintenance decisions regarding navy ship coatings, a previously proposed model[14] for infrastructure maintenance was adapted to the vessel case[15].

Firstly, the coating degradation is modelled with a non-stationary gamma process. The parameters of this process have been derived from information provided by maintenance experts (data type A1 and A2). As the state of the coating is only assessed periodically (i.e. once a year) and visually (i.e. the % of the hull area that has a degraded coating), the resulting gamma process still contains quite some uncertainty. The left-hand plot in Figure 3.6 shows the obtained degradation over time, which depends on the maintenance option that is applied. A *full replacement* of the coating (with the ship docked) restores the condition to the as new situation. The options *repainting* and *spot-repair* only provide a slight improvement of the condition, and also yield a higher degradation rate after the maintenance action (due to the imperfect treatment). The magnitude of these effects is also derived from expert interviews.

Once the degradation behaviour of the coating is properly captured by the gamma process, the second step is to optimise the maintenance decisions over time. Given a threshold for the amount of degradation allowed and the costs associated with each of the maintenance options, the model returns the sequence of maintenance activities that minimises the total expected life cycle costs. The results for a specific case are shown in Figure 3.6 (right), indicating at each age of the vessel what the best maintenance option is (if maintenance must be executed at that moment in time). One of the boundary conditions is that the ship will be docked every five years, which makes full replacement the best option at those moments. At all other moments in time, the high costs of (additional) docking and full replacement do not outweigh the resulting lower degradation rate, and either repainting or spot repair are the best options.

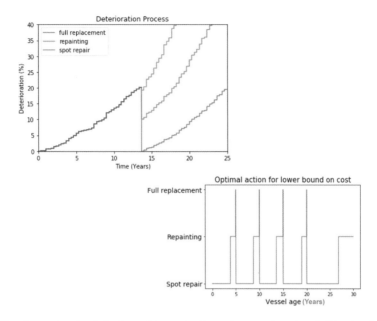

Fig. 3.6. Deterioration behaviour of the coating for various repair options (left) and optimal actions proposed by the model (right).

This case study shows that even low quality, human entered data can be used for long-term maintenance optimisation (LCF decision type III and IV). However, the low frequency of inspections (once a year) and visual assessment make the data very crude, which introduces a large amount of uncertainty in the degradation curves, and thus also in the decisions advised by the model. One of the recommendations of this project is therefore to increase the frequency and amount of detail of these hull coating inspections. This can, for example, be achieved by using drones that on a weekly basis scan the vessel hull surface and apply image processing techniques to automatically calculate the fraction of the coating surfaces that has degraded.

3.5. Decisions based on autonomously collected control system data

A control system steers a (weapon) system to the desired state and intervenes when potential unsafe situations occur (e.g. over-speeding or over-heating). By collecting the data that flows through a control system, many details about the usage and the loads during operations may be retrieved. As usage causes a load that may deteriorate the system health, yielding a nonconformity in quality, fault signatures may be based on control data. Therefore, data-driven maintenance becomes more prosperous when control data is available in addition to CMMS data. As the generation of control data is often "designed in" in many military systems, it provides an accessible entry to mature in data-driven maintenance. Field experiments with control data enable the Dutch MoD to learn about data-driven maintenance practices that include (*i*) the original equipment manufacturer involvement, (*ii*) data governance and (*iii*) the selection of prosperous case studies.

Still, it should be realised that the control systems of weapon systems intend to support *operations* rather than *maintenance*. Dedicated health monitoring systems may provide better fault signatures of impending nonconformities in quality, as will be discussed in section 3.6. The present section introduces two realistic case studies at the Dutch MoD on the use of just control data. The first case study[16] exemplifies the reconstruction of usage and load characteristics of a vehicle corresponding with the "collect data" step in the maturity model (Figure 3.3). The second case study[17] illustrates the "fault isolation" step in the maturity model using a knowledge-based model selection (Figure 3.4).

3.5.1. Military vehicle speed registration

The control system of a military vehicle records the velocity of the vehicle which allows a reconstruction of the variation in the vehicle speed (Figure 3.7). From this signal and knowledge of the mass of the vehicle, the acceleration forces and the deceleration forces can be calculated.

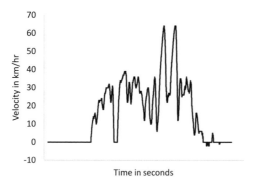

Fig. 3.7. Variation of the speed of a vehicle as registered by the control system.

For each activation of the brakes, the dispersed energy and the peak power can be estimated from the measured speed variations (Figure 3.8). The cumulative brake energy and the number of "high" peak power cycles may be better predictors of brake failures than the historic arrival rate of brake replacement notifications in the CMMS. These features may generally be helpful to distinguish rough and gentle operating regimes experienced by a specific vehicle and may be used to predict failure rates from that (LCM decision type III).

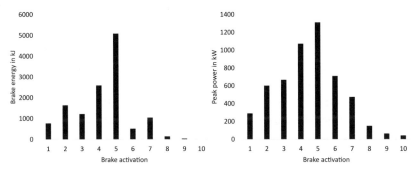

Fig. 3.8. Estimated energy and peak power at consecutive brake activations.

3.5.2. Naval vessel temperature sensor

Four installations on a naval vessel each independently measure the sea water temperature. Each of these temperature measurements has been logged by the vessel's control system, and Figure 3.9 shows the difference in measured temperature of three of the sensors (T2-T4) relative to the first sensor (T1).

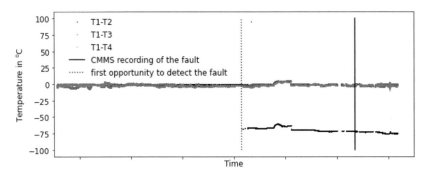

Fig. 3.9. Evolution of four independent measurements of the sea water temperature surrounding a vessel.

Figure 3.9 shows that one of the three sensors shows an offset of 75°C after some time. An expert's model that assumes equality of the four temperature measurements may then autonomously identify a faulty sensor. Figure 3.9 also shows a considerable time lag between the recording of the nonconformity in quality in the CMMS (as indicated by the solid black line) and the first opportunity to detect the fault (the dashed blue line). So, autonomous fault identification might have reduced the downtime of this sensor considerably (maintenance decision type I / II).

3.6. Decisions based on autonomously collected condition monitoring data

As an example of the use of sensors specifically aimed at condition monitoring for a complex failure mechanism, this section will address the use of corrosion monitoring for improving maintenance decisions, and the challenges this poses for their integration in data-driven maintenance.

3.6.1. Corrosion management

Military systems operate under a vast diversity of conditions. These induce a multitude of failure mechanisms, including different types of corrosion, often localised. Corrosion is estimated to cost society over 3% of the world's GDP[18], while for military systems, the consequences of corrosion account for over 20% of the total US DoD maintenance expenditures[19]. This shows that corrosion has serious cost and availability impact on military systems but can also seriously impact their safety.

Corrosion management of military assets is getting more and more demanding due to increasing design complexity. Two important materials-related developments in the industry are causing this. Firstly, the emergence of additive manufacturing processes, which significantly increase the freedom of design by allowing a high geometric complexity. And secondly, the growing use of exotic materials to meet increasingly stringent weight and strength requirements. Examples are combinations of different advanced polymer- and metal-matrix composites, often with lightweight metal alloys, also in complex geometries. These two developments cause the aforementioned increasing design complexity, making it increasingly difficult to predict its corrosion behaviour. Traditional lab-based experiments that expose simple combinations of materials to wet/dry cycles with varying temperature, salinity or pH may prove insufficient to predict specific local corrosion phenomena at micro-environments that arise throughout such geometries.

The understanding and in situ quantitative assessment of a systems response to a variety of combinations of environmental conditions and usage profiles is vital for both short term maintenance decisions (LCM decision type II) and long-term maintenance optimisation (type III). This demands for sophisticated condition monitoring techniques and prognostics, as well as verification and refinement via lab and field-based testing. Prognostic models that translate the condition, or damage accumulation, into a repeatedly updated remaining lifetime should be fed by autonomous in situ monitoring of the systems actual corrosion response to its environment, as will be discussed in section 3.6.3. This can be combined with other relevant failure parameters such as fatigue or contamination. Such an integral, real-time system status is vital for substantiated operational planning and prioritising of maintenance activities, with insight in the availability and safety consequences of e.g. mission extension (decision type I).

3.6.2. Corrosion monitoring challenges

Corrosion sensors can be based on a variety of different techniques, enabling either only fault detection or in addition fault isolation and identification. These techniques all share the same overall aim: to indicate a system's condition. However, each technique has its own disadvantages and flaws.

- Ultrasonic, (fibre-)optics, magnetism, Eddy current and corrosion coupons are all indirect techniques that only detect the corrosion damage, e.g. a wall thickness reduction or the formation of corrosion product. This may well be too late, because detection only occurs if the damage accumulation exceeds a detection threshold. Moreover, most of these techniques are not integrated in the system, and therefore require human effort in the collection (data type $A2$);
- Corrosion coupons, although widely used in the industry, are especially indirect in the sense that the coupons are assumed to be representative for the system under study, which is difficult to acknowledge in practice. Lab-based tests serve as a basis to correlate the condition of the coupon with that of the actual system. Features such as welds, crevices, alloy composition, microstructure, manufacturing process parameters and coating details can differ between the coupon and the system under investigation. These effects impede proper fault detection from corrosion coupons and reduce their validity.
- Environmental sensors are an interesting option, since this technique does not require the position of the sensor at or near the corrosion process. Instead, the detection is based on monitoring the corrosivity of the environment, in section 3.3.1 referred to as background data (type $C1$). Obviously, environmental sensors demand a high level of knowledge of the system's corrosion response to a variety of different types of operational conditions. This monitoring technique is therefore quite common in aerospace. Nevertheless, this knowledge about the system's corrosive properties results from substantial investments in research for each specific type and combination of materials.
- Since corrosion is in essence an electrochemical process, it makes sense to use electrochemistry for monitoring. Techniques such as electrochemical impedance

spectroscopy (EIS) or electrochemical noise (EN) monitor the process itself, i.e. before any significant damage has occurred[20,21,22,23]. This is an important asset, because a robust lifetime prediction model relies not only on the quality of the input, but also on timely input from corrosion sensors. EN and EIS can enable valuable fault isolation and identification as well, which is quite uncommon for a corrosion monitoring technique. Moreover, data can be collected during periodic inspections (type *A2*), but these sensors can also be integrated permanently in the system, which yields autonomously generated data of type *B2*.

Besides type-specific disadvantages, there is also a number of common issues of current corrosion monitoring techniques that require attention:
- Most importantly, all previously mentioned techniques only operate at pre-selected locations. This significantly complicates the detection of the most dangerous, localised forms of corrosion, which are also the most difficult to predict. The most promising ways of monitoring are those that potentially detect localised corrosion over larger surface areas. Unfortunately, the common monitoring techniques applied in industry, e.g. acoustic emission, have their limitation (e.g. effect of noisy environments) and have only been shown to be effective in specific use cases.
- The presence of the sensor can alter, inhibit or accelerate the corrosion process by inducing differences in the local conditions, e.g. in electrolyte composition, with respect to the area around the sensor. Taking into account these differences in a lifetime prediction model is very difficult, as this is not a constant factor, nor is its time-dependence predictable in itself.
- The detection threshold is an important parameter: considering that most non-electrochemical techniques only detect corrosion damage, their ability to serve as an early-warning technique largely depends on the damage accumulation that is required for their detection. This damage can e.g. involve the build-up of the corrosion product (e.g. in the case of fibre-optics) or the propagation of a crack (e.g. in the case of ultrasonic technique).
- The inspection interval should also be selected properly. Corrosion is a relatively slow process that may not require status updates in the millisecond range; however, the use of military systems under extreme conditions can initiate localised corrosion, which subsequently propagates within a timespan of several weeks to months. This initiation can be detected by electrochemical (EIS or EN) techniques and requires the right moment and location of monitoring. In its early stages, the corrosion process can still be mitigated in the presence of a corrosion inhibitor like Cr-VI or a suitable alternative. Once in the propagation phase, the damage accumulation becomes easier to detect but will increasingly compromise system safety.

- It is very difficult or even impossible to monitor locations inside enclosed environments or complex geometries. The often used alternative is to ascertain 'lifetime' corrosion protection, e.g. by the use of coatings with active inhibition components, of which the toxic Cr(VI) is the most widely applied.
- It is likely that the root cause of a corrosion problem is not corrosion-related. The functional degradation of an organic coating is a good example of this: mechanical, thermal or UV radiation loading of the coating can cause a loss of its corrosion protective properties. In this case, the resulting corrosion damage is merely a consequence of coating degradation. This indicates the significance of a proper root cause analysis (RCA) and of a good understanding of the prevalent failure mechanism(s). One may question whether corrosion monitoring would deliver the preferred data input for prognostic models in these cases, or whether monitoring the coating condition would provide more reliable data. Electrochemistry, such as EIS or EN, can monitor both the coating condition as well as the resulting corrosion activity.

3.6.3. Corrosion prediction

As was discussed before, the prediction of corrosion degradation is rather difficult due to the complexity of the physical processes involved. On the other hand, monitoring of corrosion only indicates the present condition, which makes it hard to plan future maintenance activities. However, if these two approaches are combined, a much stronger (hybrid) approach is obtained. Such an approach for a corrosion problem has recently been proposed[24]. A simple physical model has been used in a particle filter approach, which uses regular measurements of the actual condition to tune and adapt the model continuously.

In this work, the corrosion mass loss has been modelled using a multi-factor combination model. This type of corrosion model predicts mass loss using accelerating factors for environmental parameters. In the present model, only the ambient temperature has been included, using the Arrhenius equation. The model then allows for calculating the mass loss for a certain temperature, provided that the values of the four model parameters are known for the material and situation considered. In practice those values are often not known precisely, but the particle filter used here can utilise the periodic measurements to tune the model parameters to the actual situation.

This is shown in Figure 3.10, where the particle filter has been trained with the corrosion mass loss measurements for a certain period of time, as bounded by the dashed vertical line: 5 months for the left plot, 15 months for the right plot. From that moment in time, the physical model (with the tuned parameters) is used to predict the evolution of the corrosion mass loss (the green line) until the end of the time period (40 months), given a certain (seasonal) variation of the ambient temperature. In these plots the actual mass loss is also shown (the blue line). The left-hand plot shows that after five months the prediction of the mass loss is rather inaccurate, considerably underestimating the time to failure. But the right-hand plot shows that using the measurements up until month 15, the model can be tuned much better to the actual situation, and the resulting prediction of the time to failure is much more accurate.

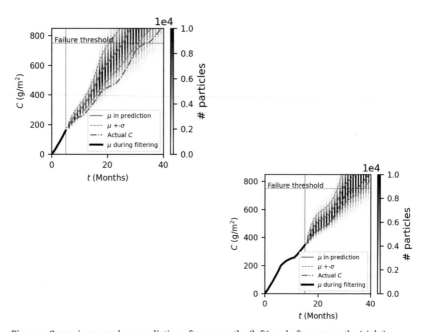

Fig. 3.10. Corrosion mass loss prediction after 5 months (left) and after 15 months (right).

This model shows that prediction of corrosion mass loss purely based on a physical model is rather difficult but strengthening the model with measurements yields a considerably improved prediction. Further, this model only considers the simple situation of mass loss of coupons due to uniform corrosion. Extension to a more realistic situation with localised corrosion on a real structure is still very challenging.

3.6.4. Final remarks on corrosion monitoring

From the discussion and cases in this section it can be concluded that data-driven maintenance for inherently dynamic, difficult to predict and spontaneously occurring processes such as corrosion is not straightforward. Looking at the cost, availability and safety figures, corrosion is an undeniably important problem for the Dutch MoD. However, to improve corrosion management, it is essential to be aware of the limitations of data-driven maintenance. The data required for the detection, identification or even prediction of corrosion is very specific, and cannot be expected to be available from standard control systems. Therefore, the Dutch MoD is currently not able to tackle an important failure mechanism such as (localised) corrosion based on data-driven maintenance. If there is an ambition to do so, investments in dedicated monitoring techniques and data collection must be done.

3.7. Conclusions

This work has discussed the potential of data-driven maintenance, as well as the challenges encountered in developing these approaches. The main conclusions regarding these two aspects are given next.

3.7.1. Potential of data driven maintenance

Data-driven maintenance potentially supports the life cycle process by (*i*) an enhanced situational awareness, (*ii*) short-term maintenance planning, (*iii*) long-term maintenance planning and scheduling and (*vi*) optimised supply chain processes. An enhanced situational awareness appeared the easiest attainable potential for an organisation in the process of maturing in data-driven maintenance, like the Dutch MoD. This finding corresponds with conventions about maturity growth in data driven maintenance that claim that autonomous fault *detection* precedes the more advanced levels of fault *isolation*, fault *identification* and *prognostics*.

3.7.2. Challenges in data-driven maintenance

The case studies predominantly covered the beginning of the maturity growth in data-driven maintenance. The following conclusions can be drawn from these cases:
- The (autonomous) registration of usage (profiles) – *maturity level 0* – does not directly lead to detection or isolation of faults, but does support the understanding of its effect on system degradation;
- The conventional use of human entered (failure) events in a CMMS – *maturity level 1 & 2* – is prone to human factors. To improve the data, (i) collecting the data with a hand-held device and (ii) using autonomously collected data from

a control system were shown to increase the possibilities for data-driven maintenance;

- The quantification of damage – *maturity level 3* – is still challenging, as is shown by the corrosion monitoring case. Although techniques are available, their application in a real operational setting, as well as the analysis and interpretation of the data, is not trivial;
- The two corrosion prediction cases (3.4.2 and 3.6.4) showed that *maturity level 4*, i.e. the prediction of failures, is theoretically already possible, and could even work with 'low level' data. However, validation and demonstration can only be done when this data is actually collected in a structured manner.

Notes

[1] CEN. EN 13306; *Maintenance terminology*. Brussels: European Committee for Standardisation, 2001.

[2] IEC. IEC 60050-191; *International Electrotechnical Vocabulary, chapter 191: Dependability and quality of service*. Geneva: International Electrotechnical Commission, 1990.

[3] ISO. ISO 9000: *Quality management systems – fundamentals and vocabulary*. Geneve: International Organisation for Standardisation, 2015.

[4] Rolf Isermann and Peter Ballé. "Trends in the application of model-based fault detection and diagnosis of technical processes". Control Engineering Practice 5, no. 5, (1997): 709-719.

[5] Wieger Tiddens, Jan Braaksma and Tiedo Tinga. *Framework for predictive maintenance method selection*. Universiteit Twente, 2022.

[6] Wieger Tiddens, Jan Braaksma and Tiedo Tinga. "Decision Framework for Predictive Maintenance Method Selection". *Applied Sciences* 13 no. 3, (2023): 2021.

[7] Wieger Tiddens, Jan Braaksma and Tiedo Tinga. "Exploring predictive maintenance applications in industry". *Journal of Quality in Maintenance Engineering* 28, no. 1, (2022): 68-85.

[8] Andrew K.S. Jardine, Daming Lin and Dragan Banjevic. "A review on machinery diagnostics and prognostics implementing condition-based maintenance". *Mechanical Systems and Signal Processing* 20, no. 7, (2006): 1483-1510.

[9] Xiao-Sheng Si, Wenbin Wang, Chang-Hua Hu and Dong-Hua Zhou. "Remaining useful life estimation; a review on the statistical data driven approaches". *European Journal of Operational Research* 213, no. 1, (2011): 1-14.

[10] Venkatasubramanian, Venkat, Raghunathan Rengaswamy and Surya N. Kavuri. "A review of process fault detection and diagnosis part II: Qualitative models and search strategies". *Computers & Chemical Engineering* 27, no. 3, (2003): 313-326.

[11] See https://www.lockheedmartin.com/en-us/products/autonomic-logistics-information-system-alis.html

[12] More cases are available on the internal network of the MoD (https://doscoportal.mindef.nl/sites/0400/010/).

[13] Rick Kunz. *Onderhoudsvalidatie; een toepassing op de maandelijkse inspecties*. BSc thesis, Den Helder: NLDA, 2019.

[14] Robin P. Nicolai, Johannes B.G. Frenk and Rommert Dekker. "Modelling and optimizing imperfect maintenance of coatings on steel structures". *Structural Safety* 31, no. 3, (2009): 234–244.

15 E. Hendriks. *Improving the maintenance strategy of naval vessels' hull coating*. BSc thesis, Eindhoven: TU Eindhoven, 2021.

16 Tom Curfs. *Predictief onderhoud met voertuigdata. Een haalbaarheidsstudie*. BSc thesis, Den Helder: NLDA, 2022.

17 Chris Rijsdijk, Willem van der Sluis and Tiedo Tinga. Autonomous fault detection and diagnostics, an enabler to control risks of military operations. *Proceedings of the 58th ESReDA Seminar: Using Knowledge to Manage Risks and Threats: Practices and Challenges* (pp. 1-8). Alkmaar: European Safety and Reliability Association, 2021.

18 Gerhardus Koch, Jeff Varney, Neil Thompson, Oliver Moghissi, Melissa Gould, and Joe Payer, *International Measures of Prevention, Application, and Economics of Corrosion Technologies Study*. NACE IMPACT Report, NACE, 2016.

19 Eric Herzberg and Rebecca Stroh, *The Impact of Corrosion on Cost and Availability of U.S. Department of Defense Weapon Systems*. STO report, NATO, 2017.

20 Axel M. Homborg, Erik P.M. van Westing, Tiedo Tinga, Gabriele M. Ferrari, Xiaolong Zhang, Johannes M.W. de Wit and Arjan M.C. Mol. "Application of transient analysis using Hilbert spectra of electrochemical noise to the identification of corrosion inhibition". *Electrochimica Acta* 116, (2014): 355-365.

21 Axel M. Homborg, Bob A. Cottis and Arjan M C Mol. "An integrated approach in the time, frequency and time-frequency domain for the identification of corrosion using electrochemical noise". *Electrochimica Acta* 222, (2016): 627–640.

22 Homborg, Axel M., Patrick J. Oonincx and Arjan M.C. Mol. "Wavelet transform modulus maxima and holder exponents combined with transient detection for the differentiation of pitting corrosion using electrochemical noise". *Corrosion* 74, no. 9, (2018): 1001–1010.

23 Axel M. Homborg, Matteo Olgiati, Paul J. Denissen and Santiago J. Garcia. "An integral non-intrusive electrochemical and in-situ optical technique for the study of the effectiveness of corrosion inhibition". *Electrochimica Acta* 403, (2022): 139619.

24 Luc S. Keizers, Richard Loendersloot and Tiedo Tinga. Atmospheric corrosion prognostics using a particle filter. *European Safety and Reliability conference* (pp. 1-8). Dublin: ESRA, 2022.

References

CEN. *EN 13306; Maintenance terminology*. Brussels: European Committee for Standardisation, 2001.

Curfs, Tom. *Predictief onderhoud met voertuigdata. Een haalbaarheidsstudie*. Den Helder: NLDA, 2022.

Hendriks, E. *Improving the maintenance strategy of naval vessels' hull coating*. BSc thesis, TU Eindhoven, 2021.

Herzberg, Eric and Rebecca Stroh. *The Impact of Corrosion on Cost and Availability of U.S. Department of Defense Weapon Systems*. NATO, 2017.

Homborg, Axel M., Bob A. Cottis and Arjan M C Mol. "An integrated approach in the time, frequency and time-frequency domain for the identification of corrosion using electrochemical noise". *Electrochimica Acta* 222, (2016): 627–640.

Homborg, Axel M., Matteo Olgiati, Paul J. Denissen and Santiago J. Garcia. "An integral non-intrusive electrochemical and in-situ optical technique for the study of the effectiveness of corrosion inhibition". *Electrochimica Acta* 403, (2022): 139619.

Homborg, Axel M., Patrick J. Oonincx and Arjan M.C. Mol. "Wavelet transform modulus maxima and holder exponents combined with transient detection for the differentiation of pitting corrosion using electrochemical noise". *Corrosion* 74, no. 9, (2018): 1001–1010.

Homborg, Axel M., Erik P.M. van Westing, Tiedo Tinga, Gabriele M. Ferrari, Xiaolong Zhang, Johannes M.W. de Wit and Arjan M.C. Mol. "Application of transient analysis using Hilbert spectra of electrochemical noise to the identification of corrosion inhibition". *Electrochimica Acta* 116, (2014): 355-365.

IEC. *IEC 60050-191; International Electrotechnical Vocabulary, chapter 191: Dependability and quality of service.* Geneva: International Electrotechnical Commission, 1990.

Isermann, Rolf and Peter Ballé. "Trends in the application of model-based fault detection and diagnosis of technical processes". *Control Engineering Practice* 5, no. 5, (1997): 709-719.

ISO. *ISO 9000: Quality management systems – fundamentals and vocabulary.* Geneve: International Organisation for Standardisation, 2015.

Jardine, Andrew K.S., Daming Lin and Dragan Banjevic. "A review on machinery diagnostics and prognostics implementing condition-based maintenance". *Mechanical Systems and Signal Processing* 20, no. 7, (2006): 1483-1510.

Keizers, Luc S., Richard Loendersloot and Tiedo Tinga. Atmospheric corrosion prognostics using a particle filter. *European Safety and Reliability conference* (pp. 1-8). Dublin: ESRA, 2022.

Koch, Gerhardus, Jeff Varney, Neil Thompson, Oliver Moghissi, Melissa Gould and Joe Payer. *International Measures of Prevention, Application, and Economics of Corrosion Technologies Study.* NACE, 2016.

Kunz, Rick. *Onderhoudsvalidatie; een toepassing op de maandelijkse inspecties.* Den Helder: NLDA, 2019.

Nicolai, Robin P., Johannes B.G. Frenk and Rommert Dekker. "Modelling and optimizing imperfect maintenance of coatings on steel structures". *Structural Safety* 31, no. 3, (2009): 234–244.

Rijsdijk, Chris, Willem van der Sluis and Tiedo Tinga. Autonomous fault detection and diagnostics, an enabler to control risks of military operations. *Proceedings of the 58th ESReDA Seminar: Using Knowledge to Manage Risks and Threats: Practices and Challenges* (pp. 1-8). Alkmaar: European Safety and Reliability Association, 2021.

Si, Xiao-Sheng, Wenbin Wang, Chang-Hua Hu and Dong-Hua Zhou. "Remaining useful life estimation; a review on the statistical data driven approaches". *European Journal of Operational Research* 213, no. 1, (2011): 1-14.

Tiddens, Wieger., Jan Braaksma and Tiedo Tinga. "Exploring predictive maintenance applications in industry". *Journal of Quality in Maintenance Engineering* 28, no. 1, (2022): 68-85.

Tiddens, Wieger, Jan Braaksma and Tiedo Tinga. "Decision Framework for Predictive Maintenance Method Selection". *Applied Sciences* 13 no. 3, (2023): 2021.

Venkatasubramanian, Venkat, Raghunathan Rengaswamy and Surya N. Kavuri. "A review of process fault detection and diagnosis part II: Qualitative models and search strategies". *Computers & Chemical Engineering* 27, no. 3, (2003): 313-326.

CHAPTER 4

Federated learning for enabling cooperation between the Royal Netherlands Navy and external parties in developing predictive maintenance

Anna C. Vriend, Wieger W. Tiddens & Relinde P.M.J. Jurrius

Abstract

Nowadays, Machine Learning is used successfully in many applications. Two important challenges are involved with its implementation: i) in many industries, data is stored in separate places; ii) the growing demand for Machine Learning that respects the privacy of the training data. Conventional Machine Learning is based on centralised data collection and is thus unable to face these challenges. Federated Learning is a solution to this problem. This work focuses on how the Royal Netherlands Navy can use Federated Learning to train a Long Short Term Memory Machine Learning model with data sets from an external party, Royal Van Oord, without sharing raw data. The Federated Learning approach exceeds the accuracy of central Machine Learning and can be used on the Ministry of Defence's intranet.

Keywords: Machine learning, Federated learning, Predictive maintenance, Government-industry collaboration

4.1. Introduction

Data from military platforms (such as sensor and failure data) is increasingly being used to develop predictive maintenance algorithms. These learning algorithms offer an opportunity to identify failures at an earlier stage, better plan maintenance and reduce the (corrective) workload aboard ships with the help of data analysis and Machine Learning (ML). ML is the use of algorithms to analyse data, learn from it and then give advice or predictions to a user. This large amount of data offers many possibilities, but data is often privacy-sensitive or not accessible. As a result, the demand for privacy-friendly ways to use data for ML is growing.

One potential way to solve the problems outlined is Federated Learning (FL). This new ML technique ensures that the owner of the data remains in control of their own data and that privacy-sensitive data does not have to be shared with external parties[1]. The term FL was coined by Google in 2017 and has had an increasing number of uses since then. Google itself uses it, for example, to improve suggestions for the Gboard keyboard and for the Health Studies app[2,3]. Users do not share their raw data, but Google does learn from this data through FL. FL can also be used for Natural Language Processing (NLP) and Predictive Maintenance, among other things.

A constant pressure on reducing the Royal Netherlands Navy's ships' crews and an increasing complexity of systems aboard naval ships creates pressure on the maintenance of future naval ships. The RNLN therefore works on a transition from planned periodic maintenance with a high corrective workload towards predictive maintenance based on advanced condition monitoring and data analysis techniques. Predictive maintenance is a preventive maintenance approach that uses analytics to inform the RNLN as owner-operator about the current, and preferably also the future state of their physical assets[4,5]. The RNLN uses ML models to predict the maintenance needs of its fleet based on (sensor) data of (machinery aboard) their ships. As an example, for a diesel engine this data may include measurements like temperatures of the bearings, fuel flow, engine speeds and power.

To facilitate this development, the RNLN needs to collaborate with knowledge institutions, industry and other external parties. When several parties work together with similar tools, there is a greater amount of data available and this potentially results in more accurate predictions on for example the required maintenance of diesel engines or pumps aboard ships. These collaborations can be made possible through the use of FL, so that the RNLN does not have to share raw data with external parties, as this data is oftentimes classified. This paper focuses on a specific case study of sharing an ML model with Royal Van Oord. Van Oord is one of the largest dredging companies in the world and has a lot of sensor data available from their ships.

In this chapter an FL architecture is set up, tested, and compared with central ML on aspects such as the model accuracy and the required data traffic. A Long Short Term Memory (LSTM) neural network (NN) from the RNLN[6] is used to test central ML and FL. Data sets from one of Van Oord's vessels, the Flexible Fall Pipe Vessel (FFPV) Stornes, were made available as training data. Experiments to train the NN federated were carried out with several laptops that served as clients and the central server. The chosen NN evaluation parameters were used to compare the results of FL and central ML. The selected LSTM network, a Recurrent Neural Network (RNN), is able to detect anomalies in ship sensor data. This model is specifically developed within the RNLN to detect anomalies in diesel engines. To train this model, various sensors can serve as input; a specific data set is not required. The outputs of this LSTM result in a warning table, Figure 4.1 shows an example. This warning table helps maintenance engineers in their day-to-day maintenance decision making (i.e., deciding whether to inspect, repair or replace an item or to continue operation).

This paper is structured as follows. Section 2 gives a description of FL. Section 3 discusses the designed system architecture of FL with one of the more flexible FL frameworks suitable for using time series data, the Flower framework. In Section 4, the reliability and validity of the study is discussed. Next, Section 5 compares our FL architecture with central ML. Finally, conclusions, limitations and general reflections will be given in Section 6.

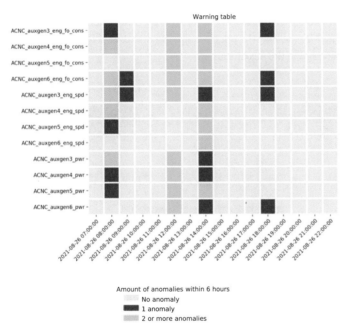

Fig. 4.1. An example of a warning table generated by the LSTM model shows the number of anomalies generated within blocks of six hours of generators 3, 4, 5 and 6 of FFPV Stornes.

4.2. Federated learning

The idea of Machine Learning (ML) is to allow systems to make data-driven decisions instead of programming the systems explicitly to execute a certain task. An early example of ML is a spam filter for email. Instead of trying to describe to the spam filter what makes a message spam, the system is provided with many examples of spam and non-spam emails and learns the difference itself. This training is done via an ML algorithm, where the result of the training is an ML model.

Nowadays, Machine Learning is used successfully in many applications. There are two important challenges involved with the implementation of this technique. The first is that in many industries, data is stored in separate places. The second challenge is the growing demand for ML that respects the privacy of the training data. Conventional ML is based on centralised data collection and is thus unable to face these challenges[7].

A solution to this increasing fragmentation and isolation of data is found in FL. This is an ML approach that makes it possible to train models with decentralised data. The idea of FL is that the algorithm is brought to the data instead of the data to the algorithm[8]. This is visualised in Figure 4.2. FL is fundamentally different in this respect from, for example, the ML technique of distributed learning, in which the data is transferred to the algorithm.

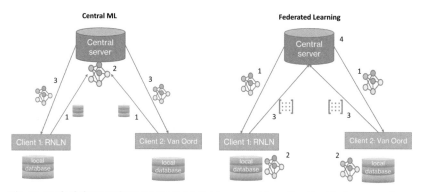

Fig. 4.2. On the left Central ML where in 1 data sets are sent from the client to the server, in 2 the server trains the model and in 3 the trained model is sent back to the clients. On the right FL where in 1 the ML model is initialized, in 2 the model is trained locally, in 3 the clients model gradients are sent to the server and in 4 the sub models are aggregated.

In FL, sub-models are trained at different parties, called *clients*, using only local data. A client can be a company, a ship, but also a mobile device. Client raw data is stored locally and not shared to ensure user privacy and data confidentiality. The various parties then share their sub-models with a central server, who *aggregates* the sub-models into a global model. The process must be carefully designed so that no party can trace the confidential data of other parties[9,10,11].

4.3. Characteristics of federated learning

FL is a relatively new concept and as a result there are different interpretations of the term in existing literature. This section examines five characteristics of FL as adopted in this study to avoid misinterpretation of terms and concepts. The characteristics are based on various definitions of FL[12,13].

4.3.1. Central orchestration

In setting up the training process, the central server has to take a number of aspects into account. The design of the server is driven by the number of clients, the size of model updates and the amount of data traffic. The number of clients can vary from tens to hundreds or millions, depending on the application of the model. The updates that are collected and communicated can also vary in size from kilobytes to tens of megabytes. Finally, the amount of data traffic from certain clients can vary greatly throughout the day.

4.3.2. Local data processing

With FL, data is generated locally at clients and remains decentralised. Raw data is not shared with other clients or with the central server. The data generated by clients is often not independently identically distributed (non-i.i.d.), hence the model may be *biased*. There are several solutions to counteract a bias in the model, for example local reweighting[14].

4.3.3. Communication

Communication is a bottleneck in *federated* networks. The communication links between clients and the server can be slow and unstable and clients can be physically distributed across the earth. It is therefore important to develop efficient communication methods that frequently send small messages or model updates for the training of the model.

Local updating methods, such as *federated averaging* (FedAvg), can reduce the number of communication rounds required for training. Furthermore, using model compression schemes significantly reduces the size of communication messages each round. Decentralised training also makes communication more efficient. It is faster than centralised training on networks with low bandwidths or high data transfer delays.

4.3.4. Privacy and confidentiality

Privacy and confidentiality are often very important for parties that use FL. FL protects local data by sharing model updates, instead of the raw data. Although this is a step towards data protection, it may still be possible for a third party or the central server to recover raw data from the model updates during the training process.

Methods that aim to improve privacy within FL, such as *Secure Multiparty Computation* (SMC) or *Differential Privacy* (DP), come at the cost of model performance or system efficiency. A balance must be struck between these two considerations when realising FL systems.

4.3.5. System heterogeneity

The communication, computing and storage capabilities of each client in a federated network often differ. This is due to a variation in hardware, network connections (3G, 4G, 5G or Wi-Fi) and computing power. In addition, the network size and system-related limitations of the clients imply that only a small part of the clients is active at the same time. It also happens that active clients drop out and disconnect.

Challenges such as FL process delays from lagging clients and fault tolerance are compounded by these characteristics. Thus, FL methods must accept heterogeneous hardware, anticipate low participation, and be robust enough to accept failed clients.

4.4. Choosing the federated learning framework

The first step in designing an FL-system is choosing a framework. A framework is an interface for the developer with built-in features to develop the systems more easily and quickly. Every framework has different characteristics, making them suitable for different applications.

In this research six frameworks have been chosen for evaluation based on popularity. The RNLN prefers the frameworks to be compatible with Python. These frameworks and their characteristics can be seen in Table 4.1[15].

Every FL framework can be used in combination with one or more ML frameworks. These ML frameworks are libraries with standard algorithms to create ML models. The mode of the framework can be local simulation on one system only or both simulation and federated mode. For the RNLN it was important the framework could be used in federated mode to be able to test the FL system with different laptops that serve as a central server and distributed clients. Also, the framework has to be compatible with the Operating System (OS) Linux.

The ship data of the RNLN is expressed in the form of time series (such as temperatures, engine speeds and pressures), so the framework must also be able to work with time series. DP stands for Differential Privacy and is a technology that ensures privacy of the individual data sets of clients by adding noise to the local gradients. Lastly, it was important for the continuation of the research to have enough examples of implementations online.

Table 4.1. Overview characteristics FL frameworks

	Federated Learning Framework					
	TensorFlow Federated 0.19.0	PySyft 0.6.0	Flower 0.17.0	FATE 1.7.1	OpenFL 1.2.1	PaddleFL 1.2.0
ML framework	TensorFlow	PyTorch	Any	TensorFlow	PyTorch and TensorFlow	PaddlePaddle
Mode	Local simulation only	Local simulation only	Simulation and federated	Simulation and federated	Simulation and federated	Simulation and federated
Operating System	Mac, Linux	Mac, Linux, Win, iOS, Android	Mac, Linux, Win, iOS, Android	Mac, Linux	Mac, Linux	Mac, Linux, Win
Data type	Time series and pictures	Pictures	Time series and pictures	Time series	Time series and pictures	Time series and pictures
DP	Yes	Yes	Yes	No	No	Yes
Algorithms	(A)NN, CNN, RNN, LogR, PR, LR	(A)NN, CNN, RNN, LogR, PR, LR	(A)NN, CNN, RNN, LogR, PR, LR	GBDT, (A)NN, CNN, RNN, LogR, PR, LR	(A)NN, CNN	(A)NN, CNN, RNN, LogR, PR, LR
Available resources	Basic examples in the documentation	Tutorials and examples in the documentation, blogs and community projects	Tutorials and examples in the documentation, blogs and boilerplate examples	Basic examples in the documentation	Tutorials and basic examples in the documentation	Basic examples in the documentation

In Table 4.2 the overview of the RNLN criteria can be seen. The FL framework Flower possesses all of these characteristics and is the most flexible framework. It can be used with every common ML framework, works on various OS and has the most literature available.

Table 4.2. Criteria for choosing the FL framework

Criteria	TensorFlow Federated 0.19.0	PySyft 0.6.0	Flower 0.17.0	FATE 1.7.1	OpenFL 1.2.1	PaddleFL 1.2.0
Python	✓	✓	✓	✓	✓	✓
Federated mode	✗	✗	✓	✓	✓	✓
OS Linux	✓	✓	✓	✓	✓	✓
Time series	✓	✗	✓	✓	✓	✓
Examples of implementations	✗	✓	✓	✗	✗	✗

4.5. Design of the architecture with the flower framework

Now that we have chosen Flower as the FL framework, a central server and associated clients must be set up. Many different aspects must be considered in this design, such as how the server communicates with clients, which parameters of the ML model are shared, and how they are secured.

This section first describes the general Flower framework architecture. Then we explain the design of the server and clients in this study and how this could be used in practice. Figure 4.3 shows an overview of our set-up.

The figure shows on the server side the strategy, the FL loop and the Remote Procedure Call (RPC). The strategy determines how clients are selected, the training configuration, how the model updates are aggregated and the evaluation of the model. The FL loop monitors the learning process and ensures that progress is made. It requires a strategy in order to configure a round in the FL process and sends those configurations to selected clients via the RPC server. The resulting client updates from the local training are sent back through the RPC server and aggregated based on the chosen strategy.

Clients can connect to the RPC server responsible for sending and receiving Flower protocol messages and checking connections. The client side of Flower is simpler, as it only manages its own connection to the server and responds to incoming messages by calling the programmed training and evaluation functions[16].

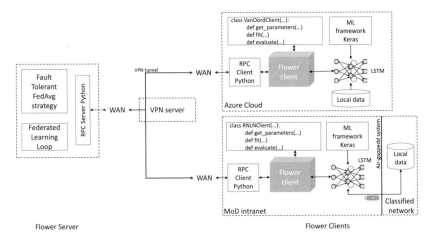

Fig. 4.3. The designed Flower framework architecture.

4.5.1. Build-in strategies

Section 2 describes the main characteristics of FL and the challenges arising from it. One of the challenges is client heterogeneity. Clients may communicate at different speeds, have different amounts of data points, or send wrong model updates to the server. Some of these challenges can be solved by choosing the right strategy depending on the amount and type of clients. Default built-in strategies within Flower are FedAvg, Fault Tolerant FedAvg, FedProx, qFedAvg and FedOptim.

Federated Averaging (FedAvg) is the standard FL strategy. It is a communication efficient algorithm in which the client first performs several updates locally and once per round the server averages, or aggregates, the model updates. Fault Tolerant FedAvg is a FedAvg variant that can deal with disconnecting or lagging clients[17].

FedProx basically works the same as FedAvg in that local updates are performed on the model and aggregated to the server once per round. It differs from FedAvg in that FedProx accepts that clients perform non-uniformly distributed amounts of work, making FedProx suitable for heterogeneous network conditions. FedOptim consists of a number of server-side optimizations for more efficient communication and consists of FedAdagrad, FedYogi and FedAdam[18].

In Section 2 the problem of bias in FL was introduced. If the aforementioned strategies are used on clients with non-i.i.d. data sets or on different clients where a certain type of device is predominant, the trained model can give a distorted picture and thus have a bias. qFedAvg is a method to solve q-Fair Federated Learning (q-FFL) and get more accurate solutions for non-i.i.d. data sets. q-FFL ensures that clients with a higher loss and higher relative weight are given less variation in the distribution of accuracy[19].

Different strategies have been tested with an LSTM model and i.i.d and non-i.i.d. data sets[20]. In case of non-i.i.d. data sets, the accuracy of qFedAvg was higher than the other strategies, but the accuracy of qFedAvg was lower than the other strategies when training with i.i.d. data sets. Van Oord's data sets have an equal number of data points and are therefore i.i.d. There is also no question of network heterogeneity in this (test) set-up. Fault Tolerant FedAvg has been chosen as the strategy because it has a higher accuracy with i.i.d. data sets and it can deal with clients that fail also by a larger number of clients.

4.5.2. Choices in the design of the federated network

The client and central server can easily communicate with each other if both are on the same Local Area Network (LAN), but if several parties, such as RNLN and Van Oord, want to train an ML model with each other, a Virtual Private Network (VPN) can be used with which they can connect remotely.

In practice, it has to be taken into account that most ship data of the RNLN is classified and stored on an air-gapped network. Two solutions are possible. Preferably, the network has to be adjusted to run Python. Once this is possible, the FL process can be set up in such a way that the model gradients are brought via a USB stick to, for example, the Ministry of Defence (MoD) intranet and then sent to the central server. Another solution is to physically transfer the data sets to the MoD intranet, thereby lowering their classification level. On the MoD intranet it is possible to train ML models with Python and connect with a VPN. Since solving this issue was beyond the scope of this research, our proof of concept only uses data from Van Oord.

At Van Oord, the data sets are not stored on an air-gapped system, but in the Azure Cloud. It is possible to train ML models in Azure Cloud and connect to a VPN. So for future cooperation, no change in infrastructure at Van Oord is required.

Wireguard was used as a VPN server in this study. Wireguard is a communication protocol with free open-source software that implements VPNs. To use this, the clients were sent configuration files, allowing them to connect to the VPN and participate in the FL process remotely.

4.5.3. Communication during the FL process

As mentioned at the beginning of this chapter, the communication of the Flower server and Flower clients goes via the RPC. It is important to think about what kind of information is being sent by each party.

Communication starts from the Flower server which selects any number of clients to run the FL process with. When this minimal number of clients is connected, the server starts FL training.

Depending on the type of strategy and the number of connected clients, client dropout and reconnection affects the FL process. Fault Tolerant FedAvg was chosen as the strategy in this study, hence client dropout and reconnection are accepted if there are enough other clients left to meet the minimum number of clients. Depending on how many parties RNLN collaborates with, this number can be adjusted.

During the FL process, the central server and the clients communicate by sending each other local and global gradients. The local gradients are the model updates from the clients and the global gradients are the server aggregated model parameters. The type of gradients and their size depend on the type of model. The LR model in this study has only one gradient (the slope of the regression line), but an NN like the LSTM in this study has many more and this means more data traffic.

In this research, the client is a Python script, where the clients themselves install the necessary libraries to run FL[21].

4.5.4. Security and privacy

FL is often used in situations where the security of data and privacy of clients is very important. Although FL solves some problems of classic ML, there are still several possibilities for attackers.

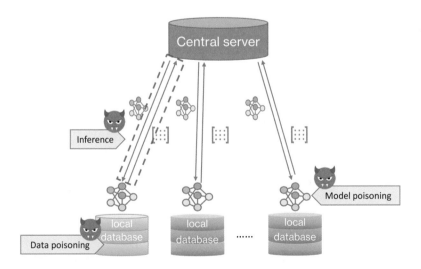

Fig. 4.4. Possibilities for attackers in the FL process[22].

Risks within FL are *poisoning* and *inference* attacks. Examples of these attacks are given in Figure 4. Poisoning can occur because a client adjusts the data set or model gradients or because the central server adjusts the global model. Inference can take place if outsiders can listen in on the data traffic between the server and clients.

Model poisoning or data poisoning is possible in FL because client data sets are not visible and clients are randomly selected. It is therefore not possible to check the quality of data sets or to check the history of a specific client. However, model poisoning or data poisoning can be prevented by regularly checking the trustworthiness of clients, for example by authentication via the VPN, and evaluating the behaviour of clients by checking for deviations.

If FL is to be used by RNLN and Van Oord, this will initially be with a small group of parties and a Trusted Third Party (TTP). This makes it easier for the central server to evaluate the reliability and behaviour of the clients. When RNLN collaborates with a larger group of external parties, random selection of clients in each round makes the effectiveness and invisibility of a poisoning attack more difficult[23,24].

In addition to poisoning, the FL process can be eavesdropped on by clients, the central server or outsiders, who thus obtain some of the local or global model gradients. With these gradients, attackers can reconstruct the data set by, for example, a Model Inversion Attack (MIA). The possibility of inference by outsiders is already reduced by using a VPN. In addition, gradient inversion methods are often unsuccessful in practice, because the architecture of the model and the global and local gradients must be known to the attacker[25,26]. If these components are all known to the attacker, *Differential Privacy* can be used to counter the attack.

4.6. Reliability and validity of the FL implementation

This section discusses the reliability and validity of the FL implementation. First, the used data sets of Van Oords's FFPV Stornes will be discussed. Next, a linear regression (LR) model was used to demonstrate the validity of the FL process. Subsequently, a script with the LSTM network was written for FL clients and a number of tests were performed to ensure the validity and reliability of the results of this model.

4.6.1. Available data of the FFPV Stornes

This section presents Van Oord's FFPV Stornes and the available data that is used in this paper for testing the FL architecture. The FFPV Stornes was designed for installation of rock on the ocean floor (in water depths up to 1500 meters) to stabilise and protect pipelines, cables and subsea templates[27]. The dataset of the

FFPV Stornes contains time series data from three of its auxiliary power generators and navigation data. Such generators are typically equipped with sensors (i.e. the engine speed, the power produced and the generator's fuel consumption) to provide for the remote monitoring and control services. The data of these generators is used for the LSTM model. The LSTM model learns the default behaviour for the generators by training on this data set. Detection of anomalies in all sensors may indicate a general problem in electrical operation, and detection of anomalies in the sensors of a specific generator may indicate a problem with this generator. The anomalies can be read in the warning table by a ship's crew or maintenance engineers ashore. This can lead to additional inspections on a generator and early detection of a problem if something is wrong with that generator. The navigation data is used to demonstrate and test whether the FL process works correctly using a LR model (see section 4.2).

The data set consists of NetCDF (NC) files, each containing data about the vessel for a time period of one day. The original frequency of the data by the monitoring system is equal to 25 points every second. In order to provide easier data handling and to train the model faster, the time interval has been reduced to 5 data points per second. No relevant information is lost in this process because the duration of anomaly is significantly longer than the new sampling rate. The available data of the vessel has been collected between 26/08/2021 and 04/09/2021. The exploited data of the FFPV Stornes is reported in Table 4.3

Table 4.3. Used sensor data of FFPV Stornes

Variable name	Unit
Timestamp	[t]
Latitude	[°]
Longitude	[°]
Heading	[°]
Heave	[°]
Roll	[°]
Pitch	[°]
Auxiliary Generators: power	[kW]
Auxiliary Generators: fuel oil consumption	[l/h]
Auxiliary Generators: engine speed	[rpm]

4.6.2. Linear Regression

The LR model was used to test the FL server and FL clients in our model. In Van Oord's data set, a linear correlation was coincidentally found between the change in latitude and longitude of FFPV Stornes, because the ship was moving more or less in a straight line. This data was therefore used as input for the LR model.

To test whether the FL process works correctly, different laptops were used with different OS (Linux, Windows and Mac) on which a data set from Van Oord and the FL client script were loaded. In the FL process, during the training of the LR model at the clients, several parameters were displayed every round on the server, and it was checked that they matched the values of the parameters at the clients.

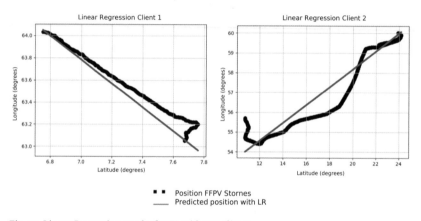

Fig. 4.5. Linear Regression results for FL with two clients.

Figure 4.5 shows an example of the output of the LR model of the two clients. These results indicate that during the FL process, the model's evaluation parameters are displayed correctly on the server and that the measurement is valid. Note that no aggregation is used here, since there was no linear correlation expected between the latitude and longitude on different days.

4.6.3. LSTM-network

An LSTM-network had already been deployed by RNLN in order to detect anomalies in sensor data of ships[28]. Our FL architecture uses an LSTM-network as an ML algorithm and uses the same hyperparameters and set-up as in previous research by Maaike Teunisse (2021)[29]. We refer there for a detailed explanation of how the LSTM-network is applied[30].

The LSTM network was first tested with a small part of the data set where there was a clear deviation in the four signals at one time point. The LSTM network is trained on the data set with the status parameters of the *heading, heave, roll* and *pitch* of FFPV Stornes. Figure 4.6 clearly shows an anomaly in the four signals at one time during the month.

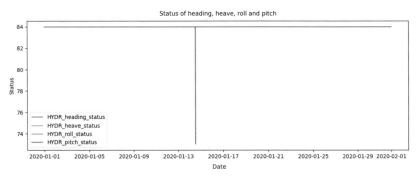

Fig. 4.6. Graph with status parameters per day of FFPV Stornes (note: dimensionless status lines are overlapping).

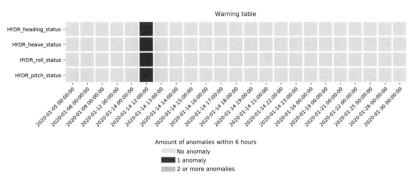

Fig. 4.7. Warning table with the detected anomalies in the data from Figure 4.6. Units as given in Table 4.3: fuel oil consumption in [l/h], engine speed in [rpm], power in [kW].

The LSTM model was tested for anomalies using the data set from Figure 4.6. The result of the training is the warning table shown in Figure 4.7. In this figure, a clear anomaly can be seen for all systems and this corresponds to the time of the deviation in Figure 4.6.

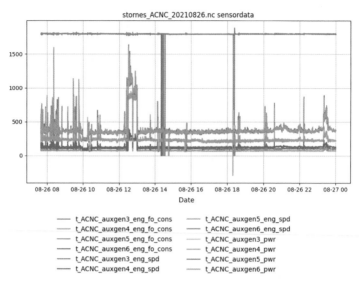

Fig. 4.8. Sensor data generators 3, 4, 5 and 6 from FFPV Stornes.

There was little variation in the sensor data in this data set, but the LSTM network successfully carried out the training. The same test was then performed on a data set with less obvious deviations. Figure 4.8 shows the signals from the data set from generators 3, 4, 5 and 6 over the period of one day. At 12:00, 14:00 and 18:00 deviations can be seen, and these deviations are reflected in the warning table in Figure 4.1.

A simpler example with one clear anomaly and an example where the anomalies were more difficult to read showed that the LSTM model is valid and the training of the model was successful.

To guarantee the validity of the study, the fit of the model was also taken into account. The appropriate parameters for the evaluation of the LSTM model have been investigated and scenarios where clients use non-i.i.d. data sets[31].

In addition to validity, a number of factors were taken into account to increase the reliability of the study. An important part of this is the repeatability of this study. During the research, the model was often trained, modified and retrained again. The functions *seed(...)* and *set_seed(...)* were used to ensure that identical results are generated for training rounds with the same software and data sets. These functions ensure that the random processes that take place in the NN can be repeated.

4.7. Results: central machine learning versus federated learning

The goal of this research is to evaluate the possibilities for RNLN to use FL for collaboration with external partners. To achieve this, we compare the LSTM model obtained from this FL set-up with central ML. For the computer code and the tables used for creating the graphs in this section, we refer to the appendices of the research of Anna Vriend (2022)[32]. To compare FL with central ML, the parameters *loss* and *accuracy* were used. In addition, the data traffic between the central server and the clients is monitored with Wireshark.

First, the loss and accuracy of a centrally trained LSTM model were obtained by training the LSTM model with one day of data from FFPV Stornes. After this, the LSTM network was trained again with FL with various laptops via a VPN, using the same data. To obtain the loss and accuracy, the global model was evaluated by the server after aggregation of the gradients. To simulate an increasing number of clients, each client had as local data a data set from FFPV Stornes of one day. The same FL process was performed in two rounds, whereby the central server thus aggregated the gradients twice. The data traffic was still below the limit of central ML.

The traffic at central ML is calculated by adding the size of the client data sets to the size of the LSTM model multiplied by the number of clients. This is done because with central ML the data sets must first be collected at a TTP and then the trained model must be distributed to all clients.

The results are summarised in two figures. Figure 4.9 shows an overview of the evaluation parameters of the LSTM model at one and two rounds of FL. The accuracy increases in both cases with a larger number of clients, and therefore more data. With one round of FL, 6 clients are needed to exceed the accuracy of central ML and with two rounds of FL, this already happens with 2 clients. The loss at two rounds of FL is lower than at one round, but in both cases the loss is not as low as at central ML.

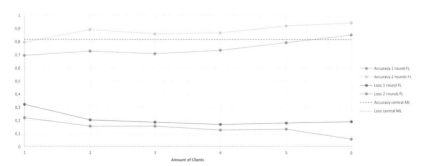

Fig. 4.9. Evaluation parameters LSTM-model 1 versus 2 rounds of FL.

In addition to the evaluation parameters for one and two rounds of FL, the data traffic between the clients and server at FL versus central ML was also examined. Figure 4.10 shows that the amount of data traffic during FL at one and two rounds remains below the amount of data traffic during central ML. The amount of data traffic during central ML increases linearly.

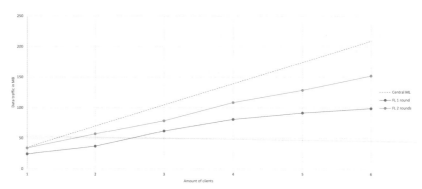

Fig. 4.10. Comparison of data traffic central ML versus FL with 1 and 2 rounds.

4.8. Conclusion

Nowadays, Machine Learning is used successfully in many applications. There are two important challenges involved with the implementation of this technique. The first is that in many industries, data is stored in separate places. The second challenge is the growing demand for ML that respects the privacy of the training data. Conventional ML is based on centralised data collection and is thus unable to face these challenges[33].

In this paper, an FL architecture has been designed using the Flower FL framework to train an LSTM model of the RNLN on data of Royal Van Oord's FFPV Stornes. Tests have been conducted in this work that show that it is possible to apply FL to the execution of this ML algorithm if the central server and the clients are laptops connected to different networks. An LR model was used to test the FL server and FL clients in the designed architecture. Next, an example with one clear anomaly and an example where the anomalies were more complex showed that the LSTM model is valid, and the training of the model was successful.

In both cases, the model accuracy increased, as expected, with a larger number of FL clients. With one round, 6 FL clients were needed to exceed the accuracy of central ML and with two rounds 2 FL clients were required. The loss at two rounds of FL was lower than at one round, but in both cases the loss was not as low as with central ML. This work herewith showed that FL with multiple clients outperforms central ML in terms of accuracy (Figure 4.9) and is therefore a valuable solution for working together on Predictive Maintenance problems. Figure 4.10 shows that the amount of data traffic during FL at one and two rounds remained below the amount of data traffic during central ML. This effect will be exaggerated when large data sets are used for training the ML model, as is often the case for Predictive Maintenance problems.

If a TTP is appointed by the RNLN and Van Oord, it is technically possible to train the discussed LSTM network using FL. The RNLN should however take into account the defence security policy before the FL client can be deployed on the MoD intranet. This security policy has not been taken into account in this study. For air-gapped or disconnected networks the FL process can be set up in such a way that the model gradients are brought via a data carrier (i.e., a USB drive) to the central server. This can be relevant for networks such as the highly classified MoD networks, but also networks aboard warships.

Following the results of this study it can be concluded that it is relevant for the RNLN and Royal Van Oort to use FL. FL is a way to achieve using more data for training their ML models without the need to share their raw data. This leads, in the generic sense, to better predictions, as the tests carried out in this work have confirmed. The RNLN's LSTM model, which was able to detect anomalies in sensor data of Van Oord's FFPV Stornes, has been trained with data sets from only one ship. In practice, value lies in collaboration with multiple parties that have different (types of) ships. It should be taken into account that anomalies in the sensors used in this study may manifest themselves differently on other ships. If the model from this research is to be used, it is important that equivalent data is available and sensor data from similar types of components is used.

Finally, opportunities for FL are recognised in applying FL for one of the parties' own fleet, the ships will be the clients and a server on shore is the central server.

Notes

[1]   Peter Hulsen, "Wat is (het verschil tussen) Artificial Intelligence, Machine Learning en Deep Learning," last modified August 7, 2017, https://cqm.nl/nl/nieuws/wat-is-het-verschil-tussen-artificial-intelligence-machine-learning-en-deep-learning.

[2]   H. Brendan McMahan, Eider Moore, Daniel Ramage, Seth Hampson and Blaise Agüera y Arcas, "Communication-Efficient Learning of Deep Networks from Decentralized Data," last modified February 28, 2017, https://arxiv.org/pdf/1602.05629.pdf.

[3]   Cem Dilmegani, "What is Federated Learning?" last modified February 9, 2022, https://research.aimultiple.com/federated-learning/.

4   Wieger Tiddens, Jan Braaksma and Tiedo Tinga. "Exploring predictive maintenance applications in industry." Journal of quality in maintenance engineering (2020).

5   Tiedo Tinga, Axel Homborg and Chris Rijsdijk. "Data-driven maintenance of military systems – potential and challenges." Netherlands Annual Review of Military Studies (2022).

6   Maaike Teunisse, "Anomalie detectie voor voorspelbaar onderhoud" (bachelor's thesis, Hogeschool van Amsterdam, 2021).

7   Yong Cheng, et al. "Federated learning for privacy-preserving AI," Communications of the ACM 63, no. 12 (2020): 33-36.

8   Keith Bonawitz, Hubert Eichner, Wolfgang Grieskamp, Dzmitry Huba, Alex Ingerman, Vladimir Ivanov, Chloé Kiddon, Jakub Konecny, Stefano Mazzocchi, H. Brendan McMahan,Timon Van Overveldt, David Petrou, Daniel Ramage and Jason Roselander, "Towards Federated Learning At Scale: System Design," last modified on March 22, 2019, https://arxiv.org/pdf/1902.01046.pdf.

9   Keith Bonawitz, Hubert Eichner, Wolfgang Grieskamp, Dzmitry Huba, Alex Ingerman, Vladimir Ivanov, Chloé Kiddon, Jakub Konecny, Stefano Mazzocchi, H. Brendan McMahan,Timon Van Overveldt, David Petrou, Daniel Ramage and Jason Roselander, "Towards Federated Learning At Scale: System Design," last modified on March 22, 2019, https://arxiv.org/pdf/1902.01046.pdf.

10   Tim d'Hondt, "Federated Learning over Local Learning: an Opportunity for Collaboration" (master's thesis, Eindhoven University Of Technology, 2020), 16-24.

11   Yong Cheng, et al. "Federated learning for privacy-preserving AI," Communications of the ACM 63, no. 12 (2020): 33-36.

12   Ibid.

13   Tian Li, Ameet Talwalkar, Anit Kumar Sahu and Virginia Smith, "Federated Learning: Challenges, Methods, and Future Directions," last modified August 21, 2019, https://arxiv.org/pdf/1908.07873.pdf.

14   Annie Abay, Yi Zhou, Nathalie Baracalo, Rajamoni Shashank, Ebube Chuba and Heiko Ludwig, "Mitigating Bias in Federated Learning," last modified on December 4, 2020, https://arxiv.org/pdf/2012.02447.pdf.

15   Anna Vriend, "Federated Learning: Onderzoek naar de toepassing van privacy vriendelijke Machine Learning voor de Koninklijke Marine" (bachelor's thesis, Netherlands Defence Academy, 2022).

16   Daniel J. Beutel, Taner Topal, Akhil Mathur, Xinchi Qui, Titouan Parcollet, Pedro Porto Buarquede de Gusmao and Nicolas D. Lane, "Flower: A Friendly Federated Learning Framework," last modified on April 7, 2021, https://akhilmathurs.github.io/papers/beutel_flower2020.pdf

17   Daniel J. Beutel, Taner Topal, Akhil Mathur, Xinchi Qui, Titouan Parcollet, Pedro Porto Buarquede de Gusmao and Nicolas D. Lane, "Flower: A Friendly Federated Learning Framework," last modified on April 7, 2021, https://akhilmathurs.github.io/papers/beutel_flower2020.pdf

18   Sahank J. Reddi, Zachary Charles, Manzil Zaheer, Zachary Garrett, Keith Rush, Jakub Konecny, Sanjiv Kumar and H. Brendan McMahan, "Adaptive Federated Optimization," last modified September 8, 2021, https://arxiv.org/pdf/2003.00295.pdf.

19   Tian Li, Maziar Sanjabi and Virginia Smith, "Fair Resource Allocation in Federated Learning," last modified May 25, 2019, https://arxiv.org/pdf/1905.10497v1.pdf.

20   Ajinkya Mulay, Baye Gaspard, Rakshit Naidu, and Santiago Gonzalez-Toral, Vineeth S., Tushar Semwal and Ayush Manish Agrawa, "FedPerf: A Practioners' Guide to Performance of Federated Learning Algorithms," last modified December 11, 2020, https://proceedings.mlr.press/v148/mulay21a/mulay21a.pdf.

21   The Python version used is 3.9 and the Flower version is 0.17.0.

22 Pengrui Liu, Xiangrui Xu, Wei Wang, "Threats, attacks and defenses to federated learning: issues, taxonomy and perspectives," last modified February 2, 2022, https://cybersecurity.springeropen.com/articles/10.1186/s42400-021-00105-6.

23 Pengrui Liu, Xiangrui Xu, Wei Wang, "Threats, attacks and defenses to federated learning: issues, taxonomy and perspectives," last modified February 2, 2022, https://cybersecurity.springeropen.com/articles/10.1186/s42400-021-00105-6.

24 Xingchen Zhou, Ming Xu, Yiming Wu and Ning Zheng, "Deep Model Poisoning Attack on Federated Learning," last modified March 14, 2021, https://www.mdpi.com/1999-5903/13/3/73/htm.

25 Briland Hitaj, Giuseppe Ateniese, and Fernando Perez-Cruz, "Deep Models Under the GAN: Information Leakage from Collaborative Deep Learning," last modified September 14, 2017, https://arxiv.org/pdf/1702.07464.pdf.

26 Ali Hatamizadeh, Hongxu Yin, Pavlo Molchanov, Andriy Myronenko, Wenqi Li, Prerna Dogra, Andrew Feng, Mona G. Flores, Jan Kautz, Daguang Xu and Holger R. Roth, "Do Gradient Inversion Attacks Make Federated Learning Unsafe?" last modified February 14, 2022, https://arxiv.org/pdf/2202.06924.pdf.

27 Van Oord, "Flexible fallpipe vessel", accessed April 26, 2022, https://www.vanoord.com/en/equipment/flexible-fallpipe-vessel/.

28 Maaike Teunisse, "Anomalie detectie voor voorspelbaar onderhoud" (bachelor's thesis, Hogeschool van Amsterdam, 2021).

29 Ibid.

30 Ibid.

31 Anna Vriend, "Federated Learning: Onderzoek naar de toepassing van privacy vriendelijke Machine Learning voor de Koninklijke Marine" (bachelor's thesis, Netherlands Defence Academy, 2022).

32 Anna Vriend, "Federated Learning: Onderzoek naar de toepassing van privacy vriendelijke Machine Learning voor de Koninklijke Marine" (bachelor's thesis, Netherlands Defence Academy, 2022).

33 Yong Cheng, et al. "Federated learning for privacy-preserving AI," Communications of the ACM 63, no. 12 (2020): 33-36.

References

Abay, Annie, Yi Zhou, Nathalie Baracalo, Rajamoni Shashank, Ebube Chuba, and Heiko Ludwig, "Mitigating Bias in Federated Learning." Last modified on December 4, 2020. https://arxiv.org/pdf/2012.02447.pdf.

Beutel, Daniel J., Taner Topal, Akhik Mathur, Xinchi Qui, Titouan Parcollet, Pedro Porto Buarquede de Gusmao, and Nicolas D. Lane. "Flower: A Friendly Federated Learning Framework." Last modified on April 7, 2021. https://akhilmathurs.github.io/papers/beutel_flower2020.pdf

Bonawitz, Keith, Hubert Eichner, Wolfgang Grieskamp, Dzmitry Huba, Alex Ingerman, Vladimir Ivanov, Chloé Kiddon, Jakub Konecny, Stefano Mazzocchi, Brendan H. McMahan, Timon Van Overveldt, David Petrou, Daniel Ramage, and Jason Roselander. "Towards Federated Learning At Scale: System Design." Last modified on March 22, 2019. https://arxiv.org/pdf/1902.01046.pdf.

Cheng, Yong, Yang Liu, Tianjian Chen, and Qiang Yang. "Federated learning for privacy-preserving AI." Communications of the ACM 63, no. 12 (2020): 33-36.

D'Hondt, Tim. "Federated Learning over Local Learning: an Opportunity for Collaboration." Master's thesis, Eindhoven University Of Technology, 2020.

Dilmegani, Cem. "What is Federated Learning?" Last modified February 9, 2022. https://research. aimultiple.com/federated-learning/.

Hatamizadeh, Ali, Honxu Yin, Pavlo Molchanov, Andiry Myronenko, Wenqi Li, Prerna Dogra, Andrew Feng, Mona G. Flores, Jan Kautz, Daguang Xu, and Holger R. Roth. "Do Gradient Inversion Attacks Make Federated Learning Unsafe?" Last modified February 14, 2022. https://arxiv.org/ pdf/2202.06924.pdf.

Hitaj, Briland, Giuseppe Ateniese, and Fernando Perez-Cruz. "Deep Models Under the GAN: Information Leakage from Collaborative Deep Learning." Last modified September 14, 2017. https://arxiv. org/pdf/1702.07464.pdf.

Hulsen, Peter. "Wat is (het verschil tussen) Artificial Intelligence, Machine Learning en Deep Learning." Last modified August 7, 2017. https://cqm.nl/nl/nieuws/wat-is-het-verschil-tussen-artificial-intelligence-machine-learning-en-deep-learning.

Li, Tian, Maziar Sanjabi, and Virginia Smith. "Fair Resource Allocation in Federated Learning." Last modified May 25, 2019. https://arxiv.org/pdf/1905.10497v1.pdf.

Li, Tian, Ameet Talwalkar, Anit Kumar Sahu, and Virginia Smith. "Federated Learning: Challenges, Methods, and Future Directions." Last modified August 21, 2019. https://arxiv.org/pdf/1908.07873. pdf.

Liu, Pengrui, Xiangrui Xu, and Wei Wang. "Threats, attacks and defenses to federated learning: issues, taxonomy and perspectives." Last modified February 2, 2022. https://cybersecurity.springeropen. com/articles/10.1186/s42400-021-00105-6.

McMahan, H. Brendan, Eider Moore, Daniel Ramage, Seth Hampson, and Blaise Agüera y Arcas. "Communication-Efficient Learning of Deep Networks from Decentralized Data." Last modified February 28, 2017. https://arxiv.org/pdf/1602.05629.pdf.

Mulay, Ajinkya, Baye Gaspard, Rakshit Naidu, Santiago Gonzalez-Toral, Vineeth S. Tushar Semwal, and Ayush Manish Agrawa. "FedPerf: A Practioners' Guide to Performance of Federated Learning Algorithms." Last modified December 11, 2020. https://proceedings.mlr.press/v148/mulay21a/ mulay21a.pdf.

Phi, Michael. "Illustrated Guide to LSTM's and GRU's." Last modified September 24, 2018. https:// towardsdatascience.com/illustrated-guide-to-lstms-and-gru-s-a-step-by-step-explanation-44e9eb-85bf21.

Reddi, Sahank J., Zachary Charles, Manzil Zaheer, Zachary Garrett, Keith Rush, Jakub Konecny, Sanjiv Kumar, Brendan H. McMahan. "Adaptive Federated Optimization." Last modified September 8, 2021. https://arxiv.org/pdf/2003.00295.pdf.

Van Oord. "Flexible fallpipe vessel." Accessed April 26, 2022. https://www.vanoord.com/equipment/ flexible-fallpipe-vessel/.

Vriend, Anna. "Federated Learning: Onderzoek naar de toepassing van privacy vriendelijke Machine Learning voor de Koninklijke Marine." Bachelor's thesis, Netherlands Defence Academy, 2022.

Teunisse, Maaike. "Anomalie detectie voor voorspelbaar onderhoud." Bachelor's thesis, Hogeschool van Amsterdam, 2021.

Tiddens, Wieger, Jan Braaksma, and Tiedo Tinga. "Exploring predictive maintenance applications in industry." Journal of quality in maintenance engineering (2020).

Tinga, Tiedo, Axel Homborg and Chris Rijsdijk. "Data-driven maintenance of military systems – potential and challenges." Netherlands Annual Review of Military Studies (2022).

Zhou, Xingchen, Ming Xu, Yiming Wu, and Ning Zheng. "Deep Model Poisoning Attack on Federated Learning." Last modified March 14, 2021. https://www.mdpi.com/1999-5903/13/3/73/htm.

CHAPTER 5

Information- and data-driven organisations from promise to practice?

Reflecting on maturity dynamics in a defence sustainment organisation

Gert Schijvenaars, André J. Hoogstrate, Ton van Kampen, Gerold de Gooijer & Paul C. van Fenema

Abstract

In today's increasingly digitised world, data is the common denominator in firms and public organisations like the Ministry of Defence (MoD). The Netherlands MoD (NLMoD) declared in its long-term vision that it wants to be an information- and data-driven organisation. How realistic is this vision, and what is necessary to really become information- and data-driven? Current literature provides limited answers to these questions. This chapter elaborates on the use of data-driven maturity models to be able to assess the current situation of the NL MoD sustainment organisation, and to explore what should be done to become a more information- and data-driven organisation. Our results contribute to literature on strategising on maturity evolution, and they offer professionals actionable insights for changing their organisation.

Keywords: Data-driven, Maturity model, Data maturity assessment, Defence sustainment, Sustainment organisation

5.1. Introduction

In today's increasingly digitised world, data is the common denominator. Leading organisations are turning data into valuable insight and powerful capabilities, not just enabling but driving strategy and decision-making.[1] However, even though leading organisations increasingly rely on data, others struggle to capture value from data and fail to fulfil this data-driven 'dream.'[2] We note that various communities refer to 'data- and/or information-driven.' Data per se does not improve organisations but it provides a required foundation for information. Data can be defined as 'a symbol set that is quantified and/or qualified', while information is 'a set of significant signs that has the ability to create knowledge.'[3] We use in this chapter 'data- and/ or information-driven' interchangeably to relate to various academic literatures and professional communities, noting that conceptual clarity will be required as the field matures.

Researchers and consultants tend to believe that strategy development and execution based on pivotal elements like an organisation's own mission and goals, analysis of the environment, and intra-organisational optimisation and adjustment are of significant importance for business continuity. Based on tools like McKinsey's 7S model,[4] organisations have invested for decades tremendous resources into strategy development and execution, the renewal of business processes and the improvement of their information systems to enact competitive advantage. Correct and in-time business decisions were and are crucial to remain competitive for which reliable, accurate and punctual information needs to be provided. Information processing has been acknowledged since the seventies as the lifeline of organisations.[5]

Especially over the last decade, the societal arena has changed significantly because of the *"continuous development and penetration of digital technologies that lead to disruptive changes in the economy and society that enable novel ways of leveraging data for optimizing business processes and design innovative data-driven business models"*[6]. These trends have increased pressure on organisations to become more data driven. In literature, these organisations are called Data-Driven Organisations (DDOs), with several definitions in use. Hupperz et al. (2021), for instance, present a high-level description of DDOs using five key elements to describe such an organisation: Digital Transformation, Data Science, Data-Driven Business Model, Data-Driven Innovation and Data Analytics. Decision-making is a pivotal concept in DDO, with data-driven decision making described as 'multiple forms of data (that) are first turned into information via analysis and then combined with stakeholder understanding and expertise to create actionable knowledge.'[7]

The wish to become a data-driven organisation is often expressed. Yet many organisations fail to set steps to become a data-driven organisation because they often do not realise that DDO-transformation requires change of all relevant aspects of the organisation in a balanced way, as for example was illustrated by the 7S model of McKinsey or five key elements of DDOs.[8] Therefore, there is a gap between on the one hand expressed intention, and on the other hand acting. Organisations often lack the knowledge to determine their direction (what are we aiming for?), and to determine how to transform the organisation in order to materialise 'potential for improvement.'[9] *The objective of this chapter is to examine the relationship between organisational goal setting pertaining to becoming data-driven, maturity in this realm, and process conceptualisation of actionable-balanced change.* We draw on digital maturity theory and examine a Defence sustainment organisation as an illustration. This organisation includes maintenance, logistics, supply chain management, transport management, purchasing management, asset management and lifecycle management.

Maturity models have been used extensively to guide organisations toward domain maturity.[10] With a maturity model, we can establish a desired to-be situation based on known experience and best practices. A maturity assessment then positions an organisation on the maturity model and determines the as-is situation of the organisation. When we have established the current state and the future direction, we can provide relevant information about the current challenges and opportunities. A maturity report provides information that can be used as the fundamentals of a strategic plan to become a mature data-driven organisation. Next, we examine theory on data-driven maturity, followed by an illustration concerning a Defence sustainment organisation.

5.2. Data-driven maturity models

5.2.1. Conceptual clarity across communities

Multiple professional and academic communities use concepts like 'intelligence' and 'data-driven,' representing at least a major hermeneutical problem. The (evolving) meaning of these – very same – words seems to differ between organisations and practices/communities. We therefore briefly clarify our understanding of the concepts and communities mentioned to make the positioning of our chapter more precise.

'Multilevel decision making' is a framework that resonates across communities; it therefore serves as our starting point. Organisations and the value chains in which they are embedded require decision making at strategic, tactical, and operational levels (in military thinking the latter two terms should be exchanged) (Figure 5.1). Note that multilevel thinking and organisation-vs-chain thinking represent two different units of analysis.

At the operational level, public organisations are responsible for security operations in a particular environment; it is their primary task. We can then distinguish two broad communities that span military business and war studies (depicted with grey and orange boxes).

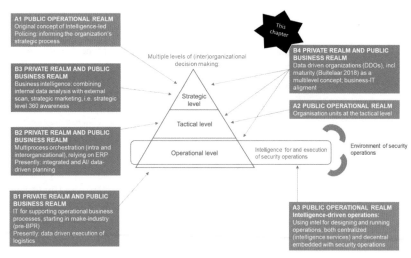

Fig. 5.1. Clarifying intelligence and data-driven concepts across communities.

–   On one hand, community A (grey) is geared towards security. Within this community, A1 depicts original strategic Intelligence-led Policing thinking that has focused on strategically orienting an organisation's activities: 'holding out the promise of a more objective basis for deciding priorities and resource allocation'.[11] At the tactical level, MoD's HQ and service branches have central-ised organisational units (A2). Shifting to the operational level, intelligence – defined as actionable information – increasingly plays a significant role in enabling or even initiating-steering operations,[12] including operational 'IGO' (Informatiegestuurd Optreden) at the NL Military Police[13] (A3).

–   On the other hand, community B concerns commercial organisations as well as business-like aspects of public organisations (including those focused on secu-rity). The advent of business computers spurred in the 1990s a new interplay of business process redesign and information technology (B1), starting bottom up rather than top down as mentioned above for A.[14] Advances in computer and software capabilities expanded this approach to tactical (e.g. Enterprise Resource Planning, control towers, managing multiple business processes using dashboards,[15] B2) and strategic levels (updating organisational position-ing, business model, B3). Intelligence in the B community is predominantly reserved for strategic level decision-making.[16] We note that at present, data-driven is still mainly focused on operational level topics like process mining (B1), with B2 showing gradual advances towards multi-process analytics and performance management 'on top of' B1. This has to do with the availability of data and algorithms. Recently, a multilevel approach is emerging embracing the notion of information and data-driven as an organisational transformation concept, building on earlier work on business-IT alignment[17] (B4).

Our chapter is positioned in the B-tradition, focusing on B4 for military sustainment organisations. At the same time, we are aware of the 'A-tradition' and carefully use resources emanating from different communities. We note that strategic level documents from the NL MoD tend to combine operational A3 innovations with future-oriented ideas on strategic level change[18] (A1 and B4) and sometimes micro innovations (B1).[19]

5.2.2. Data-driven decision making: why, how

Organisations increasingly seek to improve their insight in operational-level processes, shifting from independent (sub)processes towards multiple interdependent processes. These efforts resonate with a simultaneous shift in analytics from description, towards prediction and prescription.[20]

An important aspect of being data-driven as an organisation is that decision-making is supported by data-backed intelligent information instead of intuition, and that data-driven activities are fully integrated into business processes not only on the production level, but also on tactical and maybe in the future, strategic business levels. Data-driven decision-making increases the ability to understand patterns from the past and to predict future demand and operations, and therefore to optimise these in terms of, for instance, resource.[21] It extends traditional ideas on organisation science on information processing.[22]

Research has shown that companies that emphasise data-driven decision-making show higher performance.[23] Decision-making is a multi-level concept in modern organisations and therefore most employees, not just business analysts or data scientists, must be empowered to understand the possibilities of exploring and exploiting data to develop data-driven decision-making capabilities.[24]

Though one might expect that faster access to, and implementation of, better and broader information positively affects business decisions, this is not easily determined. This effect seems more straightforward in a military operational context as epitomised by the OODA loop (A3 in Figure 5.1).[25] For other contexts, it is difficult to assess and/or measure the influence of information access on business results. This cannot be directly measured so far. As an intermediate step, we can use maturity models to examine the extent to which organisations are or are becoming more data driven.

5.2.3. Maturity models

Maturity models define levels of definition, efficiency, manageability, and measurement of the monitored environment. A universally used maturity model is Gartner's Maturity Model for Business Intelligence and Performance Management.[26] This model offers a non-technical view and discusses maturity in terms of business-technical aspects that can help identify bottlenecks, encourage discussions between departments and thus help improve maturity. Such improvement may concern the general maturity level of the organisation, the maturity level of individual departments and/or business units, and/ or one or more areas defined by maturity models.

Guidelines are represented as a framework, defining layers and components that need to be integrated and aligned to generate a better-defined strategic vision and plan for implementing it.[27]

Buitelaar[28] redesigned Gartner's maturity model and developed a new maturity model geared towards organisations interested in becoming more data driven. For this maturity model, the design approach described by Hevner, March, and Park (2004)[29] is followed. It is also in accordance with the research of Becker, Knackstedt, and Pöppelbuß (2009)[30] who published a paper on the development of maturity models for IT management and proposed a procedural model for the development.

Dimensions. Most maturity models are focused on analytics as an isolated activity within the organisation. The relation between analytics and the organisation is often left out or implicitly included in some dimensions. Therefore, Buitelaar's model includes two more dimensions: Integration & Empowerment. These dimensions are aimed to position analytics as an activity within the organisation as a whole. Integration is the concept of integrating analytically produced insight into business processes. Empowerment is the concept of empowering members and products of the organisation using analytical techniques and data.

The complete list of dimensions then becomes as follows:

- **Data.** The fuel for all data-driven activities. How do you source and manage your data?
- **Metrics.** The key to measuring output and managing performance. How do you use, collect, and enrich your KPIs?
- **Skills.** Essential for operating a data-driven organisation. Do you hire and educate the right people?
- **Technology.** The foundation for a data-driven organisation. What technology do you need to build an analytical process?
- **Leadership.** The cornerstone for a successful analytical transformation. How does leadership successfully steer the transformation?

- **Culture.** The driving force behind a data-driven organisation. How does culture affect and promote data-driven adoption?
- **Strategy.** The plan for success. What role does analytics have in your plans and vision of the future?
- **Agility.** The ability to adapt and deliver. How well are your roles and processes organised to change and deliver?
- **Integration.** The integration of analytical insight into processes. How is the organisation using and integrating analytical output?
- **Empowerment.** The empowerment of the organisation. How is data analytics helping your employees and products to succeed?

Data-driven organisations do not isolate analytics in a certain process or department. The entire organisation should be better equipped for success through the power of data and analytics.

Maturity stages. While the dimensions help organisations distinguish areas available for interventions, stages are required to conceptualise process. Five maturity stages pertaining to becoming data-driven are distinguished in a normative sense, i.e. they are believed to provide a developmental pathway for organisations (Figure 5.2).

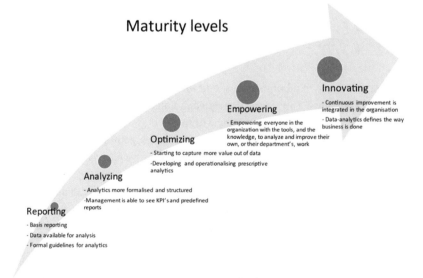

Fig. 5.2. Different maturity levels based on Buitelaar (2018).

*Table 5.1. Maturity stages related to the **data** dimension of Buitelaar's (2018) model (modified)*

		REPORTING	ANALYZING	OPTIMIZING	EMPOWERING	INNOVATING
		Visualize existing data and create the foundation for an analytical future.	Dive deeper into the data to achieve insight into why things happened.	Optimize business processes by bringing analytical insight to operations.	Empowering employees by providing the tools and knowledge to perform analytical activities.	Use data and experiments to innovate in products and transform the organization.
Data		Determine which data needs to be visible. Source and report critical data.	Determine which data is relevant for deeper analyses. Collect and analyze relevant data. Monitor and ensure data quality. Define a strategy for keeping data secure and private.	Collect data like user behavior on a bigger scale. Use customer data to personalize the experience. Focus on data quality.	Look for new data opportunities, including from third party sources. Create a clear data governance strategy that scales	Continue to look for new data sources from new products. Leverage alternative unstructured data sources, such as voice and images.

These stages apply to all dimensions, and they convey both business value and analytical capability. The stages also reflect a holistic and step by step developmental view on data-driven activities within the organisation. For instance, in stage three, the importance of *optimising* processes with analytical insight is stressed (a business process angle), while in stage four *empowering* members of the organisation shifts the emphasis to employees. These two stages then culminate in an *innovating* organisation based on data and analytics, with a workforce capable and willing to keep on experimenting and improving a data-driven approach to organisational work. Table 5.1 shows that Buitelaar's (2018) maturity model is useful for analysing the situation within (mostly large) organisations or partnerships. However, he seems to be less clear about (1) the desired maturity depending on the type of business/ industry, and (2) the extent to which an organisation has partially or fully reached a maturity level.[31] In a normative sense, maturity improvement applies to an organisation as a social system at large, not specific organisation units or processes possibly advancing more rapidly than the rest of the organisation.

Following the model, one can state that the ideal situation is that an organisation has reached a certain maturity level across its entire 'width,' but it is also conceivable that a certain independent part of an organisation has a much higher maturity level than the rest of the organisation. Hence, different normative conceptualisations of maturity dynamics can be distinguished.

Next, we examine the usefulness of these concepts in an empirical setting: the Dutch MoD sustainment organisation.

5.3. Long-term Netherlands "Defensievisie 2035" on becoming information and data-driven

The Dutch Defence organisation (NLMoD) presented in 2020 a new long term, 15-year strategy to address the envisioned national and international security threats called "Defensievisie 2035" (Defence Vision): "Vechten voor een veilige toekomst" ("Fighting for a secure future").[32] This vision describes the envisioned characteristic properties of the organisation in 2035 and ten guiding principles that will embody these organisational properties. The envisioned organisational properties and guiding principles are given in Table 5.2.

Table 5.2. NL MoD Defensievisie 2035: envisioned organisational properties and guiding principles[33]

Organisational properties		
Technologically advanced	**Information-driven**	**Reliable partner and protector**
Principle	**Principle**	**Principle**
Unique individuals and labour extensive capacities	Authoritative information position	Transparent and visible in an engaged society
Flexible operations, rapid response, scalable, independent	Multidomain and integrated operations	Strive for a stronger self-reliant Europe
Strong innovative capacity		Strive for specialisation within NATO and Europe
Escalation dominance (in cooperation with partners)		Strategic capabilities for a resilient society

A further and refined interpretation of the vision can be found in the NLMoD research agenda[34] that focuses on future, vision driven knowledge and innovation: "The Defence Vision places a strong emphasis on strengthening the innovative capacity of the Ministry of Defence. The research agenda focuses on relevant (disruptive) technologies and research needed to support the goals of the NLMoD Defensievisie 2035. In the two mentioned documents as well as not publicly published MoD-documents we notice an understanding that there is a difference between an information-driven *organisation* and information-driven *action*. The difference indicates that in the NLMoD Defensievisie 2035 there is a distinction between:

- Managing and operating the organisation, i.e., creating policy and business management aspects, and
- Operational forces and their primary processes.

Public appearances by leading officers confirm this vision. Commander of the Netherlands Defence Forces General Eichelsheim stated during the MoD strategic meeting (8th October 2021, translation by the authors):

'Focus points information-driven and digitised. Within strategic competition, digitisation is a rallying theme. Data is the foundation for our business operations and operations. This calls for modern and secure means of communications and IT, both in The Netherlands and abroad. On the battlefield, our people can make the difference because in addition to their human side (abilities), they bring something else: data. Data works for them; it gives them an edge in the strategic competition and the power to make decisions with important consequences.'

The interpretation of the information-driven organisation, in the philosophy of fact-based management, is directly tied with the data-driven philosophy (the existence, definition, conceptualisation, methods, knowledge possibilities, truth standards, and practices of working with data and information)[35] in the private and non-defence government sector (see also Figure 5.1). The information-driven action is much more tied to concepts as "intelligence led" in for instance policing (Information/Intelligence Led Policing) (A1 and A3 in Figure 5.1). This becomes clear given the interpretation of information-driven action: "Achieving the right effect in the right place by applying information smarter, faster and more powerful than our opponents."[36]

In the analysis in this chapter, we focus on the developments concerning sustainment and being or becoming data-driven in sustainment. We give a brief description of the developments since 2003, the current status and comment on some of the proposed or already started actions to reach the goals set in the NLMoD "Defensievisie 2035".

5.4. Dutch defence sustainment: from information management towards information and data-driven

The NLMoD sustainment organisation took its first steps into a more information and data-oriented approach with the decision to introduce an ERP-system in 2003 (B2 in Figure 5.1)[37]. The main goals of the transformation were: (1) more effective equipment logistics support for (joint) operations; (2) support the new Defence management model; 3) more efficiency; (4) improve the management of the information provision by replacing numerous legacy systems with one ERP system.

After the implementation in 2013, an evaluation of the ERP-migration project SPEER (Strategic Process Enabled ERP Re-engineering)[38] took place in 2017. The evaluation concluded:

"We are therefore not surprised that Defence is still at the beginning of a growth curve towards maturity in working with ERP. In various places in the organization an upward trend is visible as well as a rising enthusiasm in the use of ERP."

The report also concluded that most of the main goals that were formulated when the SPEER project started were not yet fully realised at that time. However, the implementation of the ERP-system in 2013 and the subsequent system development and migration to the ERP-system of almost all article categories and main sustainment processes within the NLMoD lead to the conclusion that its ERP-system is the leading system guiding the business process support within the sustainment organisation. Looking at today's situation, NLMoD is preparing itself for the transformation to a more information- and data-driven organisation[39].

While the new data-driven strategy was formulated, three important building blocks of the future sustainment data infrastructure that are going to dominate the next 15 years or more have been (partially) designed. The building blocks are projects to rebuild the aging IT-Infrastructure, the implementation of a new iteration of the ERP-system, and a newly to be designed HRM-architecture which will be integrated in the earlier mentioned new ERP -system. For the NLMoD sustainment organisation, this implies that most of the sustainment processes will be supported by the new S4/HANA ERP[40]-system from 2025 onwards.

Considering the efforts needed in time, capacity, knowledge, and experience to transform these building blocks during the next five years into fully deployed IT-systems that can support the goals set in the NLMoD Defensievisie 2035, the question whether there is a foreseen transformation growth path to reach the demands of the NLMoD Defensievisie 2035 should be answered. Have we learned from previous experiences and other organisations? Is the ambition of NLMoD to become an information and data-driven organisation, considering the understanding that successful organisations transform holistically in well-defined (in terms of goals, time, and budget) steps to a certain data-driven maturity level, realistic without a well-defined transformation plan? While the full transformation trajectory plan is not ready yet, one may wonder whether the building blocks presently defined suffice for reaching the next levels of maturity.

In relation to the ambitions to migrate the current ERP system SAP R3 to the new generation ERP system SAP S4/HANA the following questions are relevant. Is the S4/HANA ERP-system enabling the organisation to reach information data-driven maturity and to what extent? Is the transition path sufficiently clear? Are the ambitions realistic considering the mandatory development phases, i.e., different maturity stages? Finally, what would be realistic steps to climb the ladder? To answer these questions,

- We analysed the final targets as laid out in the NLMoD Defensievisie 2035,[41]
- Mapped these targets to the maturity model of Buitelaar (2018),
- Derived the maturity stage of being "data-driven" of the sustainment organisation,[42,43]
- Considered the steps to be taken to get from the current level (and phase) to the end goal, and
- Compared the current actions defined to climb to the next level with the path set out by the NLMoD documents at our disposal.

5.5. The maturity assessment of the current sustainment organisation

We start out by assessing the current maturity phase of the NLMoD sustainment organisation as DDO. The model by Buitelaar, as previously discussed, has ten different dimensions, and recognises five distinct stages of development. To be able to assess a current or future data-driven maturity stage of an organisation, Buitelaar has specified the ten dimensions in two or more relevant factors. For instance, the dimension DATA is constructed from the factors "data analytics storage," "source variety" and "governance." By scoring the level of use, integration, being up-to-date comparative to competitors etc. and combining the weighted scores leads to the total dimension score[44]. To derive the score for the NLMoD sustainment organisation we used the tool by Buitelaar[45]

The answers to the questions to assess the factor and dimension scores were scored based on the available documents, and experience of the research team within the NLMoD sustainment organisation. Many of the answers are related to the recognised stages of development of the implementation of the current NLMoD ERP-system, processes and other systems that are used to store and analyse data in the sustainment organisation. In this way, it is possible to estimate the data-driven maturity of our sustainment organisation.

The assessment of the current level of maturity of the NLMoD sustainment organisation shows us the following diagram and is in line with the evaluation report of the ERP-migration in 2017:

Fig. 5.3. Maturity assessment of the NL MoD sustainment organisation by the authors based on Buitelaar's (2018) assessment model.

The overall score is 1.8 out of 5. Looking at Figure 5.3, NLMoD is in the "analysing" phase implying that NLMoD is currently considering diving deeper into the data to achieve insight into why things happened and tries to make a start by optimising business processes by integrating analytical insights. In analytics terms, the organisation is at the crossroad of going from descriptive to analysis. The recently developed "Material and Personal Readiness" dashboard used by the Director Control of Operational Readiness at the Defence Staff is a good example of this maturity phase.

5.6. Maturity level of the future sustainment organisation

There are a lot of similarities between the NLMoD sustainment organisation and large public and commercial organisations. This becomes apparent when analysing the long-term goals as mentioned in the NLMoD "Defensievisie 2035" and the functionality requirements of the new S4/HANA ERP-system. This implies that we may assume – we cannot derive it from current sources – that the main underlying driving factor is algorithmic data-driven model evaluation incorporated in all or almost all decision-making processes within the sustainment organisation. Realising this and mapping the envisioned end-state of 2035 to the Buitelaar maturity model, the implication is that all ten dimensions should attain maturity level "innovating", or "empowering" at least.

Given the result of the analysis for the current state, "analysing," a transformative growth path must be formulated for all ten dimensions. It is known that an organisation must go through a development of all the different lower maturity levels to be able to reach a higher level of maturity. Experience shows that to transform from one level to another usually takes several years as learning takes time and stages cannot be skipped.[46]

This points to the risk of taking a correct step at the wrong time, which does not lead to the proposed effect and can be very costly. For example, in the "innovation" maturity level, one needs innovative oriented individuals in the workforce, while in the level "analysing" these individuals cannot do more or less than routinely executing their tasks, so in the analysing phase you "just build the damn thing." Training and hiring the wrong type of individuals at the wrong time can turn out to be a costly and ineffective affair.

This leaves us with the question: can we already find a transformative strategy in communications of the NLMoD and to what extent? And further, what should that strategy entail? Could the maturity model enable design of a transformative strategy?

Some indications can be found in NLMoD "Defensievisie 2035":

"The Defence Vision wants to make room for experimentation and scaling up new technologies and working methods in the organisation by means of short-cycle innovation. Partly against this background, the Defensievisie 2035 stimulates the further development of innovative defence ecosystems of companies, universities, and knowledge institutions."[47]

These developments are needed to define objectives and may drive organisational developments.

This statement creates problems with our current state on the maturity ladder if this innovation must be widely implemented within the organisation. This is also recognised: current successful innovative experiments cannot be embedded in the current organisation due to legal and cultural challenges, eliciting frustration with the innovators.[48] This again implies that before stimulating the innovative ecosystems, the 'regular' (routine-oriented) part of the organisation must be in order to be able and willing to support the innovators. Therefore, stimulating the innovation too much in this phase of the transformation again could have adverse effects.

In earlier NLMoD policies and statements,[49] the most important action program was to get the organisation in order and in line with its budget. The above analysis confirms that to be able to transform to a situation as defined in the NLMoD "Defensievisie 2035", the organisation should be in line with its actual goals and budget.

5.7. Discussion

This chapter's objective was to examine the relationship between organisational goal setting pertaining to becoming data-driven, maturity in this realm, and process conceptualisation of actionable-balanced change. Given the considerable number of dimensions and the size of the gaps as identified before and the necessary integrated approaches to data-driven working in the sustainment organisation, several major achievements must be realised: the implementation of the new ERP-system. Working data-driven means not only implementing the IT but also a redesign of the "business operations" side, what is required, what should the new working processes look like, and how the IT systems can support this. This in turn implies that the business side is aware of what is possible. This is called the classical business – IT alignment. But now, packaged in newer and more dynamic context, the data, information, and software rules in the SAP S/4 HANA ERP-system are even more determining the operational possibilities and efficiency than ever before.

This observation leads to the conclusion that the classic way of bringing in new IT functionalities is not only a matter of "change the business through a short-term project" or implement a new technical tool that delivers the required functionality but also needs a long-term perspective on how the organisation can be permanently transformed to a higher maturity level needed to be able to operate and work with the new functionality as a Data-driven Organisation. This means a long-term perspective not only on IT but also on all the ten mentioned dimensions.

Another consequence of maturity level awareness within a data-driven organisation should be the goal setting within (strategic) vision papers and requirement documents. An often-observed pattern in IT projects is that "the business" is asked what kind of functionality they require without completely realising the possibilities and limitations of the software and the IT department is often not fully aware what the business consequences are after implementation of the IT-system. The new SAP S/4 HANA standard, which will be the new ERP/SAP functionality from 2025 onwards, has a lot of new Business Process Management (BPM) possibilities. The project business case of the migration project, called "Roger", shows us the advanced functionalities NLMoD will be able to perform. However, the necessary maturity-steps that must be taken to be able to use the functionalities in an optimal way seem not well defined.

Not only should goal setting in strategic vision papers and requirements be looked at. In addition, the governance of "running the business" is a prominent issue. As already mentioned, to be able to use new complex functionalities on a high maturity level, all ten maturity dimensions should ideally function at the same level. This means for instance, at level X for data, that they are stored in one database, accessible through IoT, have high quality (complete, unique, actual), are well protected and managed. For the dimension skills, it means that the sustainment organisation needs high-level computer science and artificial intelligence skills to create analytical products and empower employees. It needs knowledge on how to rapidly create new analytical models for different applications and ventures to capture the maximum value out of our data. All the employees in the organisation have the mindset and the knowledge to continuously improve their work and processes, using a variety of data sources and methods.

The alignment of IT with business in a data-driven context is in the current organisation very much tied to the IT, while in successful companies the data are more tied to the business operations. To achieve a good alignment, it is necessary to bring this in balance and give IT a more integrated position in business. In the traditional strategical alignment model,[50] this situation is referred to as the competitive potential situation. NLMoD quite often seems to struggle with this problem.

5.8. Conclusion

To become a really successful information and data-driven organisation as mentioned in the NLMoD "Defensievisie 2035", a broad, long-term and transformation encompassing multiple dimensions approach must be made before this end state can be reached and innovation permeates the blood vessels of the people and the organisation, for instance procurement processes for future weapon and IT-systems and their life cycle support. It is a difficult and time-consuming transformation process that can only be successfully achieved through the transparent and structural use of an implementable "maturity development strategy".

'Implementable' concerns the ability of an extant organisation to take the next step in terms of maturity development. With respect to the extant Defence sustainment organisation, one can point at the present challenges to achieve a 'normal' level of operations when looking at for instance around 600 vacancies, stock problems, and data immaturity. Addressing these challenges calls for not only budget but also a systematic fixing of foundational capabilities. Moving from sustainment towards the MoD organisation at large, we can conceptually generalise our findings. As becoming data-driven has developed into a strategic choice for the Dutch MoD, future research topics include refinement of a maturity model across all levels and across the military operational and non-operational sides. For instance, researchers may evaluate the extent to which military operational units are data-driven. They can examine operationalisation and use of fact-based/ data-driven strategic management of military operations and organisational performances.

Notes

[1]   Stefaan G. Verhulst, "Unlock the Hidden Value of Your Data," *Harvard Business Review*, no. March 2020.

[2]   McKinsey, "The Age of Analytics: Competing in a Data-Driven World," *McKinsey Global Institute* https://www.mckinsey.com/~/media/mckinsey/industries/public%20and%20social%20sector/ our%20insights/the%20age%20of%20analytics%20competing%20in%20a%20data%20driven%20 world/mgi-the-age-of-analytics-full-report.pdf 2016.

[3]   Chaim Zins, "Conceptual Approaches for Defining Data, Information, and Knowledge," *Journal of the American Society for Information Science and Technology* 58, no. 4 2007.: 480

[4]   Robert H. Waterman Jr, "The Seven Elements of Strategic Fit," *The Journal of Business Strategy* 2, no. 3 1982.

[5]   Jay R. Galbraith, *Designing Complex Organizations* (Reading, Massachusetts: Addison-Wesley, 1973).

[6]   Marius Hupperz et al., "What is a Data-Driven Organization?," *Proceedings Twenty-Seventh Americas Conference on Information Systems, Montreal* 2021.: 1

[7]   Gina S. Ikemoto and Julie A. Marsh, "Cutting through the "Data-driven" Mantra: Different Conceptions of Data-driven Decision Making," *Teachers College Record* 109, no. 13 2007.: 108

[8]   Hupperz et al., "What is a Data-Driven Organization?."

[9]   Mert O. Gökalp et al., "Data-Driven Manufacturing: An Assessment Model for Data Science Maturity," *Journal of Manufacturing Systems* 60 2021.

[10] Harmen S Wijbenga, Paul C. van Fenema, and Nynke Faber, "Diagnosing Recurrent Logistics Problems: A Combined SCM Disciplines and Maturity Perspective," *Supply Chain Management: An International Journal* Forthcoming 2021; Maximilian Roglinger, Jens Poppelbuß, and Jörg Becker, "Maturity Models in Business Process Management," *Business Process Management Journal* 18, no. 2 2012.

[11] Jerry H. Ratcliffe, *Intelligence-led Policing* (Milton Park: Routledge, 2016).

[12] https://www.nisa-intelligence.nl/sites/default/files/2019-10/Redeboekje%20Rietjens%20-%20definitief.pdf

[13] https://magazines.defensie.nl/kmarmagazine/2021/07/02_samenwerking_nlda_07-2021

[14]   Thomas H. Davenport and James E. Short, "The New Industrial Engineering: Information Technology and Business Process Redesign," *https://dspace.mit.edu/bitstream/handle/1721.1/48613/newindustrialengoodave.pdf* 1990.

[15]   Jan vom Brocke and Michael Rosemann, *Handbook on Business Process Management 2: Strategic Alignment, Governance, People and Culture* (Heidelberg: Springer, 2010).

[16]   Gert Laursen and Jesper Thorlund, *Business Analytics for Managers: Taking Business Intelligence beyond Reporting*, vol. 40 (Hoboken, NJ: Wiley, 2010).

[17]   Jeanne Ross, Peter Weill, and David C. Robertson, *Enterprise Architecture as Strategy: Creating a Foundation for Business Execution* (Boston, MA: Harvard Business Press, 2006).

[18]   https://www.defensie.nl/downloads/publicaties/2020/11/25/strategische-kennis--en-innovatie-agenda-2021-2025, https://www.defensie.nl/onderwerpen/defensievisie-2035

[19]   https://magazines.defensie.nl/vliegendehollander/2020/12/06_robotics

[20]   Mustafa Emirbayer and Ann Mische, "What Is Agency?," *American Journal of Sociology* 103, no. 4 1998; Yingfeng Zhang et al., "A Big Data Analytics Architecture for Cleaner Manufacturing and Maintenance Processes of Complex Products," *Journal of Cleaner Production* 142 2017.

[21]   Zhang et al., "A Big Data Analytics Architecture for Cleaner Manufacturing and Maintenance Processes of Complex Products."

[22]   Galbraith, *Designing Complex Organizations*; Prem Premkumar, Katikireddy Ramamurthy, and Carol S. Saunders, "Information Processing View of Organizations: An Exploratory Examination of Fit in the Context of Interorganizational Relationships," *Journal of Management Information Systems* 22, no. 1 2005.

[23]   Erik Brynjolfsson, Lorin M. Hitt, and Heekyung H. Kim, "Strength in Numbers: How Does Data-Driven Decision Making Affect Firm Performance?," *https://ide.mit.edu/sites/default/files/publications/2011.12_Brynjolfsson_Hitt_Kim_Strength%20in%20Numbers_302.pdf* 2011.

[24]   Nicolaus Henke, Jordan Levine, and Paul McInerney, "Analytics Translator: The New Must-Have Role," *Harvard Business Review* https://www.mckinsey.com/~/media/McKinsey/Business%20Functions/McKinsey%20Analytics/Our%20Insights/Analytics%20translator/Analytics-translator-The-new-must-have-role.pdf 2018.

[25]   Frans P.B. Osinga, "On Boyd, Bin Laden, and Fourth Generation Warfare as String Theory," in *On New Wars*, ed. J. Olson (Oslo, Norway: Oslo Files on Defence and Security no 4/2007, 2007).

[26]   John Hagerty and Bill Hostmann, "ITScore Overview for Business Intelligence and Performance Management," *Gartner Report, Article ID G, 205072, https://www.gartner.com/en/documents/1433813* 2010.

[27]   Bill Hostmann, Nigel Rayner, and Ted Friedman, "Gartner's Business Intelligence and Performance Management Framework," *Gartner, Stamford, CT* 2006.

[28]   Ruben Buitelaar, "Building the Data-Driven Organization: A Maturity Model and Assessment," *Master thesis, Leiden University* 2018.

[29]   Alan R. Hevner et al., "Design Science in Information Systems Research," *MIS Quarterly* 28, no. 1 2004.

[30]   Jörg Becker, Ralf Knackstedt, and Jens Pöppelbuß, "Developing Maturity Models for IT Management," *Business & Information Systems Engineering* 1, no. 3 2009.

[31]   This concerns not only the organisation itself but – in a world increasingly demanding organisations to function as open systems – also strategic partnerships required for executing its core tasks. In fact, some level of symmetry of digital maturity is becoming a precondition for such partnerships to work. Kemal Ö. Yılmaz, "Mind the Gap: It's About Digital Maturity, Not Technology," in *Managerial Issues in Digital Transformation of Global Modern Corporations*, ed. Thangasamy Esakki (Hershey, PA: IGI Global, 2021).

32 MoD, "Defensievisie 2035 – Vechten voor een veilige toekomst," *https://www.rijksoverheid.nl/documenten/rapporten/2020/10/15/defensievisie-2035-vechten-voor-een-veilige-toekomst* 2020.

33 Translated from Dutch as mentioned in: MoD, "Defensievisie 2035 – Vechten voor een veilige toekomst."

34 MoD, "Strategische Kennis en Innovatie Agenda 2021-2025," *https://www.google.com/url?sa= t&rct=j&q=&esrc=s&source=web&cd=&cad=rja&uact=8&ved=2ahUKEwjbyrr-hLD3 AhVDPOwKHZ6TCVMQFnoECAYQAQ&url=https%3A%2F%2Fwww.defensie.nl%2Fdownloads%2F- publicaties%2F2020%2F11%2F25%2Fstrategische-kennis--en-innovatieagenda-2021-2025&usg=AOv- Vaw1RgmwAUctsty6JpRKoCWEo* 2020.

35 See earlier on data-driven decision making, Ikemoto and Marsh, "Cutting through the "Data-driven" Mantra: Different Conceptions of Data-driven Decision Making."

36 Translated by the authors from: MoD, "Defensie Strategie Data Science en AI 2021-2025: Werken aan een slimme krijgsmacht," *Internal document* 2021.

37 PBLQ, "Eindrapportage Programma SPEER," *https://www.binnenlandsbestuur.nl/whitepaper/eindrapportage-programma-speer* 2013; Jan-Bert Maas, Paul C. van Fenema, and Joseph M.M.L. Soeters, "ERP as an Organizational Innovation: Key Users and Cross-boundary Knowledge Management," *Journal of Knowledge Management* 20, no. 3 2016.

38 ADR, "Evaluatie basisimplementatie ERP M&F bij Defensie," *https://www.google.com/ url?sa=t&rct=j&q=&esrc=s&source=web&cd=&ved=2ahUKEwjytbfgg7D3AhWQ2qQKHWAx- DwEQFnoECAgQAQ&url=https%3A%2F%2Fzoek.officielebekendmakingen.nl%2Fblg-830047. pdf&usg=AOvVaw2Z5GApdPZRiub9yDhDtVYv* 2018.

39 MoD, "Defensievisie 2035 – Vechten voor een veilige toekomst."

40 For more information: https://blogs.sap.com/2017/04/03/what-is-sap-s4hana-in-simple-terms/

41 MoD, "Defensievisie 2035 – Vechten voor een veilige toekomst."

42 PBLQ, "Eindrapportage Programma SPEER."

43 ADR, "Evaluatie basisimplementatie ERP M&F bij Defensie."

44 Buitelaar, "Building the Data-Driven Organization: A Maturity Model and Assessment."

45 Available at https://data-driven.rubenbuitelaar.com/.

46 Ross, Weill, and Robertson, *Enterprise Architecture as Strategy: Creating a Foundation for Business Execution.*: 86-87

47 MoD, "Strategische Kennis en Innovatie Agenda 2021-2025."

48 Paul C. van Fenema et al., "Sustaining Relevance: Repositioning Strategic Logistics Innovation in the Military," *Joint Forces Quarterly* 101, no. April 2021; Feng Li, "Leading Digital Transformation: Three Emerging Approaches for Managing the Transition," *International Journal of Operations & Production Management* 40, no. 6 2020.

49 MoD, "Investeren in onze mensen, slagkracht en zichtbaarheid," *https://www.google. com/url?sa=t&rct=j&q=&esrc=s&source=web&cd=&cad=rja&uact=8&ved=2ahUKEwiQ-Yz- RhbD3AhWngfoHHcbXDSEQFnoECAIQAQ&url=https%3A%2F%2Fwww.rijksoverheid. nl%2Fbinaries%2Frijksoverheid%2Fdocumenten%2Frapporten%2F2018%2F03%2F26%2Fdefensieno- ta-2018-investeren-in-onze-mensen-slagkracht-en-zichtbaarheid%2Fdefensienota-2018-investeren-in- onze-mensen-slagkracht-en-zichtbaarheid.pdf&usg=AOvVaw3076nLd7OvV5q9V7WBfogP* 2018.

50 John C. Henderson and N. Venkatraman, "Understanding Strategic Alignment," *Business Quarterly* 56, no. 3 1991.

References

ADR. "Evaluatie Basisimplementatie Erp M&F Bij Defensie." *https://www.google.com/ url?sa=t&rct=j&q=&esrc=s&source=web&cd=&ved=2ahUKEwjytbfgg7D3AhWQ2qQKHWAx-DwEQFnoECAgQAQ&url=https%3A%2F%2Fzoek.officielebekendmakingen.nl%2Fblg-830047. pdf&usg=AOvVaw2Z5GApdPZRiub9yDhDtVYv*, 2018.

Becker, Jörg, Ralf Knackstedt, and Jens Pöppelbuß. "Developing Maturity Models for It Management." *Business & Information Systems Engineering* 1, no. 3, 2009: 213-22.

Brynjolfsson, Erik, Lorin M. Hitt, and Heekyung H. Kim. "Strength in Numbers: How Does Data-Driven Decision Making Affect Firm Performance?". *https://ide.mit.edu/sites/default/files/publications/2011.12_Brynjolfsson_Hitt_Kim_Strength%20in%20Numbers_302.pdf*, 2011.

Buitelaar, Ruben. "Building the Data-Driven Organization: A Maturity Model and Assessment." *Master thesis, Leiden University*, 2018.

Davenport, Thomas H., and James E. Short. "The New Industrial Engineering: Information Technology and Business Process Redesign." *https://dspace.mit.edu/bitstream/handle/1721.1/48613/newindustrialengoodave.pdf*, 1990.

Emirbayer, Mustafa, and Ann Mische. "What Is Agency?". *American Journal of Sociology* 103, no. 4, 1998: 962-1023.

Galbraith, Jay R. *Designing Complex Organizations.* Reading, Massachusetts: Addison-Wesley, 1973.

Gökalp, Mert O., Ebru Gökalp, Kerem Kayabay, Altan Koçyiğit, and P. Erhan Eren. "Data-Driven Manufacturing: An Assessment Model for Data Science Maturity." *Journal of Manufacturing Systems* 60, 2021: 527-46.

Hagerty, John, and Bill Hostmann. "Itscore Overview for Business Intelligence and Performance Management." *Gartner Report, Article ID G, 205072, https://www.gartner.com/en/documents/1433813*, 2010.

Henderson, John C., and N. Venkatraman. "Understanding Strategic Alignment." *Business Quarterly* 56, no. 3, 1991: 72-78.

Henke, Nicolaus, Jordan Levine, and Paul McInerney. "Analytics Translator: The New Must-Have Role." *Harvard Business Review* https://www.mckinsey.com/~/media/McKinsey/Business%20Functions/ McKinsey%20Analytics/Our%20Insights/Analytics%20translator/Analytics-translator-The-new-must-have-role.pdf, 2018.

Hevner, Alan R., Salvatore T. March, Jinsoo Park, and Sudha Ram. "Design Science in Information Systems Research." *MIS Quarterly* 28, no. 1, 2004: 75-105.

Hostmann, Bill, Nigel Rayner, and Ted Friedman. "Gartner's Business Intelligence and Performance Management Framework." *Gartner, Stamford, CT*, 2006: 1-6.

Hupperz, Marius, Inan Gür, Frederik Möller, and Boris Otto. "What Is a Data-Driven Organization?". *Proceedings Twenty-Seventh Americas Conference on Information Systems, Montreal*, 2021.

Ikemoto, Gina S., and Julie A. Marsh. "Cutting through the "Data-Driven" Mantra: Different Conceptions of Data-Driven Decision Making." *Teachers College Record* 109, no. 13, 2007: 105-31.

Laursen, Gert, and Jesper Thorlund. *Business Analytics for Managers: Taking Business Intelligence Beyond Reporting*. Vol. 40, Hoboken, NJ: Wiley, 2010.

Li, Feng. "Leading Digital Transformation: Three Emerging Approaches for Managing the Transition." *International Journal of Operations & Production Management* 40, no. 6, 2020: 809-17.

Maas, Jan-Bert, Paul C. van Fenema, and Joseph M.M.L. Soeters. "Erp as an Organizational Innovation: Key Users and Cross-Boundary Knowledge Management." *Journal of Knowledge Management* 20, no. 3, 2016.

McKinsey. "The Age of Analytics: Competing in a Data-Driven World." *McKinsey Global Institute* https://www.mckinsey.com/~/media/mckinsey/industries/public%20and%20social%20sector/our%20insights/the%20age%20of%20analytics%20competing%20in%20a%20data%20driven%20world/mgi-the-age-of-analytics-full-report.pdf, 2016.

MoD. "Defensie Strategie Data Science En Ai 2021-2025: Werken Aan Een Slimme Krijgsmacht." *Internal document*, 2021.

———. "Defensievisie 2035 – Vechten Voor Een Veilige Toekomst." *https://www.rijksoverheid.nl/documenten/rapporten/2020/10/15/defensievisie-2035-vechten-voor-een-veilige-toekomst*, 2020.

———. "Investeren in Onze Mensen, Slagkracht En Zichtbaarheid." *https://www.google.com/url?sa=t&rct=j&q=&esrc=s&source=web&cd=&cad=rja&uact=8&ved=2ahUKEwiQ-YzRhbD3AhWngfoHHcbXDSEQFnoECAIQAQ&url=https%3A%2F%2Fwww.rijksoverheid.nl%2Fbinaries%2Frijksoverheid%2Fdocumenten%2Frapporten%2F2018%2F03%2F26%2Fdefensienota-2018-investeren-in-onze-mensen-slagkracht-en-zichtbaarheid%2Fdefensienota-2018-investeren-in-onze-mensen-slagkracht-en-zichtbaarheid.pdf&usg=AOvVaw3076nLd7OvV5q9V7WBfogP*, 2018.

———. "Strategische Kennis En Innovatie Agenda 2021-2025." *https://www.google.com/url?sa=t&rct=j&q=&esrc=s&source=web&cd=&cad=rja&uact=8&ved=2ahUKEwjbyrr-hLD3Ah-VDPOwKHZ6TCVMQFnoECAYQAQ&url=https%3A%2F%2Fwww.defensie.nl%2Fdownloads%2Fpublicaties%2F2020%2F11%2F25%2Fstrategische-kennis--en-innovatieagenda-2021-2025&usg=AOvVaw1RgmwAUctsty6JpRKoCWEo*, 2020.

Osinga, Frans P.B. "On Boyd, Bin Laden, and Fourth Generation Warfare as String Theory." In *On New Wars*, edited by J. Olson. Oslo, Norway: Oslo Files on Defence and Security no 4/2007, 2007.

PBLQ. "Eindrapportage Programma Speer." *https://www.binnenlandsbestuur.nl/whitepaper/eindrapportage-programma-speer*, 2013.

Premkumar, Prem, Katikireddy Ramamurthy, and Carol S. Saunders. "Information Processing View of Organizations: An Exploratory Examination of Fit in the Context of Interorganizational Relationships." *Journal of Management Information Systems* 22, no. 1, 2005: 257-94.

Ratcliffe, Jerry H. *Intelligence-Led Policing*. Milton Park: Routledge, 2016.

Roglinger, Maximilian, Jens Poppelbuß, and Jörg Becker. "Maturity Models in Business Process Management." *Business Process Management Journal* 18, no. 2, 2012: 328-46.

Ross, Jeanne, Peter Weill, and David C. Robertson. *Enterprise Architecture as Strategy: Creating a Foundation for Business Execution*. Boston, MA: Harvard Business Press, 2006.

van Fenema, Paul C., Ton van Kampen, Gerold Gooijer, Nynke Faber, Harm F. Hendriks, Andre J. Hoogstrate, and Loe Schlicher. "Sustaining Relevance: Repositioning Strategic Logistics Innovation in
the Military." *Joint Forces Quarterly* 101, no. April, 2021: 59-68.

Verhulst, Stefaan G. "Unlock the Hidden Value of Your Data." *Harvard Business Review*, no. March, 2020:
https://hbr.org/2020/05/unlock-the-hidden-value-of-your-data.

vom Brocke, Jan, and Michael Rosemann. *Handbook on Business Process Management 2: Strategic
Alignment, Governance, People and Culture.* Heidelberg: Springer, 2010.

Waterman Jr, Robert H. "The Seven Elements of Strategic Fit." *The Journal of Business Strategy* 2, no. 3,
1982: 69-73.

Wijbenga, Harmen S, Paul C. van Fenema, and Nynke Faber. "Diagnosing Recurrent Logistics Problems:
A Combined Scm Disciplines and Maturity Perspective." *Supply Chain Management: An International Journal* Forthcoming, 2021.

Yılmaz, Kemal Ö. "Mind the Gap: It's About Digital Maturity, Not Technology." In *Managerial Issues in
Digital Transformation of Global Modern Corporations*, edited by Thangasamy Esakki. Hershey,
PA: IGI Global, 2021.

Zhang, Yingfeng, Shan Ren, Yang Liu, and Shubin Si. "A Big Data Analytics Architecture for Cleaner
Manufacturing and Maintenance Processes of Complex Products." *Journal of Cleaner Production*
142, 2017: 626-41.

Zins, Chaim. "Conceptual Approaches for Defining Data, Information, and Knowledge." *Journal of the
American Society for Information Science and Technology* 58, no. 4, 2007: 479-93.

PART II

Data Driven Support to Decision Making

CHAPTER 6

The effect of big data and AI on forecasting in defence and military applications

André J. Hoogstrate

Abstract

In this chapter we analyse the effects Big Data and AI have and will have on the practice of forecasting for defence and military applications. Big Data brings an abundance of heterogeneous data from many sources to the table and AI and its subfield Machine Learning brings a whole host of new models, algorithms, and forecasting methods with it. Combining both concepts, it is foreseen that forecasting and foresight development will be greatly influenced by these two developing trends. The impact on applications at the strategic, operational as well as tactical level is considered.

Keywords: Forecasting, Big Data, Artificial Intelligence, Military, Defence.

6.1. Introduction

Forecasting is an important tool in defence and military strategy, operations, management, and sustainment. The applications vary from direct use in budgeting, planning, logistics and numerous sustainment functions and weather forecasting to more varying applications in strategy and foresight development[1], intelligence, operational and tactical aspects of predicting enemy forces and friend and enemy civilian behaviour.

Forecasting is founded on the assumption that current and past knowledge can be used to make predictions about events in the future. It is not expected though that the forecasts match future values exactly but are close in some sense. Common forecasts to use are the expected value or the most likely value but it might also be a forecast interval or the entire probability distribution of possible forecasts.[2] Several relevant aspects of forecasting are discussed in more detail below.

Artificial Intelligence (AI) and big data (BD) are topics that have attracted the interest of business and military strategists alike. Both concepts are projected to become disruptive technologies[3]. While Big Data and AI are used extensively by several very successful early adopters in the private sector, the most dramatic/

influential impact is still to materialise, certainly in the military context. AI is still in an early development phase, but most nations have recognised its importance and setup large scale scientific research programs in cooperation with the private sector and the military-industrial complex[4].

Big data is considered important in analytics and forecasting as it provides the option to include more and other types of data in the forecasting process than traditionally available. Available means having in place a data registration, storage, and collection process such that data from the own organisation such as registry, ERP-systems, sensors, etc as well as third party data be it structured, un-structured, semi-structured and/or heterogeneous can be included in the forecasting process without much effort from the analyst or forecaster. Getting this infrastructure organised and working is a precondition to enable an organisation to use BD successfully but requires in most organisations an expensive and intensive reengineering of most business and operational processes[5].

Artificial Intelligence could informally be defined as, given that no precise single commonly accepted definition exists, "the capability of a computer system to perform tasks that normally require human intelligence such as visual perception, speech recognition, and decision-making"[6] (Cummings, 2018). DARPA[7] gives a somewhat more focused definition "Artificial intelligence is the programmed ability to process information." Below, we return to the DARPA vision on AI, as this characterisation comes closest to the perception of the military practice.

In the remainder of this chapter, we first give introductions to BD, AI and the theory of forecasting. Thereafter we introduce a framework to access the impact of BD and AI on forecasting methods and processes. Finally, we illustrate the current and foreseen effects by discussing several defence and military applications.

6.2. Big data

Big data in business is often characterised by the 5 V's: volume, velocity, variety, veracity, and value. Volume indicating the large amounts of data, velocity the speed at which the data are generated and become available for analysis, variety indicating the heterogeneity of the data, from users-log to images, from simple spreadsheets to complex relational data. These three V's are in a sense always positive; they create analytical opportunities whether we seize the opportunity or not. The same analysis holds for defence and military applications.

Veracity and value are a bit different: veracity is whether the data are consistent and "truthful" enough to be used; it is about data quality and its relation to the real-world. In business there are applications where the veracity of data is necessarily high, for instance in most transactional data the importance consistency is paramount, think about your bank statement at the end of the month. For other applications veracity is not so important; in a marketing campaign, if observations do not have the correct GPS coordinates for 20 percent of the potential clients it is usually not problematic.

In defence, intelligence, and military applications the percentage of applications where a high veracity is of utmost importance is much greater than in business applications. Having one wrong GPS coordinate can lead to many lost lives. The fact that most data are not first-hand data and stakes are high means veracity always must be established. This is particularly important in intelligence where information must be handled by multiple analysts. Therefore, several grading systems have been developed[8] and will for BD be implemented in some form, but this discussion is outside the scope of this research.

Value in national security and military applications also has a different meaning than in business or for instance health applications. The value of BD for intelligence and military applications lies in the ability to derive actionable information from the data. It is envisaged that having large sets of historical and current data consisting of as many observables as possible increases the probability of deriving actionable information and aids the decision-making processes to a certain extent. Indeed, extrapolating the results of BD in business, health, and science it is projected that BD has a predictive power in terms of containing value in historical and actual information, which can potentially enrich the inputs of forecasting models and in turn also the quality of the forecasts[9], which in turn will create actionable information for military decision making.

However, BD for forecasting does not come for free. Different types of big data contain specific information, have unique characteristics but are usually stored in various formats. This means that the actual forecasting step in the forecasting process is in effect often being dominated by different data preparation and analysis techniques to first process the data and extract the hidden predictive knowledge[10]. Usually only after this analysis can it be determined whether the data are suitable for the forecasting tasks at hand.

6.3. Artificial intelligence

There are several characterisations of the level of sophistication of AI systems in mimicking human information processing behaviours. An often-used categorisation[11] is given by "narrow AI", "strong AI" and "superintelligence". Narrow AI indicates that a system equals or surpasses human performance in one task; strong AI indicates that a system equals human capabilities on ay task, while superintelligence surpasses human intelligence on any task. Several more refined characterisations of human information processing exist, but in this analysis the characterisation given in "The DARPA perspective on artificial intelligence"[12] is used. It characterises AI based on the level of ability to process information in perceiving, learning, abstracting, and reasoning, by three phases: handcrafted knowledge, statistical learning, contextual adaptation.

The phase "handcrafted knowledge" is characterised by humans creating sets of rules to represent knowledge in well-defined domains. The structure of the knowledge is defined by humans, the specifics are explored by the machine. The expert-systems developed in the 70's and 80's are typical examples of this. Another example is most rule-based systems.

The "statistical learning" phase is characterised by humans creating statistical models for specific problem domains and training them on big data. Observe that the statistical models can either be model-driven or data-driven although most well-known methods i.e., Machine Learning, Deep Learning are data-driven. The definitions of and distinctions between data-driven and model-driven are deferred until later in this chapter. The current scientific state is that we are in the middle of this phase.

The "contextual adaptation" phase is characterised by systems constructing contextual explanatory models for classes of real-world phenomena. This implies that systems can not only detect, observe, and classify objects and phenomena in their environment, often using techniques from the statistical learning phase, but can also reason about them using conceptual models. This phase has not been reached yet; research is just starting.

In Table 6.1 below the different phases and their associated level of ability to process information in perceiving, learning, abstracting, and reasoning, is visualised. The more crosses the more the specific ability is addressed.

Table 6.1. Notional Intelligence scale, after Launchbury[13]

	Perceiving				Learning				Abstracting				Reasoning			
Handcrafted knowledge	x												x	x	x	x
Statistical learning	x	x	x	x	x	x	x	x	x				x			
Contextual adaptation	x	x	x	x	x	x	x	x	x	x	x		x	x	x	x

6.4. Forecasting

As stated before, forecasting is based on the premise that current and past knowledge is useful to make predictions about events in the future. To make discussions more precise, it is necessary to distinguish between the concepts of estimation, prediction, and forecasting which are frequently used interchangeably; however, they differ from each other in the following way. Estimation is about guessing the value of a parameter of interest. Prediction is concerned with guessing the outcomes of unseen data. Forecasting is the sub-field of prediction concerned with guessing the outcomes of future and thus unseen data.

A forecasting procedure or process tries to predict the future value of a stochastic or uncertain event or observation, while a forecasting method is defined here to be a predetermined sequence of steps that produces forecasts of future points in time[14]. This implies that a forecasting procedure might entail several forecasting methods and the result being a combination of the outcomes of the used methods.

Whatever the goal of the forecasting process, four steps can always be distinguished: data collection, data processing, prediction improvement and the actual forecasting. For different applications these steps can differ[15]. Also note that in the design-stage or ad-hoc analysis, steps might be handcrafted and time consuming while in "production" mode the steps will be automated as far as possible.

In interpreting, decision making and forecasting models play a crucial role. In this context it suffices to define a model as a simplified description, often but not necessarily a mathematical one, of a system or process to assist analysis and predictions. For a more extensive introduction into forecasting in general see for instance Armstrong[16] or the overview article by Petropoulos et al[17].

6.4.1. Characterisation of forecasting methods

Traditionally[18] forecasting methods are characterised within two categories based on whether they use implicit or explicit models. The implicit characterisation is often referred to as "judgmental" the second as "statistical". The first category consists of methods where one elicits forecasts from experts, laymen based on their experience of opinion. No model for the phenomenon under consideration is made explicit, only a secondary model on how to collect, combine and interpret the derived forecasts. The second, "statistical" category consists of methods that explicitly use mathematical models, stochastic or deterministic. The term "statistical" might be a bit confusing as the forecast might be based on a deterministic model, but the interpretation is still probabilistic as measurement errors and unaccounted variables still influence the actual future outcome.

For the current analysis, we recognise that the statistical model's category entails two different philosophical approaches and we will characterise these. The first category is characterised by models that are in simulacra, (within likenesses) that is they bear some likeness to the real-world, be it physical, social, economic, or otherwise, and are constructed to reflect certain aspects that are essential for the analysis or prediction at hand. This category of models will from here on be denoted as model-driven. This category consists of models that are created such that the theoretical behaviour of the models mimics the theorised real-world behaviour of the object or phenomenon. Most of the models in this category are not only used for prediction or forecasting but also for causal reasoning.

The second category, denoted here as data-driven, is based on directly mimicking the observed data of the real-world object or phenomenon. This implies that the model does not have the intention to match theorised real-world behaviour. This makes it very hard to interpret these types of models in a causal fashion; they mainly try to mimic correlations or non-linear relations apparent in the data. Currently most machine learning algorithms fall into this class. Also, some traditional statistics, for instance strict time series analysis, fall into this class.

In Table 6.2 some forecasting methods are presented based on the introduced characterisation. Two remarks are in order. First, the list is not complete; secondly, the characterisation of forecasting methods concerns the way in which one derives a single forecast or forecast distribution, but it has been demonstrated that combining the results of several forecasting methods can greatly improve the final forecasting results. Combining forecasts based on different methods is often denoted as hybrid forecasting methods. Currently hybrid methods that make use of a combination of judgemental, model-driven and data-driven methods are starting to emerge in applications[19].

Table 6.2. Several forecasting methods grouped by the characterisation derived in the text

Judgmental	Model-driven	Data-driven
Forecast is based on – Expert intuition – Analogy – Reference classes	Model is derived from theory about the object or phenomenon to forecast	The model is defined by the method used and the data, not by theory about the object or phenomenon to forecast
Aggregated judgment: – expert survey – Delphi method – Reputation base prediction – Forecasting platforms – Super forecasters – Prediction markets Conjoint analysis Automated judgement – Judgmental bootstrapping – Expert system Gaming	Regression Causal modelling Probabilistic modelling – Markov Chains – Econometric models Deterministic modelling Structural equation modelling Simulation	Statistical methods – Time Series – fuzzy modelling – Data reduction methods – Heuristic methods – Machine learning – supervised – unsupervised – reinforcement – Deep learning – LSTM, GRU, RNN Simulation

6.4.2. Representation of the forecast and evaluation criteria

As pointed out in the introduction, forecasts can come in many shapes and are a "guess" of a future event based on current and past knowledge and information. In that sense a point forecast comes down to trying to guess the outcome of one single realisation of a random draw from all possible outcomes. Truly an almost impossible task certainly if the distribution of the possible outcomes is almost infinite as is often the case. Therefore, in practice one settles for a forecast that is expected to be close to that single draw in some sense. It also implies that a point forecast without measure of uncertainty of how close the prediction is likely to be, can hardly be interpreted and is in most cases not useful. With the rise of BD and AI and the increasing computing power, the probabilistic forecast is gaining practical use. Instead of giving a point forecast with some measure of uncertainty, like a forecast interval, one estimates the whole probability distribution of the future event. This distribution is called the prediction or forecasting distribution. This distribution can be used in decision-making.[20]

Evaluation of the quality of forecasting methods is hard. There are two important reasons. First, although the idea is that past and current information conveys information about the future, for forecasting procedures to be consistent a certain stability of the mechanisms generating the data should be present. This is in practice almost always only by assumption. So, determining why one forecasting method is good or suddenly very bad is hard. The second problem is that most forecasting methods only work for a specific set of models. And whether the data for the application under consideration are generated by a model included in that set of models is never certain.

In practice the quality of a forecasting method for a certain application is assessed as follows: split the available data into a training and test set, first estimate or train the proposed forecasting method on the training set, thereafter the performance is assessed using the test set. This method works better with more data and for model driven as well as data-driven methods. For judgmental methods one can assess performance by recordkeeping and setting up experiments that mimic this setup. BD facilitates these approaches as more data are available.

6.5. The effect of AI and big data on forecasting

Before analysing the effect of AI and BD on forecasting we first analyse the relation between BD and AI as they influence each other. Without the success of Machine Learning there would be less of a business case for BD and without BD Machine

Learning would be less successful in the current state of knowledge within the AI field. Table 6.3 shows the most important factors that BD brought to AI.

Table 6.3. The effect of Big Data on Artificial Intelligence, after Deshpande and Kumar[21]

AI without Big Data	AI with Big Data
Availability of limited data sets, in size and granularity, as well as number of variables.	Availability of increasingly detailed and wide datasets
Limited sample size.	Massive sample size resulting in increased prediction accuracy in several specific applications.
Batch oriented.	More and more real-time, on-line.
Slow learning curve.	Accelerated learning curve.
Single and homogenous data sources.	Multiple and heterogeneous data sources.
Based mostly on traditionally structured data.	Based on structured, semi- and unstructured data.

Repeating the analysis in our framework we combine the three DARPA AI phases with the 3 V's of the big data characterisation and obtain the results in Table 6.4.

Table 6.4. Order of effects of big data on the DARPA levels of ability to process information

		Big Data Characteristic		
		Velocity	Volume	Variety
DARPA levels of ability to process information	Handcrafted knowledge	+	++	+++
	Statistical learning	++	+++	+
	Contextual adaptation	++	+	+++

In the handcrafted knowledge phase, which started before the BD introduction, the entrée of BD introduced a greater variety of data that could be used. This influenced AI the most. Volume helped in the sense that various estimates and predictions could be improved by removing the limitations of small sample size, sample data type and out-of-date information, a problem forecasting often had before the BD era. In the statistical learning phase, the phase we are currently in, it's the sheer volume of data that influences the development of Machine Learning but basically all data-driven methods. Also, velocity influenced statistical learning and with new applications, even in (near) real-time, online forecasting and nowcasting. In the contextual adaption phase, the gains from the first two phases will be used as well; the additional contribution in this phase is reasoning with the results of the previous phases. In that sense variety will bring the most, velocity will still help to improve reaction times, volume will have less additional influence in this phase.

To assess the effect of BD and AI on forecasting we combine in a similar fashion as above the characterisations of BD and AI with those of forecasting methods. The results are presented in Table 6.5. The strength of the effect, indicated by the number of "+", only indicates the relative order of the foreseen effect, not the absolute magnitude of the effect.

Table 6.5. The relative order of the effect of BD and AI on forecasting methods in the three different categories

		judgmental	Model driven	Data driven
Big data	velocity	+	++	+++
	volume	+	++	+++
	variety	+++	++	+
Artificial Intelligence	Handcrafted knowledge	++	+++	+
	Statistical learning	+	++	+++
	Contextual adaptation	+++	++	+

The velocity and volume characteristics of BD have a dramatic effect on the rise of data-driven methods; they certainly also benefitted model-driven methods with increased sample sizes and more up-to-date models. This led for both types of forecasting methods to increased accuracy in general. Although, velocity and volume affect the judgemental methods, it is less pronounced. Several judgmental methods benefit mostly from the rise of online communication which makes aggregated judgment methods easier to apply. In contrast the variety of new data gives forecasters the option to make associations that benefit forecasting accuracy, while the mathematical formal models, and data driven models even less so, are not yet that flexible in combining heterogenous data.

The handcrafted knowledge phase gives a boost to model-driven methods and judgemental forecasts while it impacts the data-driven methods the least. Model driven is more impacted as to automate and operationalise the methods in this phase means building on explicitly formulated formal mathematical and logical models. In the statistical learning phase AI is all about data-driven methods, which in turn result in additional model driven methods, further strengthened by the increased date availability. The judgemental methods are least effected by statistical learning, although we see the rise of hybrid methods which in most cases imply the use of some judgemental component. Finally, the contextual adaptation phase will take off when AI systems can conceptually understand and reason based on the model-driven and data-driven results. Therefore, the judgemental method will be greatly impacted, as the conceptual models can enrich to a great extent the judgemental methods.

6.6. Effect of AI and BD on forecasting in military and defence applications

Based on the discussion in the previous sections the expected impact of the develop-ments in BD and AI on several national security, defence, and military applications where forecasting plays a role is discussed. It should be noted that AI and BD also have effects on numerous applications not related to forecasting but more related to prediction and analysis. Important examples include, facial recognition, API/PNR data for risk-based border patrol, detecting objects on imagery and many more. Often these methods will be combined in one application. Consider the case of autonomous moving vehicles where forecasting in the current statistical learning phase is mostly limited to navigational and manoeuvring issues concerning its own position and behaviour and the behaviour of vehicles and objects, possibly incom-ing missiles, in its surroundings. The detection and classification of the objects and vehicles is usually also done by BD and AI methods, but this is not forecasting.

6.6.1. Autonomy in defence: systems, weapons, decision-making

High on the priority list of most countries is the development and use of systems that have a certain degree of autonomy. Whether armed or not. Three subcategories can be distinguished. First, autonomous systems that have the capability to operate in the physical world, i.e., autonomous unmanned vehicles. Second, autonomous weapons, i.e., systems that can launch weapons autonomously and third, decision- making systems, i.e., systems that give autonomous advice to entities on how to act. These later systems are usually also part of the first mentioned categories at a lower level but can also have the C2 role for multiple entities, being machines and/or humans.

Currently all three types of system are still based on handcrafted knowledge combined with statistical learning, i.e., sensor data is analysed by statistical learning methods, objects recognised, then based on a rule based expert system actions are devised. Most systems are currently still semi-autonomous, higher order decisions are still made by humans. The level of autonomy of the systems will grow when the methods that show contextual adaptation capabilities have arrived.

In this phase there is the question of whether humans are 'in the loop', when they retain a great degree of control over robotic autonomous systems, 'on the loop', when the system can autonomously take actions, but humans retain the ability to abort these actions, or 'off the loop', if they are neither asked to confirm action nor can abort such actions. The ability to forecast with high confidence without human support will be an important factor on whether, without touching upon the ethical dimensions, in, on or off the loop is feasible.

Near- and real-time forecasting is necessary for navigation and manoeuvring to prevent collisions with obstacles and other vehicles. If and/or when autonomous vehicles are used in conflict situations it should be possible to forecast near- and longer-term enemy behaviour automatically in real-time. This type of forecasting very quickly requires techniques from the contextual adaptation phase. Also, for the C2 forecasting at the contextual adaptation level is necessary. An important example thereof is the need for forecasting in combination with strategic games.

6.6.2. Strategic intelligence and foresight

BD and AI will affect the analytical, predictive as well as operational roles in national security and military intelligence environments.[22] Analytical roles detect, collect, describe, analyse, and try to model and explain why things happen. This field is currently booming, typical applications are monitoring surveillance, situational awareness and understanding. In this context BD and AI have already impacted the way intelligence organisations are operating[23]. Most techniques currently used are judgemental methods and to a smaller extent model- and data-driven methods from the statistical learning phase. The prominence of judgemental methods holds even stronger for predictive roles, so considerable improvements might be on the horizon when the contextual adaptation phase arrives. These methods might improve the forecasts of the outcome of complex negotiations and situations where numerous factors can influence outcomes of political or military processes and conflicts.[24,25]

Foresight, the process of analysing how changes in the current situation might affect us in the future[26], plays an important role in military contexts, especially strategic foresight and greatly influences and impacts national security and relative military strength and capabilities[27,28]. Forecasting plays a big role in foresight development, particularly when actions and plans based on foresight analysis are being developed. Considering the tools to develop foresight: horizon scanning, trend analysis, scenario analysis, modelling, simulation, and forecasting[29,30], and observing that some of these tools coincide with the tools used in forecasting, therefore foresight is expected to be impacted by BD and AI in a similar fashion as forecasting. Not only will BD and AI probably change the used methods and possibly the accuracy, but also new applications are developed. Examples thereof are Conflict Early Warning Systems, systems that forecast when conflicts will erupt, but also whether and when an irregular leadership change is to be expected[31,32]. Also, in these applications a mix of BD and statistical learning is the current driving factor. However, it is also here that probably only when the conceptual adaptation phase starts can AI and BD rival the current practice of judgemental and model-driven forecasts if ever[33].

6.6.3. Supply chain management, business operations, budgeting, planning, human resources

In fields as supply chain management, predictive maintenance, business operations, budgeting, planning, human resources, in short, the fields that keep the
military organisation daily business running, forecasting is used much more like in
any other government and larger private organisation. It is expected that military
organisations will follow suit. Therefore, we will not further discuss this group of
applications, just make the remark that the importance of the forecasting and decision making at this level is often undervalued: it might even be the most important
factor of determining military performance at a strategic level as it determines
for military organisations the condition in which they will enter conflict. Once in
conflict, repairs are usually expensive and might not be ready in time.

6.6.4. Forecasting at the operational and tactical level, military logistics (sustainment)

We start with a sobering thought: "Given the unpredictability of the future, in
military strategic, operational and tactical planning foresight and forecasting play
a role but commanders typically favour adaptivity and agility"[34].

The adaptivity and agility is traditionally interpreted by planners to make
sure that commanders can handle anything, any time at any place which leads in
many cases to overly conservative estimates of means needed[35]. This is very costly
and will in the future no longer be a viable manner of operating given the limited
means (human, platforms, financial) available. Also, the expanding use of more
small units makes it necessary to incorporate forecasting processes in planning and
operations, for instance by embedding forecasting teams[36].

In operations, situational awareness and understanding are already improved
by BD and statistical learning enhanced forecasting: Nowcasting, again a product
of combining statistical learning and BD is when one makes predictions of the very
recent past, the present and forecasts of the very near future. This is done as the most
recent data are not always fully available, correct or verified. Starting in macro-eco-
nomic forecasting, it has developed into a tool that improves weather forecasts and
SA in tactical operations by, for instance, forecasting crowd flows in cities.

In military logistics forecasting is a longstanding practice and procedures are
well developed, mostly based on judgemental and model-driven procedures. To get
some idea the reader is referred to Cap. Johnson and Lt. Col. Coryell's[37] description.
The procedural legacy is focused on robustness (implied by the requested operational adaptivity and agility); instead adaptivity makes it harder to implement
more agile forecasting methods as it also requires the real-world organisation to be
able to cope with the needed agility. In general, it also implies a less resilient supply

chain. It is envisioned that BD together with a combination of model and data-driven methods can improve logistics efficiency and effectiveness considerably by improved forecasting.

6.7. Conclusion

In this chapter a categorisation of forecasting methods into three instead of two categories "judgemental" and "statistical" methods is introduced by subdividing the "statistical methods" category into "model-driven" and "data-driven" methods. This is to enable the analysis of the effects of Big Data and Artificial Intelligence on forecasting practices and applications. Using DARPA's characterisation of Artificial Intelligence by considering the level of information processing capabilities "handcrafted knowledge", "statistical learning" and "contextual adaptation", the direction and relative order of the expected effects of AI and BD on forecasting are obtained. Finally, the consequences of results have been discussed for some national security, defence, and military applications.

Forecasting will become an ever more important tool for the military as the still standing military robustness practice to make sure to "be able to handle anything, any time at any place" is becoming less and less viable. This is due to costs and new ways of operating with smaller more specialist units. For the smaller units to be effective, good situational understanding is of utmost importance. Forecasting will be one of the most important instruments in the situational understanding toolbox of the future, and Big Data and Artificial Intelligence will enhance forecasting greatly.

Notes

[1]   Matthew Enderlein, "Foresight in decision making: improving intelligence analysis with probabilistic forecasting." *Small Wars Journal.* December 8, 2018. Accessed April 18, 2022. https://smallwarsjournal.com/jrnl/art/foresight-decision-making-improving-intelligenceanalysis-probabilistic-forecasting.

[2]   Fotios Petropoulos, et al., "Forecasting: theory and practice", *International Journal of Forecasting,* 2022. https://doi.org/10.1016/j.ijforecast.2021.11.001.

[3]   James Johnson, "Artificial intelligence & future warfare: implications for international security," *Defense & Security Analysis* 35, no. 2 (2019).

[4]   Roxana Radu, "Steering the governance of artificial intelligence: national strategies in perspective," *Policy and society* 40, no. 2 (2021).

[5]   Kerem Kayabay et al., "Data science roadmapping: An architectural framework for facilitating transformation towards a data-driven organization," *Technological Forecasting and Social Change* 174 (2022).

6   Mary L Cummings, "Artificial Intelligence and the Future of Warfare", In Artificial Intelligence and International Affairs: Disruption Anticipated (eds.: Cummings, Mary L. et al.), Chatham House, 2018.

7   John Launchbury, "A DARPA perspective on artificial intelligence." *Retrieved April* 6 (2017): 2022.

8   John Joseph and Jeff Corkill, "Information evaluation: how one group of intelligence analysts go about the task," (2011).

9   Ling Tang et al., "Big Data in Forecasting Research: A Literature Review," *Big Data Research* 27 (2022).

10 Tang et al., "Big Data in Forecasting Research: A Literature Review."

11 De Spiegeleire, Stephan, Matthijs Maas, and Tim Sweijs. *Artificial intelligence and the future of defense: strategic implications for small-and medium-sized force providers.* The Hague Centre for Strategic Studies, 2017.

12 John Launchbury, "A DARPA perspective on artificial intelligence." *Retrieved April* 6 (2017): 2022.

13 John Launchbury, "A DARPA perspective on artificial intelligence," *Retrieved November* 11 (2017).

14 Fotios Petropoulos, et al., "Forecasting: theory and practice", *International Journal of Forecasting*, 2022. https://doi.org/10.1016/j.ijforecast.2021.11.001.

15 Tang, Ling, Jieyi Li, Hongchuan Du, Ling Li, Jun Wu, and Shouyang Wang. "Big data in forecasting research: a literature review." *Big Data Research* 27 (2022): 100289.

16 Jon Scott Armstrong, *Principles of Forecasting: a Handbook for Researchers and Practitioners*, vol. 30 (Springer, 2001).

17 Fotios Petropoulos et al., "Forecasting: theory and practice," *International Journal of Forecasting* (2022), https://doi.org/10.1016/j.ijforecast.2021.11.001.

18 Armstrong, *Principles of Forecasting: a Handbook for Researchers and Practitioners*, 30.

19 Tang et al., "Big Data in Forecasting Research: A Literature Review."

20 Matthew Enderlein, "Foresight in decision making: improving intelligence analysis with probabilistic forecasting," *Small Wars Journal*, https://smallwarsjournal.com/jrnl/art/foresight-decision-making-improving-intelligenceanalysis-probabilistic-forecasting.

21 Anand Deshpande and Manish Kumar, *Artificial Intelligence for Big Data: Complete Guide to Automating Big Data Solutions Using Artificial Intelligence Techniques.* Packt Publishing Ltd, 2018.

22 Mary L. Cummings, "Artificial Intelligence and the Future of Warfare", in *Artificial Intelligence and International Affairs: Disruption Anticipated* (eds.: Cummings, Mary L. et al.), Chatham House, 2018.

23 Giliam de Valk, "Analytic black holes: a data-oriented perspective.," *National Security and the Future* 23, no. 1 (2022), https://doi.org/10.37458/nstf.23.1.1.

24 David R Mandel and Alan Barnes, "Accuracy of forecasts in strategic intelligence," *Proceedings of the National Academy of Sciences* 111, no. 30 (2014).

25 David R Mandel, "Accuracy of intelligence forecasts from the intelligence consumer's perspective," *Policy Insights from the Behavioral and Brain Sciences* 2, no. 1 (2015).

26 Kevin Kohler, "Strategic Foresight: Knowledge, Tools, and Methods for the Future." *CSS Risk and Resilience Reports* (2021).

27 North Atlantic Treaty Organization (2017). Strategic Foresight Analysis Report. https://www.act.nato.int/application/files/1016/0565/9725/171004_sfa_2017_report_hr.pdf

28 S. Babst, NATO's strategic foresight: Navigating between Black Swans, Butterflies and Elephants [Statement] (2018). https://securityconference.org/news/meldung/natos-strategic-foresight-navigating-between-black-swans-butterflies-and-elephants/

29 Kevin Kohler, "Strategic Foresight: Knowledge, Tools, and Methods for the Future." *CSS Risk and Resilience Reports* (2021).

30 Angela Wilkinson. "Strategic foresight primer." *European Political Strategy Centre* (2017).

31 Tim Sweijs and Joris Teer, "Practices, Principles and Promises of Conflict Early Warning Systems," (2022).

32 Anna L Buczak et al., "Crystal cube: forecasting disruptive events," *Applied Artificial Intelligence* (2021).

33 Hykel Hosni and Angelo Vulpiani, "Forecasting in light of big data," *Philosophy & Technology* 31, no. 4 (2018).

34 Enderlein, "Foresight in Decision Making: Improving Intelligence Analysis with Probabilistic Forecasting."

35 Aaron F Anderson, *Forecasting Approaches in Operations Desert Shield and Desert Storm*, US Army School for Advanced Military Studies (2020).

36 Enderlein, "Foresight in Decision Making: Improving Intelligence Analysis with Probabilistic Forecasting."

37 Michael Johnson and Brent Coryell, "Logistics forecasting and estimates in the brigade combat team," *Army Sustainment Magazine* (2016).

References

Anderson, Aaron F. "Forecasting Approaches in Operations Desert Shield and Desert Storm". US Army School for Advanced Military Studies (2020).

Armstrong, Jon Scott. *Principles of Forecasting: A Handbook for Researchers and Practitioners.* Vol. 30: Springer, 2001.

Buczak, Anna L, Benjamin D Baugher, Christine S Martin, Meg W Keiley-Listermann, James Howard, Nathan H Parrish, Anton Q Stalick, Daniel S Berman, and Mark H Dredze. "Crystal cube: forecasting disruptive events." *Applied Artificial Intelligence* (2021): 1-24.

de Valk, Giliam. "Analytic black holes: a data-oriented perspective.". *National Security and the Future* 23, no. 1 (2022): 21-48. https://doi.org/10.37458/nstf.23.1.1.

Deshpande, Anand, and Manish Kumar. *Artificial Intelligence for Big Data: Complete Guide to Automating Big Data Solutions Using Artificial Intelligence Techniques.* Packt Publishing Ltd, 2018.

Enderlein, Matthew "Foresight in Decision Making: Improving Intelligence Analysis with Probabilistic Forecasting." Small Wars Journal. https://smallwarsjournal.com/jrnl/art/foresight-decision-making-improving-intelligenceanalysis-probabilistic-forecasting.

Hosni, Hykel, and Angelo Vulpiani. "Forecasting in light of big data." *Philosophy & Technology* 31, no. 4 (2018): 557-69.

Johnson, James. "Artificial intelligence & future warfare: implications for international security." *Defense & Security Analysis* 35, no. 2 (2019): 147-69.

Johnson, Michael, and Brent Coryell. "Logistics Forecasting and Estimates in the Brigade Combat Team." Army Sustainment Magazine (2016).

Joseph, John, and Jeff Corkill. "Information Evaluation: How One Group of Intelligence Analysts Go About the Task." (2011).

Kayabay, Kerem, Mert Onuralp Gökalp, Ebru Gökalp, P Erhan Eren, and Altan Koçyiğit. "Data science roadmapping: an architectural framework for facilitating transformation towards a data-driven organization." *Technological Forecasting and Social Change* 174 (2022): 121264.

Launchbury, John. "A Darpa Perspective on Artificial Intelligence." Retrieved November 11 (2017): 2019.

Mandel, David R. "Accuracy of intelligence forecasts from the intelligence consumer's perspective." *Policy Insights from the Behavioral and Brain Sciences* 2, no. 1 (2015): 111-20.

Mandel, David R, and Alan Barnes. "Accuracy of Forecasts in Strategic Intelligence." Proceedings of the National Academy of Sciences 111, no. 30 (2014): 10984-89.

Petropoulos, Fotios, Daniele Apiletti, Vassilios Assimakopoulos, Mohamed Zied Babai, Devon K. Barrow, Souhaib Ben Taieb, and Christoph Bergmeir. "Forecasting: theory and practice." *International Journal of Forecasting* (2022). https://doi.org/10.1016/j.ijforecast.2021.11.001.

Radu, Roxana. "Steering the governance of artificial intelligence: national strategies in perspective." *Policy and Society* 40, no. 2 (2021): 178-93.

Sweijs, Tim, and Joris Teer. "Practices, Principles and Promises of Conflict Early Warning Systems." (2022).

Tang, Ling, Jieyi Li, Hongchuan Du, Ling Li, Jun Wu, and Shouyang Wang. "Big data in forecasting research: a literature review." *Big Data Research* 27 (2022): 100289.

CHAPTER 7

Military helicopter flight mission planning using data science and operations research

Roy Lindelauf, Herman Monsuur & Mark Voskuijl

Abstract

Military helicopter flight mission planning consists of many aspects ranging from for instance route selection, helicopter configuration design, opponent modelling to personnel to platform allocations. In this chapter we will survey recent algorithmic techniques from the fields of operations research, data science and aircraft trajectory optimisation that can aid in military flight mission planning automation and optimisation. We will present examples and describe potential uses with a specific focus on the Dutch Ministry of Defence. As such this chapter provides a first approach to military helicopter mission optimisation.

Keywords: *Helicopter performance, Mission planning, Route selection, Search games, Trajectory optimization*

7.1. Introduction

The nature of modern conflict varies among four archetypes of fight, ranging from counterterrorism, grey-zone and asymmetric fights to high-end conflict fights.[1] Air power provides unique opportunities to create a wide range of effects at all possible levels, ranging from the strategic to the tactical level. In the Nicaraguan intervention of 1927, aircraft supported ground troops at close range in such a manner that they were being actively coordinated toward their targets whilst at the same time being deconflicted from fire and movement of friendly ground forces. Offensive Air Support was born. Initially coordination was done by ground troops only, later however airborne platforms also started to incorporate this coordinating function.[2]

Even though we have come a long way since 1927, offensive air missions are in large part still being planned manually, e.g., battle positions are picked by hand and analysed by a large group of experienced aviators, load planning is done by load masters who do weight and balance calculations (using simple calculation tools) and route allocations are determined by visual inspection of maps and terrain data. In this chapter we introduce a methodology that can help to optimise components of helicopter mission (planning), using the key elements of the strike coordination and reconnaissance (SCAR) mission. SCAR is a typical counter-land mission that gained prominence during the Gulf War and in theatres such as Iraq and Afghanistan.

7.1.1. Strike coordination and reconnaissance

Offensive Air Support (OAS) consists of a variety of missions and can be divided into deep air support missions (DAS) that prematurely disrupt the enemy's operational cycle at a distance from friendly forces and close air support missions (CAS) providing flexible and responsive fire support against hostile targets in close proximity to friendly forces (see Figure 7.1). One key DAS mission, the use-case of this chapter on helicopter flight mission optimisation, is the so-called strike coordination and reconnaissance (SCAR) mission. It is flown in a specific geographic area with the purpose of locating, reporting and coordinating the attack of targets of opportunity and to perform battle damage assessments (BDA). See Figure 7.1 for an overview of the various types of OAS missions and SCAR's position in this taxonomy.

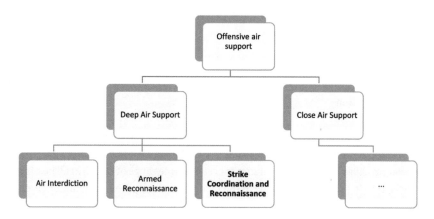

Fig. 7.1. A taxonomy of offensive air operation mission types according to Marine Corps Doctrinal Publication 3-23 Offensive Air Support.[3]

SCAR missions differ from Air Interdiction (AI) in the sense that they do not respond to targets that are known and briefed in advance. SCAR platforms discover and mark targets and/or verbally talk-on other attacking flights through the target area. Additionally, they provide prioritised targeting guidance including the location, description and threat type of targets. SCAR missions prevent redundant targeting and assist in ensuring the flow of aircraft through an assigned area of operations. Typically, a SCAR mission consists of searching for targets, target discovery, identification and tracking and finally attack coordination and execution. The key elements of SCAR type missions can therefore be summarised as follows[4]:

- Targets are not known in advance.
- Conducted by combat aircraft able to detect, mark and interdict targets including the neutralisation of enemy air defences.
- Conducted in a specified geographic zone (so-called surface kill boxes).
- SCAR aircraft are part of command-and-control interface and can cycle multiple attacking flights through the target area.
- SCAR aircraft can provide battle damage assessments (BDA).

SCAR missions are normally part of the command and control (C2) interface to coordinate multiple flights, detect and strike targets, neutralise enemy air defences, and provide battle damage assessment. Typical tasks include cycling multiple attacking flights through the target area and providing prioritised targeting guidance to maximise the effect of each sortie.

Platforms like remotely piloted aircraft can perform specific SCAR tasks such as locating, verifying, and cross-cueing other assets to positively identify targets and pass target updates. These platforms may also be able to engage targets on their own, buddy lase for manned aircraft, and provide BDA for the same mission.

7.1.2. Overview of the chapter

We illustrate how scientific methods can be used to optimise elements in helicopter mission planning in general with the SCAR mission as a particular use-case. After the collection and analysis of data concerning the mission, the geographical situation (data analysis) and the defence options of the enemy (data analysis, machine learning (ML) about locations of for example surface to air missiles (SAMs)), one enters the deployment phase, consisting of routing toward the kill-box and neutralising enemy air defence, prioritising detected targets and assigning platforms to those targets (operations research), trajectory optimisation (flight mechanics and human factors) and finally the realisation and execution of the mission (Military, data science, ML, flight guidance, scheduling). Each of these elements will be taken up in the sections below. In Section 7.2 a general overview and decomposition of helicopter mission planning is given. Optimal routing towards the kill-box and trajectory optimisation is discussed in Section 7.3. Section 7.4 deals with the methodology to optimise search patterns for target discovery inside the kill-box together with target prioritisation and allocation to platforms. Finally, we present our conclusions in Section 7.5.

7.2. Helicopter mission planning

The current helicopter mission planning process consists of several steps. It starts with a mission order. Next, information is gathered, and a plan is made. Once the plan is complete, the mission is prepared. If permitted by time and resources, pilots can practice the mission. The process ends with the take-off and the execution of the mission. This process has several shortcomings. First of all, it is time consuming since it is largely a manual process. As a consequence, it is also prone to errors and it will most likely lead to suboptimal solutions. Furthermore, it can be an inflexible process if the situation changes during the mission. In case of relatively straightforward missions, the above-mentioned shortcomings are acceptable. However, when complex missions are planned with multiple helicopters in a contested environment it would be highly beneficial to have a mission planning process in place that automatically creates optimised mission plans from which the end user can select the most appropriate one. A new mission planning process for future helicopter operations is therefore proposed. This process is summarised in Figure 7.2.

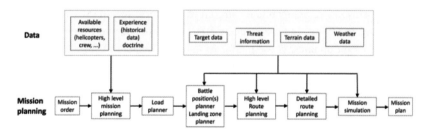

Fig. 7.2. Proposed future helicopter mission planning process based on multi-level optimization.

Just as in the current process, future mission planning starts with a mission order. In the high-level mission planning step, it is decided how many helicopters and of which type are required. This step also includes the selection of weapons and payload. Next, the payload must be distributed over the helicopters, taking into account weight and balance constraints. The load planner phase includes the determination of the required amount of fuel. In case of an attack mission, feasible locations for battle positions must be identified and from this set the most optimal location should be selected. This step can also be replaced by a landing zone planner in case the objective is to transport payload or personnel to a specific location. With the battle position known, the route planning can start. The different steps in the process require data, ranging from information about the available resources to threat information and detailed digital terrain maps. Regardless of the mission, it will consist of different phases. A part of a mission is given as an example in Figure 7.3. In this example three phases can be identified. The helicopter first performs a nap of the earth terrain following flight in order to evade enemy air defences. This flight will follow several pre-determined waypoints. Near the final destination, a manoeuvre is executed to decelerate the helicopter. Finally, the helicopter performs its intended manoeuvres at the destination.

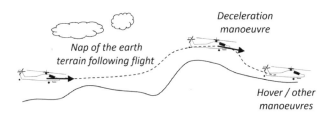

Fig. 7.3. Example elements of a military mission.

The route planning for a mission can be decomposed in two levels. High level route planning in order to determine waypoints with for example the objective to maximise survivability and minimise flight time or fuel burn. The second level pertains to the trajectory optimisation between waypoints. In this second step a more refined analysis of survivability, mission effectiveness and flight-time and fuel burn can be made, taking into account helicopter limitations and external factors such as wind. Cheng and Menon and Kim decomposed helicopter missions in a similar fashion and identified a third level designated near field guidance which is intended to provide a least expected deviation path from the nominal flight path.[5] A third level could be useful for future operations with unmanned aircraft. The route planning is described in detail in Section 7.3. Finally, if permitted by time and resources, pilots can practice the mission in the mission simulation phase.

7.3. Route planning

7.3.1. Route towards the area of operation (high level planning)

Given a mission, the helicopter has to find an optimal route towards the area of operation, and once arrived has to determine an optimal search pattern inside the kill-box. *En route* it may encounter several threats in the form of (mobile) SAM installations, interceptors/radar. A chosen route has to balance time (duration of the route) and survivability (non-detection). When an aircraft is required to penetrate air defences, it is necessary to select an optimum route that will ensure a minimum exposure to enemy fire and achieve the maximum probability of survival in accomplishing the specified mission.

To this end, the region that has to be traversed to reach the area of operation is represented by a network, where the nodes form the set of possible waypoints. For this it is necessary that (electronic) maps and terrain data are available. Intelligence may provide information about positions of SAM installations, radar, etc. Given such information, one may use simulation techniques to assign the probability of intercepting a traversing helicopter on a path between two consecutive nodes. This enables optimising a route from entry point A to target B that has the largest probability of survival and non-detection.

One may use linear programming techniques to find this route: given the probability of detection p_{ij} on a path from node (waypoint) i to node (waypoint) j, the following optimisation problem finds the route with minimum detection probability. The route can be identified from the optimal solution in terms of the variables x_{ij}, which is 1 if the optimal route contains the path from i to j.

$$\min \sum_{i \to j} -\ln\left(1 - p_{ij}\right) x_{ij}$$

subject to

$$\sum_{j:i \to j} x_{ij} - \sum_{j:j \to i} x_{ji} = 0, i \neq A, B$$

$$\sum_{i:A \to i} x_{Ai} = 1$$

$$\sum_{i:i \to B} x_{iB} = 1$$

with all variables binary.

The airspace between points A and B is defended against any penetrating aircraft and the probability of kill over any segment of flight within that space will be assumed to be known, as in the linear program above, either in terms of the actual probabilities or probability densities per unit distance or unit time. To ensure a minimum exposure to enemy fire and achieve the maximum probability of survival in accomplishing the specified mission for aircraft, one may also use this linear program to identify on which paths a reduction of probability of detection is most rewarding. In other words, one may try to identify the installation whose destruction or degradation is most rewarding in terms of survivability of a penetrating aircraft. Analogous to this problem, the enemy may try to maximise this probability of interception by placing its installations (mobile or fixed) at optimal locations from a set of possible locations. These two problems then become a so-called max-min or min-max problem, for which several solution procedures exist.[6] One may also use game-theory to devise optimal interdiction plans for a flow of helicopters that try to penetrate the area.[7] In many cases, one may use genetic algorithms to solve these problems. Results of these analyses provide the waypoints that are input to the trajectory optimisation phase.

7.3.2. Trajectory optimisation (detailed planning)

The waypoints determined in the high-level route planning process provide the global path which the helicopter has to follow in order to optimise a combination of survivability and flight time. The geometric distance between the nodes (waypoints) in the high-level route planning process should be small enough to capture the most important features of the terrain and the threat. At the same time, the geometric distance should not be too small for two reasons. First of all, a denser grid increases the computational effort. Second, if the grid is too dense, solutions may be found which are infeasible when taking into account the dynamics and physical limitations of the helicopter. In practice, waypoints are expected to have a distance which typically requires flight times ranging from 15 seconds up to 5 minutes. At this stage, the optimal path of the helicopter in three-dimensional space between the waypoints is not yet determined. This optimal path consists not only of positions in space but also of the groundspeed of the helicopter as a function of time. The performance of a helicopter in terms of manoeuvrability and fuel burn depends primarily on airspeed, flight altitude and the current weight of the helicopter. External factors such as wind and atmospheric conditions also play an important role. For the analysis of optimal trajectories, one has to take into account the limits of the helicopter such as engine and drivetrain limitations, the maximum load factor and airspeed limits. The technique to solve this type of problem is called trajectory optimisation. It is essentially a problem within the field of optimal control theory which has a wide variety of applications in the aerospace domain such as space flight, missile guidance and the calculation of environmentally friendly trajectories for transportation aircraft.

1. Problem definition and numerical methods

When applied specifically to helicopters, Bottasso defines trajectory optimisation as:[8]

> "The process of computing the optimal control inputs and the resulting response of a model of a helicopter which minimizes a cost function (or maximizes an index of performance) while satisfying given constraints which specify for example, the vehicle flight envelope boundaries and/or safety and procedural requirements for a manoeuvre of interest."

An example path of a helicopter between two waypoints is visualised in Figure 7.4. This finite-time transition from condition 1 to condition 2 is called a manoeuvre.[9]

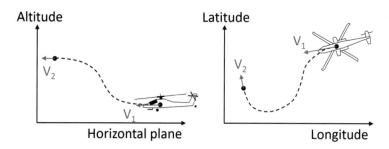

Fig. 7.4. Example manoeuvre.

2. Numerical methods to solve the trajectory optimisation problem.

In mathematical terms the optimal control problem can be defined as follows:

$$\begin{array}{c} \min \\ \underline{u}(t) \in U \end{array} \int_{t_0}^{t_1} L[\underline{x}(t), \underline{u}(t), t] dt$$
$$subject\ to : \underline{\dot{x}}_i = f_i[\underline{x}(t), \underline{u}(t), t]$$

Where the objective function depends on the control usage $\underline{u}(t)$, the state of the helicopter $\underline{x}(t)$ and the time (t). The control usage $u(t)$ is subject to rate limits and maximum deflections. It therefore belongs to a closed and bounded set U. The performance index is expressed as an integral of a function (L) of the state variables, control usage and time. Hence, survivability, mission effectiveness, fuel burn and flight time can be incorporated in the objective function. The set of first order nonlinear differential equations represents the dynamics of the helicopter which is a function (f) of its state variables and control variables. There is a large variety of numerical methods which can be used to solve the trajectory optimisation problem. The methods which are most widely used for trajectory optimisation problems in the aerospace domain are summarised in Table 7.1. This summary is based on the studies by Betts, Rao and Ben-Asher.[10]

The primary two classes of numerical methods are direct and indirect methods. Indirect methods rely on the analytical derivation of adjoint equations and their gradients. This can be very challenging if the physical model of the vehicle is complex and nonlinear as is typically the case for helicopter simulation models. The analytical derivation is not required for direct methods. Direct methods are based on a parameterisation of the control function. For sake of completeness, 'other' methods are also included in the table. Both dynamic programming and heuristic methods have been applied successfully to trajectory optimisation problems, but they are computationally more expensive than direct and indirect methods.

Table 7.1. Numerical methods to solve the trajectory optimisation problem

Direct methods	Indirect methods	Other methods
Single shooting	Single shooting	Dynamic programming
Multiple shooting	Multiple shooting	Heuristic (usually genetic algorithms)
Collocation (transcription)	Collocation (transcription)	
Pseudospectral		

3. Helicopter flight simulation models

In order to solve the trajectory optimisation problem, a simulation model of the helicopter is required as well. The helicopter can be modelled with different levels of fidelity, ranging from basic point mass models to high fidelity flight dynamics models. Padfield proposed a classification of different levels of model fidelity.[11] An overview of the classification, extended with a *Level 0* model is presented in Table 7.2. The *level 0* model is a point mass model of the helicopter. Performance charts such as those listed in a helicopter flight manual can serve as an aerodynamic database for such a model. Alternatively, simple aerodynamic models based on momentum theory can be used in conjunction with a point mass model.

Table 7.2. Classification of helicopter models used for trajectory optimisation (extension of classification by Padfield[12])

Model designation	Equations of motion (fuselage)	Rotor dynamics	Aerodynamics
Level 0	Point mass model or perfor-mance constraints	Not included	Look-up tables with power required for complete helicopter or momentum theory combined with look-up tables for fuselage aerodynamics. Aerodynamics are not included in case the helicopter is modelled based on performance constraints.
Level 1	Rigid body	Quasi steady or rigid blades with flap and possibly lag and torsional degrees of freedom	Linear 2-D dynamic inflow or local momentum theory. Analytically integrated loads

Model designation	Equations of motion (fuselage)	Rotor dynamics	Aerodynamics
Level 2	Rigid body	Rigid blades with flap and possibly lag and torsional degrees of freedom	Nonlinear (limited 3-D) dynamic inflow or local momentum theory. Local effects of blade vortex interaction. Unsteady 2-D compressibility. Numerically integrated loads
Level 3	Rigid body or elastic body	Elastic blades	Nonlinear 3-D full wake analysis, unsteady 2-D compressibility numerically integrated loads

Hence, different types of simulation models and numerical methods for optimisation can be used. The overarching question is which combination of numerical method and helicopter simulation model is most suitable in the context of helicopter mission planning considering all operational constraints. In the next section, a comprehensive literature survey is presented of helicopter trajectory optimisation applications in scientific literature.

4. Survey of helicopter trajectory optimisation applications
Research in the field of helicopter trajectory optimisation with military relevance dates back to the early 1980s. Falco and Smith investigated the influence of manoeuvrability on helicopter combat effectiveness.[12] This method was intended to be used for short term manoeuvres in the context of helicopter conceptual design. Slater and Hertzberger used an indirect method to optimise a complete mission of a civil helicopter with the aim to minimise fuel burn.[13] Menon and Kim performed a variety of studies where the objective was to maximise a linear combination of terrain masking and flight time. Terrain masking is defined as the integral of the altitude above the terrain over the complete trajectory. An indirect method was used in combination with a level 0 model of a military helicopter (AH-1S Cobra).[14] Dynamic programming techniques were later used by Nikiforova to find optimal terrain following trajectories for attack missions.[15] More than 10 years later, a similar method was used by Toupet and Mettler to optimise trajectories of an unmanned military helicopter in obstacle avoidance flight.[16] A key difference in the latter study is that the algorithm was used for online computation on-board the unmanned helicopter. The studies described above all made use of low fidelity (level 0) helicopter simulation models. Bottasso performed various studies in which higher fidelity simulation models were successfully used to optimise specific short-term manoeuvres both in the civil and military domain.[17] Higher fidelity simulation models are typically black box models developed by specialists. For this reason, direct methods were applied

in the studies by Bottasso. Since noise and emissions are becoming more and more important to the helicopter community, several studies were performed in the last years with the aim to minimise helicopter noise for approach and departure trajectories of civil helicopters.[18] It is a challenge to accurately predict helicopter noise. A key aspect in these latest research efforts was the inclusion of accurate noise models within the helicopter simulation model. The literature survey of trajectory optimisation specifically for helicopters is summarised in Table 7.3.

5. Proposed trajectory optimisation approach for the Royal Netherlands Air Force
The Royal Netherlands Air Force operates various helicopter types. The mission planning process should therefore accommodate missions in which different helicopter types are used simultaneously. In the future, one could even include (unmanned) fixed wing aircraft and rotorcraft in this process. Of each helicopter type, the performance and its limitations are known in detail based on flight manuals and technical manuals. Detailed flight simulation models are available for pilot training. However, these models are typically black box models provided by helicopter manufacturers in combination with the flight simulator hardware. Therefore, these models cannot easily be integrated within an automated mission planning and optimisation process. Altogether, it is recommended to create dedicated *level 0* or *level 1* helicopter simulation models for the mission planning process. These models should be combined with direct methods for trajectory optimisation as demonstrated in the studies by Hartjes and Visser.[19]

7.4. Detect, mark and interdict targets

In this paragraph we discuss techniques from operations research and game theory that can help to optimise other elements of helicopter missions, in particular regarding the strike coordination and reconnaissance mission.

7.4.1. Prioritising targets in areas of operation / situation awareness / coordinating multiple flights

Intelligence reports, based on historical analysis of data using common data scientific techniques, might indicate (or predict) the presence of specific targets in the kill-box designated to SCAR platforms prior to their mission. This a priori information can be used even though targets are not known in advance in a typical SCAR mission, e.g., the platforms conducting the SCAR mission in a designated geographic area are searching for unknown targets that may be static or dynamic in nature. A research field of particular use in this domain is the field of search games

Table 7.3. Survey of trajectory optimisation studies for helicopter applications

Helicopter category	Manoeuvre/mission phase	Timescale (seconds)	Objective(s)	Numerical method	Flight simulation	Special features	References
Military utility	Hover depart	15	Flight time	Direct collocation	Level 1		Bottasso et al. 2008
	Lateral reposition	20					
	Pirouette	60					
	Slalom	30					
Civil tiltrotor	Continued and rejected take-off	20	Safety and flight time	Direct multiple shooting and direct single shooting	Level 2		Bottasso et al. 2012
Military utility	Minimum time turn	15	Flight time	Direct collocation	Level 1		Bottasso et al. 2010
	Pirouette	60					
	Slalom	30					
	Continued take-off	20	Safety and flight time				
Civil tiltrotor	Minimum time turn	15	Flight time	Direct multiple shooting	Level 2		Bottasso et al. 2010
Military utility	Lateral reposition	20	Flight time	Direct multiple shooting and direct collocation	Level 1	Inclusion of pilot model	Bottasso et al. 2009
	Pirouette	60					
	Slalom	30					
	Rejected take-off	20	Safety and flight time				
	Engine off landing	10	Safety				
Civil	Descent	150	Noise	Genetic algorithm	Level 0	Noise source and propagation models	Cruz et al. 2012
Military attack	Air-to-air missile avoidance	30	Survivability	Stochastic learning method	Level 0	Missile and gun models	Falco 1982
	Air-to-air gun combat	30					

Civil utility	Flight over populated area	1000	Noise	Heuristic	Only performance constraints	Acoustic database	Greenwood 2019
Civil utility	Approach trajectories, Climb to cruise, Course change	300	Minimal accelerations (smooth flight)	State Dependent Ricatti Equation technique	Level 0 and Level 2	Automatic flight path vector generation and tracking	Halbe et al. 2018
Civil utility	Approach trajectories	300	Noise and fuel burn	Pseudospectral	Level 0	Basic noise model	Hartjes et al. 2009
Civil utility	Approach trajectories	300	Flight time, fuel burn and NO_x emissions	Pseudospectral	Level 1	Basic noise model	Hartjes et al. 2011
Civil utility	Approach trajectories	400	Noise, flight time and fuel burn	Pseudospectral	Level 1	Noise source, noise propagation and nonstandard atmosphere models	Hartjes & Visser 2019
Military attack	Nap of the earth terrain following flight	60	Altitude and flight time	Various indirect methods	Level 0	Wind included	Menon and Kim, 1990
Civil small unmanned	Obstacle avoidance in urban terrain	60	Flight time	Various	Level 0	Standardised approach for comparing guidance systems	Mettler & Kong 2010
Military attack	Nap of the earth terrain following flight	300	Altitude and flight time	Dynamic programming	Level 0		Nikiforova 1995
Civil transport	Climb, cruise and descent	3600	Fuel burn and flight time	Indirect method based on energy height approach	Level 0	Helicopter performance based on flight manual	Slater & Ertzberger 1984
Military unmanned	Obstacle avoidance	variable	Flight time	Dynamic programming	Level 0	Online trajectory optimisation for obstacle free elements of the flight	Toupet &Mettler 2006

which stems from search theory.[20] The focus of search theory is to find optimal ways to search for targets in a plethora of settings that occur between searchers and targets (often called 'hiders'). If one also has information in terms of heat maps or has intelligence of how targets move around in the area, specialised search and detection theory techniques may be employed to find a target as soon as possible and with the highest probability. These methods are based on meta-heuristics and discrete Markov chains for instance.[21] Table 7.4 presents an overview of the most common search games (as found in literature).[22]

Table 7.4. Search games classification[23]

		Searcher's strategy		
		Special	**Moving**	**Resource distribution**
Hider's strategy	**Special**	Smuggling game		Inspection game
	Stationary	Binary search game	Linear search game Hide-search game	Hide-allocation game
	Moving	Path constrained search game	Evasion-search game Princess-monster game	
	Resource distribution			Search-search game Blotto game Attack-defence game

Consider for instance platforms observing a fixed location known for possible illicit activity. If due to resource constraints (budget, personnel, platforms) the platforms could observe the location only for m times in a fixed time-window whereas the opponent has n opportunities ($n > m$) to conduct its illicit activity (smuggle weapons for instance), then this situation can be modelled as an inspection game and optimal strategies for both players determined. On the other hand, if the opponent is dynamic and aware of being searched for in a continuous space and time, the evasion-search game is more suited to model and find optimal strategies (continuous search paths for the helicopters).

Once targets are within sensor range of SCAR platforms, one may use neural networks and data science to classify and mark the objects using techniques from image recognition.[24] Multiple SCAR platforms active in a given kill-box can use their datalink to communicate about found targets and use onboard target prioritisation algorithms to rank them. Techniques like the orienteering problem can then be used to plan and coordinate a route that visits the set of targets that maximise the total profit, while also giving a final classification.[25] There are several OR techniques that can be used in this regard. These include methods for online planning and planning with time windows. Online planning may be very important to take into consideration new targets that appear. In that case the sequence of visiting the targets can be interrupted and a new optimal route has to be computed. To minimise the effect of these interruptions, one uses meta-heuristics and robust planning techniques to also take into consideration these online events in advance.

Additionally, network models can be used to assess the situation awareness in coordinated attacks. The intelligence reports regarding the targets (combined with the results of neural networks) may be combined with availability and capability of individual helicopters, resulting in an assignment problem at the operational level that can be represented as an integer linear programming problem. It gives a deployment plan that may serve as input at the tactical level of a mission.

7.4.2. Battle position planner

Even though battle positions are generally not an element of a typical SCAR mission we include them in this paragraph as they are important to (attack) helicopter missions in general. To determine battle positions and route to a mission essential enemy targets in an area of operation, one may utilise game theory, that models the situation of conflicts with opposing interest. It has, for example, been applied to attacking or defending a ship that coordinates and executes a mission in a coastal area by a submarine. The essential ship has frigates, helicopters with dipping sonar to protect itself. In such a situation, movements of some friendly assets are detectable by the enemy (like the movement of a frigate), and some are only detectable when it is too late (like the dipping sonar of a helicopter). In these situations, one may find an optimal plan for the deployment of friendly forces, as well as an optimal attack plan. The same approach may be used to determine optimal battle positions and attack plans.[26] An optimal configuration of, say four helicopters to attack an enemy wave of aircraft can be found using war game simulations, neural networks and heuristics.[27] If the configuration consists of unmanned aircraft, the results become even more interesting, as some of the aircraft in an optimal configuration serve as decoy and are attacked. Such configuration with a very risky position for one aircraft is less likely to be flown in the case of manned aircraft.

7.5. Conclusions

In this chapter, we investigated how Operations Research, Trajectory Optimisation and Data Science may be utilised in helicopter mission planning. In particular, we focused on SCAR missions to identify where optimisation techniques may improve the planning of such missions. The blue part of figure 7.2 (terrain and target data) lends itself for applications of modern AI techniques and image recognition. In the figure below, which illustrates a part of figure 7.2, we highlight the multi-level optimisation approach that we explored in more depth in this chapter, using OR techniques, including artificial intelligence and machine learning.

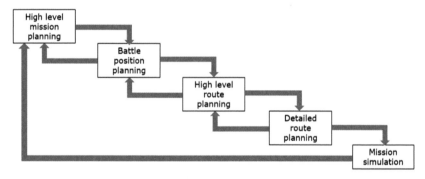

Fig. 7.5. Multi-level optimization approach.

As stated in the introduction, offensive air missions are in large part still being planned manually, e.g., battle positions are picked by hand and analysed by a large group of experienced aviators, load planning is done by load masters who do weight and balance calculations (using simple calculation tools) and route allocations are determined by visual inspection of maps and terrain data. In this approach experienced military personnel play a significant role.

Valuable as this is, we propose to add to this the application of OR techniques. As explained, the battle position planning, the route planning and the prioritising of targets in the area of operation can be supported by already existing scientific methods. The same holds for coordinating multiple flights. Applying these methods and using the outcomes of these analyses can be used as decision support. It serves as a platform for discussion to improve the situation awareness of the whole mission with all its aspects, tests tacit assumptions, and reveals the various subjective opinions. We conjecture that, using field testing in a simulated environment, one may already prove the added value of applying existing OR tools as described in the previous three sections. This may hold for the various aspects of a mission in isolation, as well as for planning and evaluating the mission as a whole. All the different elements can be automated to avoid a time-consuming manual process. It also avoids operator fatigue, especially in the case of a SCAR helicopter mission consisting of more than one wave, generating additional tactical situations.

Given the complexity of the interacting processes (terrain information, weather, routing, neutralising, prioritisation, coordinating) as shown in the figure above, classical OR techniques may be used in an iterative manner to generate an optimal or even suboptimal solution that takes into account all these interacting processes in a balanced way. Combined with modern image recognition techniques regarding terrain and target data, and AI/ML, one may develop a comprehensive approach to mission planning.

Tactical problems (like battle position planning, prioritising targets, etc.) become (only) difficult when there is a scarcity of helicopters and other means. This scarcity may be due to operational considerations that assigns just a few helicopters to the mission but may also be the result of maintenance issues, operational reserve and logistic reserve. Given the number of targets and the influence of enemy air defence, this scarcity poses several other problems in the planning process, from planning for battle positions to prioritising targets. For example, a tactical issue may be: do we distribute our means evenly over the targets or do we create excess situations for just a few targets. Here a game-theoretic approach may be beneficial as it is able to generate and identify dominating strategies.

Notes

[1] Raphael Cohen, Nathan Chandler, Shira Efron, Bryan Frederick, Eugeniu Han, Kurt Klein, Forrest Morgan, Ashley Rhoades, Howard Shatz and Yuliya Shokh. *The future of warfare in 2030: project overview and conclusions*. Santa Monica, California: RAND Corporation, 2020.

[2] Michael Bergerud. *Two heads are better than one: the need for a two-seat aircraft for strike coordination and reconnaissance missions and airborne forward air control missions*. Quantico, Virginia: Master of Military Studies Thesis, United States Marine Corps, Command and Staff College, 2001.

[3] In Dutch Air Force doctrine SCAR is a specific mission type within the counter-land operations task in the air attack role.

[4] See for instance NATO Allied Joint Doctrine for Close Air Support and Air Interdiction – AJP 3.2.2(A).

[5] Victor H.L. Cheng. "Obstacle-avoidance automatic guidance." *AIAA Guidance, Navigation and Control Conference*. Minneapolis, 1988; Padmanabhan Menon and E. Kim. *Optimal helicopter trajectory planning for terrain following flight*. Contractor Report 177607, NASA, 1990.Menon and Kim, *Optimal Helicopter Trajectory Planning*.

6   David Alderson, Gerald Brown and Matthew Carlyle. "Assessing and improving operational resilience of critical infrastructure and other systems." *INFORMS Tutorials in Operations Research* (2014): 180-215.

7   Corine Laan, Tom van der Mijden, Ana Barros, Richard Boucherie and Herman Monsuur. "An interdiction game on a queueing network with multiple intruders." *European Journal of Operational Research* 260, no. 3, (2017): 1069-1080.

8   Bottasso, Carlo, Giorgio Maisano and Francesco Scorcelletti. "Trajectory optimization procedures for rotorcraft vehicles, their software implementation, and applicability to models of increasing complexity." *Journal of the American Helicopter Society* 55, no. 3, (2010): 0320101-03201013.

9   Bottasso, Carlo, Francesco Scorcelletti, Giorgio Maisano and Andrea Ragazzi. "Trajectory optimization strategies for the simulation of ADS-33 mission task elements." *34th European Rotorcraft Forum*. Liverpool, 2008.

10 John Betts. "Survey of Numerical Methods for Trajectory Optimization." *Journal of Guidance, Control and Dynamics* 21, no. 2, (1998): 193-207; Anil Rao. "A survey of numerical methods for optimal control." *Advances in Astronautical Sciences* 135, no. 1, (2010): 497-528; Joseph Ben-Asher. *Optimal control theory with aerospace applications*. Reston, Virginia: AIAA Education Series, American Institute for Aeronautics and Astronautics, 2010.

11 Gareth Padfield. *Helicopter flight dynamics*. Oxford, United Kingdom: Blackwell Publishing, 2007, p. 90.

12 Michael Falco and Roger Smith. "Influence of maneuverability on helicopter combat effectiveness." *38th Annual Forum of the American Helicopter Society*. Annaheim, California, 1982.

13 Gary Slater and Heinz Erzberger. "Optimal short range trajectories for helicopters." *Journal of Guidance, Control and Dynamics* 7, no. 4, (1984): 393-400.

14 Menon and Kim, *Optimal Helicopter Trajectory Planning*. 1990.

15 Nikiforova, Lidia N. "Nap-of-the-Earth flight optimization using optimal control techniques." 21st European Rotorcraft Forum. Saint Petersburg, 1995.

16 Berenice Mettler, Goerzen Zhaodan, Chad Kong and Matthew Whalley. "Benchmarking of obstacle field navigation algorithms for autonomous helicopters." *66th Annual Forum of the American Helicopter Society*. Phoenix, Arizona, 2010.

17 Bottasso et al. *Trajectory optimization strategies...*, 2008; Bottasso, Maisano and Scorcelletti, *Trajectory optimization procedures ...*, 2009; Bottasso, Maisano and Scorcelletti, *Software implementation*, 2010; Bottasso, Luraghi and Maisano, *Efficient rotorcraft trajectory optimization...*, 2012.

18 Sander Hartjes, Hendrikus Visser and Marilena, D. Pavel. "Optimization of simultaneous non-interfering rotorcraft approach trajectories." *35th European Rotorcraft Forum*. Hamburg, Germany, 2009.; Sander Hartjes and Hendrikus Visser. "Environmental optimization of rotorcraft approach trajectories." *37th European Rotorcraft Forum*. Vergiate and Gallarate, Italy, 2011.; Luis Cruz, Andrea Massaro, Stefano Melone, and Andrea D'andrea. "Rotorcraft multi-objective trajectory optimization for low noise landing procedures." *38th European Rotorcraft Forum*. Amsterdam, The Netherlands, 2012; Eric Greenwood. "Dynamic replanning of low noise rotorcraft operations." *75th Annual Forum of the Vertical Flight Society*. Philadelphia, Pennsylvania, 2019.; Sander Hartjes and Hendrikus G. Visser. "Optimal control approach to helicopter noise abatement trajectories in nonstandard atmospheric conditions." *Journal of Aircraft* 56, no. 1, (2019): 43-52.

19 Sander Hartjes and Hendrikus G Visser. "Optimal control approach to helicopter noise abatement trajectories in nonstandard atmospheric conditions." *Journal of Aircraft* 56, no. 1, (2019): 43-52.

20 Steve Alpern, Robbert Fokkink, Leszek Gasieniec, Roy Lindelauf and V. S. Subrahmanian. *Search Theory: A Game Theoretic Perspective*. New York: Springer, 2013.

21 Alan Washburn. *Search and detection*. Create Space Independent Publishing Platforms, 2014

[22]   Ryusuke Hohzaki. "Search Games: Literature and Survey." *Journal of the Operations Research Society of Japan* 59, no. 1, (2016): 1-34.

[23]   Taken from Hohzaki, "Search Games: Literature and Survey," 1-34.

[24]   Geraldo M. de Lima Filho, Felipe L.L Medeiros, and Angelo Passaro. "Decision support system for unmanned combat air verhicle in beyond visual range air combat based on artificial neural networks." *Journal of Aerospace Technology and Management*, 2021.

[25]   Evers, Lanah. *Robust and agile UAV mission planning.* Rotterdam: PhD Thesis, Erasmus University, 2012.

[26]   Herman Monsuur, Rene Janssen, and Rick Jutte. "A game-theoretic attacker-defender model for a sea-base: optimal deployment at the maritime battleground." In: *Optimal deployment of military systems,* by Patrick Oonincx and Arjan van der Wal, 179-206. Asser Press, 2014.

[27]   Geraldo de Lima Filho, Andre Kuroswiski, Felipe Medeiros, Mark Voskuijl and Herman Monsuur. "Optimization of unmanned air vehicle tactical formation in war games." *IEEE Access* 10, (2022): 21727-21741.

References

Alderson, David, Gerald Brown and Matthew Carlyle. "Assessing and improving operational resilience of critical infrastructure and other systems." *INFORMS Tutorials in Operations Research* (2014): 180-215.

Allied joint doctrine for close air support and air interdiction. Allied Joint Publication (AJP)3.3.2(A), STANAG 3736 AO, 2019.

Alpern, Steve, Robbert Fokkink, Leszek Gasieniec, Roy Lindelauf and V. S. Subrahmanian. *Search Theory: A Game Theoretic Perspective.* New York: Springer, 2013.

Ben-Asher, Joseph. *Optimal control theory with aerospace applications.* Reston, Virginia: AIAA Education Series, American Institute for Aeronautics and Astronautics, 2010.

Bergerud, Michael. *Two heads are better than one: the need for a two-seat aircraft for strike coordination and reconnaissance missions and airborne forward air control missions.* Quantico, Virginia: Master of Military Studies Thesis, United States Marine Corps, Command and Staff College, 2001.

Betts, John. "Survey of Numerical Methods for Trajectory Optimization." *Journal of Guidance, Control and Dynamics* 21, no. 2, (1998): 193-207.

Bottasso, Carlo, Fabio Luraghi and Giorgio Maisano. "Efficient rotorcraft trajectory optimization using comprehensive models by improved shooting methods." *Aerospace Science and Technology* 23, no. 1, (2012): 34-42.

Bottasso, Carlo, Francesco Scorcelletti, Giorgio Maisano and Andrea Ragazzi. "Trajectory optimization strategies for the simulation of ADS-33 mission task elements." *34th European Rotorcraft Forum.* Liverpool, 2008.

Bottasso, Carlo, Giorgio Maisano and Francesco Scorcelletti. "Trajectory optimization procedures for rotorcraft vehicles including pilot models with applications to ADS-33 MTEs CAT-A and engine-off landings." *65th Annual Forum of the American Helicopter Society.* Texas, 2009.

Bottasso, Carlo, Giorgio Maisano and Francesco Scorcelletti. "Trajectory optimization procedures for rotorcraft vehicles, their software implementation, and applicability to models of increasing complexity." *Journal of the American Helicopter Society* 55, no. 3, (2010): 0320101-03201013.

Cheng, Victor H L. "Obstacle-avoidance automatic guidance." *AIAA Guidance, Navigation and Control Conference.* Minneapolis, 1988.

Cohen, Raphael, Nathan Chandler, Shira Efron, Bryan Frederick, Eugeniu Han, Kurt Klein, Forrest Morgan, Ashley Rhoades, Howard Shatz and Yuliya Shokh. *The future of warfare in 2030: project overview and conclusions.* Santa Monica, California: RAND Corporation, 2020.

Cruz, Luis, Andrea Massaro, Stefano Melone, and Andrea D'andrea. "Rotorcraft multi-objective trajectory optimization for low noise landing procedures." *38th European Rotorcraft Forum.* Amsterdam, The Netherlands, 2012.

de Lima Filho, Geraldo, Andre Kuroswiski, Felipe Medeiros, Mark Voskuijl and Herman Monsuur. "Optimization of unmanned air vehicle tactical formation in war games." *IEEE Access* 10, (2022): 21727-21741.

de Lima Filho, Geraldo, Felipe Medeiros and Angelo Passaro. "Decision support systems for unmanned combat air vehicle in beyond visual range air combat based on artificial neural networks." *Journal of Aerospace Technology and Management* 13, (2021): 1-18.

Evers, Lanah. *Robust and agile UAV mission planning.* Rotterdam: PhD Thesis, Erasmus University, 2012.

Falco, Michael and Roger Smith. "Influence of maneuverability on helicopter combat effectiveness." *38th Annual Forum of the American Helicopter Society.* Annaheim, California, 1982.

Greenwood, Eric. "Dynamic replanning of low noise rotorcraft operations." *75th Annual Forum of the Vertical Flight Society.* Philadelphia, Pennsylvania, 2019.

Hartjes, Sander and Hendrikus G Visser. "Optimal control approach to helicopter noise abatement trajectories in nonstandard atmospheric conditions." *Journal of Aircraft* 56, no. 1, (2019): 43-52.

Hartjes, Sander and Hendrikus Visser. "Environmental optimization of rotorcraft approach trajectories." *37th European Rotorcraft Forum.* Vergiate and Gallarate, Italy, 2011.

Hartjes, Sander, Hendrikus Visser and Marilena, D Pavel. "Optimization of simultaneous non-interfering rotorcraft approach trajectories." *35th European Rotorcraft Forum.* Hamburg, Germany, 2009.

Hohzaki, Ryusuke. "Search Games: Literature and Survey." *Journal of the Operations Research Society of Japan* 59, no. 1, (2016): 1-34.

Laan, Corine, Tom van der Mijden, Ana Barros, Richard Boucherie and Herman Monsuur. "An interdiction game on a queueing network with multiple intruders." *European Journal of Operational Research* 260, no. 3, (2017): 1069-1080.

Menon, Padmanabhan and E. Kim. *Optimal helicopter trajectory planning for terrain following flight.* Contractor Report 177607, NASA, 1990.

Mettler, Berenice, Zhaodan, Goerzen, Chad Kong and Matthew Whalley. "Benchmarking of obstacle field navigation algorithms for autonomous helicopters." *66th Annual Forum of the American Helicopter Society.* Phoenix, Arizona, 2010.

Monsuur, Herman, Rene Janssen, and Rick Jutte. "A game-theoretic attacker-defender model for a sea-base: optimal deployment at the maritime battleground." In *Optimal deployment of military systems*, by Patrick Oonincx and A van der Wal, 179-206. Asser Press, 2014.

Nikiforova, Lidia N. "Nap-of-the-Earth flight optimization using optimal control techniques." *21st European Rotorcraft Forum.* Saint Petersburg, 1995.

Padfield, Gareth. *Helicopter flight dynamics.* Oxford, United Kingdom: Blackwell Publishing, 2007.

Rao, Anil. "A survey of numerical methods for optimal control." *Advances in Astronautical Sciences* 135, no. 1, (2010): 497-528.

Slater, Gary and Heinz Erzberger. "Optimal short range trajectories for helicopters." *Journal of Guidance, Control and Dynamics* 7, no. 4, (1984): 393-400.

Toupet, Olivier and Bernard Mettler. "Design and flight test evaluation of guidance system for rotorcraft." *AIAA Guidance, Navigation and Control Conference and Exhibit.* Keystone, Colorado, 2006.

Washburn, Alan. *Search and detection.* Create Space Independent Publishing Platforms, 2014.

CHAPTER 8

Applying GTSP-algorithms in maritime patrolling missions that require mutual support

Martijn van Ee, Geraldo de Lima Filho & Herman Monsuur

Abstract

The goal of a maritime patrol is to detect, locate and identify (opposing) vessels. In some of these missions, the patrol aircraft may need the support of another aircraft to counter a possible threat. This support may consist of using capabilities of the other aircraft, such as its sensors, communications or its weapons. Our work aims to optimise maritime patrol routes for two or more drones, while providing the possibility of mutual support. We show that this routing problem including mutual support can be modelled as a generalised travelling salesman problem (GTSP). We investigate the costs of requiring mutual support and compare it to the costs of using separate drones that detect and identify vessels in the area of operations.

Keywords: Generalised travelling salesman problem, Maritime patrolling, Mutual support, Unmanned aerial vehicles

8.1. Introduction

In this chapter, we consider the problem of routing two drones, while guaranteeing the possibility for mutual support at any point in time. We therefore first consider maritime patrols and discuss and illustrate the need for mutual support in modern warfare. We then discuss related routing problems that may be found in the literature, which are variants of the classical travelling salesman problem (TSP).

8.1.1. Maritime patrolling and mutual support

The Exclusive Economic Zone (EEZ) is an area beyond and adjacent to a coast or territorial sea to a limit of 200 nautical miles from the baseline. In this zone, the coastal state can practice sovereign rights over the exploration, conservation and management of resources and other economic activities, such as the production of tidal or wind energy. The coastal state is mainly responsible for the preservation of living resources in the EEZ. It has judicial and supervisory powers within its EEZ to counter dumping of waste from ships and pollution from deep-sea activity.[1] The EEZ zone was first formalised in international law. It was a compromise between maritime powers and coastal states. As part of the United Nations Convention on the Law of the Sea (UNCLOS) "deal package" it recognised coastal states' rights to economic resources within the EEZ, while still protecting (to some extent) access by other states to the area for non-economic purposes.[2] Nevertheless, military activities in the EEZ have been controversial as the UNCLOS text does not explicitly address military activities, leaving room for different interpretations. This has resulted in several diplomatic and military conflicts between some of the world's major powers and has been the subject of much academic debate. The focal point of disagreement is the level to which UNCLOS allows other states to carry out military activities, including surveillance, within a coastal state's EEZ. The concerns are high: while some see military activities as an unacceptable threat to a coastal state's sovereignty, others see them as securing and protecting maritime trade and under-sea cables, and of crucial importance to the global economy and global security.[3]

Maritime patrol is the action of employing aerospace means to detect, locate, identify, monitor, or neutralise opposing vessels. These vessels can operate in open sea, inland waters or other maritime spaces of interest to naval operations. Maritime patrol comprises several activities that require integrated and synchro-nised cooperation with friendly naval forces. It may also include the supervision of jurisdictional activities carried out by the Navy.[4] An aerial maritime surveillance (or maritime patrol) mission can be divided into five tasks: detection (locating something of interest), classification (boat, iceberg, oil slick, etc.), identification (name and nationality of a vessel), inspection (determination of vessel or object activity), and enforcement (issue warning or evidence collection).[5] There are many classes of maritime patrol missions: patrol and law enforcement, investigation and evidence collection, maritime search and rescue, monitoring of oil spilling and pollution of ship discharge, emergency action, patrolling and examination of the buoy, survey of the channel, and international maritime supervision or maritime patrol.[6] All these activities can be carried out by active (radar) and passive (thermal and electro-optical) sensors airborne on the aircraft and are sent to a control centre by a datalink system.

Several studies in the literature have shown the substitution of manned vehi-cles for unmanned vehicles in various military missions, including maritime patrol missions.[7,8,9] This replacement has been taking place due to several advantages such as: greater autonomy, performance (e.g., agility), overload durability, and stealth capacity.[10] Other advantages are the possibility to admit aircraft losses from a tactical formation in favour of the victory of a squadron in a given combat scenario[11], and improved accuracy when firing a missile.[12] However, there are a few practical downsides of this substitution. For example, drones, also known as unmanned aerial vehicles, may be more susceptible to electronic interference (for example jamming, which can reduce radar range, disrupt communication/data link, interfere with the aircraft control, and interfere with the drone's passive sensors) or mechanical interference (use of weapons). In a manned aircraft, electronic interference would be easily recognised by an experienced operator. An attack with kinetic weapons would be unlikely to occur on a manned aircraft, due to the legal implications. This illustrates that drones are more susceptible to interference.

In order to try to guarantee the execution of an effective and robust maritime patrol, which aims to collect data necessary for the control and surveillance of the maritime area, an additional drone can be used to guarantee mutual support between the two aircraft. Mutual support is a commitment within a flight of two or more aircraft that supports the mission objectives of the flight. In modern combat, mutual support is directly related to situational awareness.[13,14] But, in addition, an effective mutual support arrangement will also allow a maritime patrol to generate offensive power to neutralise hostile intentions, thereby increasing the survivability of the team of aircraft. The mutual support may also provide an operational capability in the event of the inoperability of one of them. That is, in the event of a failure of electronic equipment in one of the drones, the other drone must be at such a distance that ensures it can continue to perform the task that the failing drone cannot perform. In this paper we assume that the drones constantly maintain a maximum separation distance, while carrying out their patrol independently. This model assumption, the maximum separation distance, aggregates all the various aspects and types of mutual support like protection and sustaining operational capabilities. An example of this capability is the SAR (synthetic-aperture radar) whose functions are identification and surveillance, data link for command and control, and exchange of operational messages.

To summarise, the objective of this work is to optimise maritime patrol routes for two or more drones, providing the possibility of mutual support. We assume that the drones have been assigned a number of contacts in the area of operation which they have to investigate. We show that this routing problem including mutual support can be modelled as a generalised travelling salesman problem (GTSP). We prove that it can be turned into a regular TSP. We also investigate the costs of requiring mutual support and compare it with the costs of using separate drones that detect and identify vessels in the area of operations.

8.1.2. Related literature

Most of the works related to the routing of maritime patrols that can be found in the literature aim to optimise the flight time, the fuel consumption, or aim at maximising the amount of collected data. In the case of a swarm of drones, it seeks to optimise detection and tracking of targets.[15] A Lagrangian swarm model that has the capability of covering large areas of the sea effectively is presented by Kumar and Vunualailai.[16] The controllers derived in this work generate a linear formation which, if applied to dynamical systems, will have the ability to be a very good model for the surveillance of an Economic Exclusive Zone, as well as for search and rescue. The monitoring of trawler activities in Kuala Keda was studied by Suteris et al.[17] They create a route optimisation method for unmanned air vehicles for maritime surveillance. The aim is to find a short tour, in terms of time, to cover all locations at sea either by boat or by both a boat and a drone. Optimisation of a drone's trajectory for maritime radar wide area surveillance is discussed by Brown and Anderson.[18] The considerations of the dynamics, propulsion, and mission requirements of a fixed-wing drone, as well as a maritime surveillance radar, provide a method to obtain the fuel consumption, probability of detection, and revisit time for a given trajectory.

The main contribution of our work concerns the optimisation of maritime patrol routes while maintaining a maximum distance between two drones. This distance is based on the range of their sensors. This way, the two drones can provide mutual support throughout the route, also in a hostile environment. Comparable models are studied by Murray and Chu for example.[19] Their one considers the flying sidekick travelling salesman problem for the optimisation of drone-assisted parcel delivery. Carlsson and Song investigated routing problems where a drone provides service to customers while making return trips to a truck that is itself moving.[20] In other words, a drone picks up a package from the truck (which continues on its route), and after delivering the package, the drone returns to the truck to pick up the next package. Similarly, Agatz et al. study last-mile delivery concepts in which a truck collaborates with a drone.[21]

8.1.3. Travelling salesman problems

In this paper, we use several versions of the travelling salesman problem (TSP). In TSP, we are given a set of vertices V, with $|V| = n$, and costs $c(i,j)$ for vertices $i, j \in V$. The goal is to find a tour, that visits every vertex at least once, of minimum total cost. We discuss two variants: 2-person TSP (2TSP), and generalised TSP (GTSP).

In an instance of 2-person TSP, we additionally have a root vertex.[22] Now, a feasible solution consists of two tours, starting and ending in the root, such that every vertex is visited at least once. The goal is to minimise the maximum cost of the two tours. We will use this variant when we consider the inefficiency caused by demanding mutual support.

In the generalised TSP[23], we are given a set of node sets $S_1, \dots, S_m \subseteq V$. A feasible solution is a tour, that visits at least one vertex from each node set. Again, the goal is to minimise the tour cost. We will transform our routing problem, including mutual support, into GTSP. Then, we show that any instance of GTSP can be reduced to an instance of TSP.

8.1.4. Outline

Now that we have illustrated the need for mutual support during maritime patrolling, we provide a mathematical model in Section 8.2. There we define the mathematical problem of routing two drones, while demanding mutual support at any point in time. Our model can be described as a generalised travelling salesman problem (GTSP). In Section 8.3, we will show how to reduce any instance of GTSP into an instance of a regular, albeit asymmetric TSP. Mutual support comes at a cost compared to a patrol of two independently routed drones. Therefore, we study the inefficiency of demanding mutual support, both theoretically and based on real-life instances. In Section 8.5, we discuss how to generalise our approach to more than two drones, and what issues should be considered when doing this. Finally, we give our conclusion and present operational considerations.

8.2. Problem description and reduction to GTSP

We call our problem the 2-travelling salesman problem with mutual support, abbreviated to 2TSP-MS. We assume that the drones are identical, and that we will operate at a constant altitude. Hence, in an instance of 2TSP-MS, we are given a set of n points $P \subset \mathbb{R}^2$, and a root $r \in \mathbb{R}^2$. Let $d(x, y)$ denote the Euclidean distance between points $x \in P$ and $y \in P$. Each of the two drones flies a tour on a subset of the points, where it can vary its speed over time, with a maximum speed of S. The distance between the two drones has to be at most R at any point in time. Each point has to be visited at least once. The goal is to find two tours and corresponding speeds such that the time spent is minimised.

To keep the solution space comprehensible, we assume that the drones have to visit points from P simultaneously. This mimics the assumptions in the related literature, where drones return to the truck at delivery points only.[24,25,26] Of course, one may construct examples that this assumption comes at some extra costs. By matching the speeds of the two drones, we can guarantee that they will always be within a distance of R units apart, without explicitly tracking the drones over time (see Theorem 1). We also allow a drone to wait at a point. This implies that the drone will loiter around a location if it has to fly at some minimum speed. To accommodate the need to visit points that are at distance more than R from any other point in P, we assume that the two drones can visit the same point simultaneously. Finally, we assume that there is no limit on the endurance or range of our drones. Note that this final assumption might restrict the speed at which we can fly. This implies that we might choose a lower maximum speed S. Moreover, the maximum speed also depends on the altitude we have chosen.

An instance and a feasible solution of 2TSP-MS is illustrated in Figure 8.1. In the solution, drone 1 flies the route A, B, D, F, and drone 2 flies the route C, E, F. First, drone 1 flies at maximum speed S from r to A and B. Drone 2 flies at a lower speed such that the drones arrive at points A and C simultaneously. Then, drone 2 waits at C until drone 1 has reached point B. Both then continue at maximum speed to their next destinations D and E. From D to F, drone 1 again flies at maximum speed, while drone 2 lowers its speed from E to F to arrive at the same time at point F. Finally, both drones fly at maximum speed back to the root.

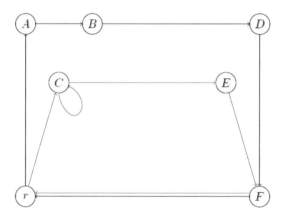

Fig. 8.1. An instance and feasible solution of 2TSP-MS.

To transform an instance of 2TSP-MS into an instance of GTSP, we use the notion of a configuration. If we use configuration (x, y), it means that drone 1 visits $x \in P$, and drone 2 visits $y \in P$ simultaneously. We only consider configurations (x, y) that satisfy $d(x, y) \leq R$. If the drones move from configuration (A_1, B_1) to configuration (A_2, B_2), both drones will travel in a straight line. The drone that has to travel the largest distance flies at maximum speed S. To make sure that the two drones arrive at their destinations A_2 and B_2 at the same time, the other drone then can adjust its speed to, for example, the speed S' that is specified in the proof of Theorem 1. This also ensures that at any point in between, the two drones are at most distance R from each other.

Theorem 1

If the two drones move from configuration (A_1, B_1) to configuration (A_2, B_2) as described above, they spend time

$$\frac{\max\{d(A_1, A_2), d(B_1, B_2)\}}{S},$$

while ensuring that the distance between the drones is at most R at any point in time.

Proof: We first show that moving from configuration (A_1, B_1) to configuration (A_2, B_2) takes $\max\{d(A_1, A_2), d(B_1, B_2)\}/S$ time. The drone that has to travel the largest distance travels at maximum speed S. It therefore arrives at time $\max\{d(A_1, A_2), d(B_1, B_2)\}/S$. The other drone can adjust its speed to

$$S' = \frac{\min\{d(A_1, A_2), d(B_1, B_2)\}}{\max\{d(A_1, A_2), d(B_1, B_2)\}} S < S$$

Moving at this speed, it will also arrive at its destination at time $\max\{d(A_1, A_2), d(B_1, B_2)\}/S$.

To show that the distance between the two drones is at most R at any point in time, let the vectors $\vec{a}_1, \vec{a}_2, \vec{b}_1, \vec{b}_2 \in \mathbb{R}^2$ denote the coordinates of points $A_1, A_2, B_1,$ and B_2, respectively. For $0 \le t \le \max\{d(A_1, A_2), d(B_1, B_2)\}/S$, the position of drone 1 can be parametrised as

$$\left(1 - \frac{t}{\max\{d(A_1, A_2), d(B_1, B_2)\}/S}\right) \vec{a}_1 + \frac{t}{\max\{d(A_1, A_2), d(B_1, B_2)\}/S} \vec{a}_2,$$

while the position of drone 2 can be parametrised as

$$\left(1 - \frac{t}{\max\{d(A_1, A_2), d(B_1, B_2)\}/S}\right) \vec{b}_1 + \frac{t}{\max\{d(A_1, A_2), d(B_1, B_2)\}/S} \vec{b}_2.$$

Note that at $t = 0$, drone 1 is at point A_1 and drone 2 is at point B_1. Similarly, at $t = \max\{d(A_1, A_2), d(B_1, B_2)\}/S$, drone 1 is at point A_2 and drone 2 is at point B_2. Moreover, both drones move at a constant speed. The distance between the drones at any time $0 \le t \le \max\{d(A_1, A_2), d(B_1, B_2)\}/S$ is at most R, as follows from the following reasoning, where we let $u = t/\left(\max\{d(A_1, A_2), d(B_1, B_2)\}/S\right)$.

$$
\begin{aligned}
\left\|(1-w)\vec{a}_1 + u\vec{a}_2 - \left((1-w)\vec{b}_1 + u\vec{b}_2\right)\right\| &= \left\|(1-w)(\vec{a}_1 - \vec{b}_1) + u(\vec{a}_2 - \vec{b}_2)\right\| \\
&\le \left\|(1-w)(\vec{a}_1 - \vec{b}_1)\right\| + \left\|u(\vec{a}_2 - \vec{b}_2)\right\| \\
&= (1-u)\left\|(\vec{a}_1 - \vec{b}_1)\right\| + u\left\|(\vec{a}_2 - \vec{b}_2)\right\| \\
&\le (1-w)R + uR \\
&= R
\end{aligned}
$$

∎

Using Theorem 1, a feasible solution to 2TSP-MS is a sequence of configurations (A_1, B_1), $(A_2, B_2), ..., (A_k, B_k)$ such that for each point $x \in P$ we have $x \in \{A_i, B_i\}$ for at least one i. In this case, we say that the configuration contains x. The sequence represents the route the two drones are flying, in which they start and end at the root r. For all $x \in P$, we allow (x, x) to be a configuration: both drones simultaneously visit x. This, for example, enables us to visit an isolated point $x \in P$ (no y exists such that $d(x, y) \leq R$). It also may occur that one of the drones keeps hovering at the same point y, while the other drone flies from point z to another point q. Therefore (y, z) and (y, q) can be two consecutive configurations in such a sequence.

We now shall formulate 2TSP-MS as a generalised travelling salesman problem (GTSP), using the following construction.

- Create a vertex (x, y) for each configuration (x, y) with $d(x, y) \leq R$. In that case, also (y, x) is a vertex.
- Create a directed arc between any pair of vertices (x, y) and (u, v).
- Define the costs of an arc from (x, y) to (u, v) as $\max\{d(x, u), d(y, v)\}/S$. The costs of the arc between the root r and (x, y) is equal to $\max\{d(r, x), d(r, y)\}/S$.
- Define for each of the n points x a nodeset S_x that contains all vertices that contain x.

This results in a complete (see Theorem 1), bi-directed graph, with overlapping nodesets. Feasible solutions for this GTSP have to take at least one vertex from each nodeset, starting and ending at r. It is easy to verify that a feasible solution to our original problem is a feasible solution of the GTSP, and vice versa. So, 2TSP-MS is equivalent to GTSP on the created instance. The reduction is illustrated in Figure 8.2.

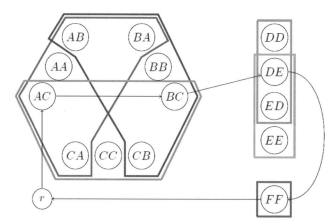

Fig. 8.2. Illustration of reduction from 2TSP-MS to GTSP, where vertex XY corresponds to configuration (x, y). The nodesets are illustrated by areas with colored borders. The black arcs depict a solution of the GTSP-instance, that corresponds to the solution of 2TSP-MS in Figure 8.1.

Although the focus of this paper is not on computational issues, we refer to research that investigates efficiency of meta-heuristics for the GTSP. Ben Nejma and Mhalla consider local search algorithms to solve the GTSP.[27] Cosma et al. present a novel Hybrid Genetic Algorithm (HGA) for solving the GTSP.[28] Finally, Wu et al. use genetic algorithms that can solve the GTSP directly without the need of intermediate transformations to TSP, the issue of the next section.[29]

8.3. Reduction to TSP

To solve our problem with two drones that require mutual support, we could transform the GTSP of the previous section into a travelling salesman problem. We provide the transformation in two steps:
- Transform the GTSP into a GTSP without overlapping nodesets.
- Apply the reduction from Noon and Bean from GTSP without overlapping nodesets to TSP.[30]

Remember that in the GTSP of the previous section, vertices (x, y), with $x, y \in P$ and $d(x, y) \leq R$ are grouped into nodesets. Our nodesets overlap. For example, S_x, as well as nodeset S_y, contains the vertex (x, y). To eliminate the overlap between two nodesets, we use the following construction.
- Make two copies of each vertex (x, y) with $x \neq y$: $(x, y)^x$ and $(x, y)^y$.
- Introduce nodesets T_z for each point $z \in P$: $(u, v)^u$ and $(v, u)^u$ are in T_z if, and only if, $u = z$. Vertex (z, z) is also included in T_z.
- Only arcs between vertices belonging to different nodesets are copied from the original GTSP, with corresponding costs. Here, the costs between $(x, y)^x$ and $(x, y)^y$ are zero. Note that, for example, an arc from (x, y) to (x, z) in the original GTSP may be transferred to an arc from $(x, y)^x$ to $(x, z)^z$.

This way we again obtain a bi-directed graph. Now, we have a GTSP without overlapping nodesets. A feasible solution of this GTSP has to take at least one vertex from each nodeset. Again, it is easy to verify that corresponding feasible solution can be found with the same total costs.

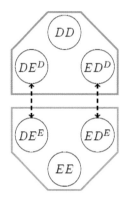

Fig. 8.3. Illustration of result of reduction from GTSP to GTSP without overlapping node sets, where vertex X Y^X corresponds to $(x, y)^x$. The dashed lines have costs zero.

To explain the reduction from GTSP without overlapping nodesets to a TSP, we use the approach of Noon and Bean.[31]

For this TSP, we introduce an arbitrary ordering of the vertices in a node-set S_i: $i_0, i_1, ..., i_{|S_i|-1}$. (Note that each i_j is a configuration (x, y).) Next, we introduce new arcs.

— Create in each nodeset a single, directed cycle according to the given ordering, i.e., $i_0, i_1, ..., i_{|S_i|-1}, i_0$, with the cost of each arc equal to zero.

— Define the costs of directed arcs between any two nodeset as follows: the cost of the arc from $i_j \in S_i$ to $k_l \in S_k$ in the created TSP-instance is set equal to the cost of the arc from $i_{j-1(\bmod |S_i|)}$ to k_l in the original GTSP-instance.

Now, take an arbitrary solution of our GTSP without overlapping nodesets, say $r, i_j, k_l, ..., p_q, r$. This then corresponds to the solution $r, i_j, i_{j+1}, ..., i_{j-1}, k_l, k_{l+1}, ..., k_{l-1}, p_q, p_{q+1}, ..., p_{q-1}, r$, where all indices of i, k, and p are modulo $|S_i|$, $|S_k|$, and $|S_p|$, respectively. This has the same total cost. To enforce that optimal solutions of the TSP are of this type (meaning that optimal tours visit all nodes of a nodeset before visiting nodes of another nodeset), we add a very large cost β to each directed arc between nodesets. This means that optimal tours will have to visit each nodeset just once and will have a total cost strictly less than $n\beta$. These solutions correspond to optimal solutions of our GTSP without overlapping nodesets. In sum, we proved the following theorem:

Theorem 2

Any instance of 2TSP-MS can be transformed into an instance of asymmetric TSP. An optimal solution of the created instance of TSP corresponds to an optimal solution of the instance of 2TSP-MS, and vice versa.

8.4. Mutual support gap

In this section, we consider the inefficiency caused by demanding mutual support. In order to quantify this inefficiency we define the mutual support gap. Here, an instance of 2TSP is a set of points $P \subset \mathbb{R}^2$ and a root $r \in \mathbb{R}^2$. An instance of 2TSP-MS in addition specifies the maximum allowed distance R between the two drones. For a given P, r, and R, let $Opt_{2TSP-MS}(P, r, R)$ be the optimal value of 2TSP-MS, and let $Opt_{2TSP}(P, r)$ be the optimal value of 2TSP. Note that, in 2TSP, both drones can fly at maximum speed at any point in time. We define $MSG(P, r, R)$, the mutual support gap, as

$$MSG(P, r, R) = \frac{Opt_{2TSP-MS}(P, r, R)}{Opt_{2TSP}(P, r)}.$$

We will first study this quantity theoretically.

Theorem 3

For any instance $P \subset \mathbb{R}^2, r \in \mathbb{R}^2, R \in \mathbb{R}$, we have

$$1 \leq MSG(P, r, R) \leq 2.$$

Proof: First, note that we can adjust any solution of 2TSP-MS to a solution of 2TSP by setting all speeds equal to the maximum speed. Since this will increase the time spent, we have

$$Opt_{2TSP}(P, r) \leq Opt_{2TSP-MS}(P, r, R).$$

On the other hand, a feasible solution for 2TSP-MS is to simultaneously fly a TSP-tour at maximum speed. Let $Opt_{TSP}(P, r)$ be the optimal value of TSP on $P \cup \{r\}$. Moreover, since the total length of the tours in the optimal solution for 2TSP is at most $2Opt_{2TSP}(P, r)$, we have

$$Opt_{2TSP-MS}(P, r, R) \leq Opt_{TSP}(P, r) \leq 2Opt_{2TSP}(P, r).$$

∎

To see that our upper bound of 2 is tight, consider the instance in Figure 8.4. Here, we have a root r, and points A and B. Furthermore, we have $d(r, A) = d(r, B) = \alpha$, and $d(A, B) = 2\alpha$, where $\alpha > R$. In the optimal solution of 2TSP-MS, the two drones will fly together from r to A, from A to B, and from B to r. The value of this solution equals 4α. On the other hand, in the optimal solution of 2TSP, one of the drones visits A, while the other visits B. The value of this solution is equal to 2α.

Fig. 8.4. Instance with mutual support gap 2.

We also investigated the magnitude of the mutual support gap for some real-life instances, provided by the Brazilian Air Force. One of these instances, and the optimal solutions for 2TSP-MS and 2TSP are illustrated in Figure 8.5 and Figure 8.6, respectively. To clarify, the solution for 2TSP-MS in Figure 8.5 uses configurations $(9,9)$, $(6,5)$, and $(6,4)$ consecutively. Hence, the first drone waits at point 6 to sustain mutual support. In Figure 8.6, the first drone primarily focuses on visiting the point farthest from the depot, to minimise the maximum time a drone is flying. It is also clear that mutual support is not provided here. The mutual support gap for this instance is approximately 1.234, i.e., the time taken by the solution of 2TSP-MS is 23.4% more than the time taken by the solution of 2TSP.

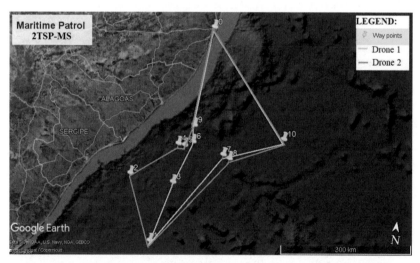

Fig. 8.5. Solution 2TSP-MS. Here, drone 2 waits at point 6, while drone 2 flies from point 5 to point 4.

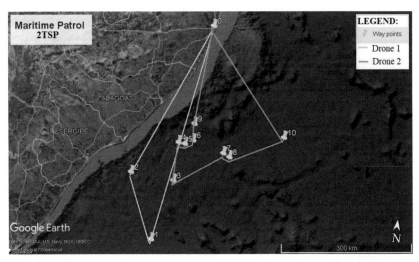

Fig. 8.6. Solution 2TSP.

Naturally, the mutual support gap of an instance relies heavily on geographical features. If the points are located in a "wide" area, like in Figure 8.4, and if the number of isolated points is large, numerical experiments show that the mutual support gap can be as large as 1.5. On the other hand, if the points are "close" to each other, and the number of configurations is large, the mutual support gap can be as low as 1.05.

8.5. Considerations for more drones

So far, we only considered our routing problem for two drones. In this section, we discuss which issues should be considered when dealing with more than two drones.

In Section 8.2, we created a configuration for each pair of points $x, y \in P$, possibly with $x = y$, with $d(x, y) \leq R$, describing the location of each of the drones. This definition naturally generalises to k drones. Whether a certain configuration is created depends on how you define the need for mutual support. For example, if you have three drones, you might define mutual support as having at least one other drone within distance R. Alternatively, you could require that all other drones are within distance R. Depending on your definition of mutual support, it is clear which configurations should be included in the reduction to GTSP. Now, we will discuss two important issues that pop up in the reduction.

A first consideration is the size of the created instance. In the original reduction, we created at most n^2 configurations. If we have k drones, this will lead to creating at most n^k configurations. Hence, the number of configurations grows exponentially in the number of drones. This limits the scalability of our approach. On the other hand, it might not be sensible to use this generalisation if the number of drones is large. We are still assuming that the drones arrive at their destinations simultaneously. Already for four drones, this assumption can increase the time spent significantly. In this case, it might be more convenient to partition the set of points and assign a pair of drones to each of the parts. However, it can be quite complicated to find the optimal partition.

Another issue is related to the result of Theorem 1. There, we showed that the drones can move between any pair of configurations spending the least amount of time possible, while guaranteeing mutual support. This is still true for more than two drones if we define mutual support as having all other drones within a distance of R. As we will show, this result does not always hold for the case with at least three drones if we define mutual support as having at least one other drone within distance R. Here, it is always possible to move the drones from points $\{A, B, C\}$ to points $\{D, E, F\}$ and guarantee mutual support, but this might not be the most efficient movement between the two sets.

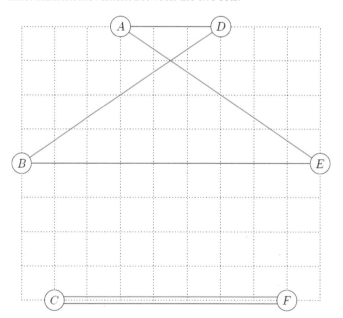

Fig. 8.7. Instance, with R = 5, showing that it might not be possible to move from any permutation of {A, B, C} to any permutation of {D, E, F} in the fastest way while guaranteeing mutual support.

To see this, consider the example in Figure 8.7, where $R = 5$. If the drones move from configuration (A, B, C) to configuration (D, E, F) (blue lines), we guarantee mutual support. This movement takes $9/S$ time. If we move from configuration (A, B, C) to configuration (E, D, F) (red lines), we only need $2\sqrt{13}/S$ time. However, this move is not allowed since the third (bottom) drone is not supported by either the first or the second drone for some period of time.

This example shows that we cannot move between any pair of configurations anymore. This will not lead to infeasible instances, since it is always possible to move between any pair of subsets of points, while maintaining mutual support. To check whether a certain movement is allowed, we have to solve three quadratic inequalities. For $k > 2$ drones, we need to solve $\binom{k}{2}$ quadratic inequalities to verify if a movement between two configurations is allowed.

8.6. Conclusion

Drones have been used for decades in maritime patrol by several countries.[32] Although many advantages are obtained from their use, drones may require special care depending on the type and risk of the mission. Hence, mutual support can be a way to ensure the fulfilment of the mission. In this paper we prove, using the notion of configurations, that a routing problem including mutual support can be modelled as a generalised travelling salesman problem (GTSP), and can be transformed into a regular TSP. We also examined the costs of operating with mutual support and compared it to the costs of using separate drones that detect and identify vessels. This gap can be as big as a factor 2. However, in instances with a large number of configurations, i.e., possibilities to simultaneously visit distinct points, the gap can be as low as a factor 1.05.

Modelling this problem with more than two drones was also addressed. With k drones, it would create at most n^k configurations, so the number of configurations grows exponentially in the number of drones. Hence, increasing the number of drones considerably increases the computational cost for calculating the optimal solution, and the total mission time (total cost). From an operational perspective, when the drone quantity is an even number and the mutual support condition can be fulfilled by a pair of drones, one should consider the possibility of planning the maritime patrol by dividing the surveillance area so that it is patrolled in pairs of drones.

Notes

[1]   P. Hoagland, J. Jacoby, and M.E. Schumacher, "Law of the sea," in *Encyclopedia of Ocean Sciences (Second Edition)*, ed. John H. Steele (Oxford: Academic Press, 2001), 432–443.

[2]   United Nations General Assembly, "Convention on the law of the sea" (vol. 1833), 10 December 1982.

[3]   Simon McKenzie, "Autonomous technology and dynamic obligations: Uncrewed maritime vehicles and the regulation of maritime military surveillance in the exclusive economic zone," *Asian Journal of International Law* 11, no. 1 (2021):146–175.

[4]   Brazilian Air Force, "Doutrina básica da Força Aérea Brasileira, Vol. II," 2020.

[5]   Siu O'Young and Paul Hubbard, "Raven: A maritime surveillance project using small UAV," *in Proceedings of the 12th IEEE Conference on Emerging Technologies and Factory Automation, Patras, 2007*, 904–907.

[6]   Gui-Jun Duan and Peng-Fei Zhang, "Research on application of UAV for maritime supervision," *Journal of Shipping and Ocean Engineering* 4 (2014): 322–326.

[7]   Robyn Hopcroft, Eleanore Burchat, and Julian Vince, "Unmanned aerial vehicles for maritime patrol: human factors issues," Technical report, Defence Science and Technology Organisation Edinburgh (Australia) Air Operations Division, 2006.

[8]   Sandeep Ameet Kumar and Jito Vanualailai, "A Lagrangian UAV swarm formation suitable for monitoring exclusive economic zone and for search and rescue," in *Proceedings of the 1st IEEE Conference on Control Technology and Applications, Kohala Coast, 2017*, 1874–1879.

[9]   Simon McKenzie, "Autonomous technology and dynamic obligations: Uncrewed maritime vehicles and the regulation of maritime military surveillance in the exclusive economic zone," *Asian Journal of International Law* 11, no. 1 (2021): 146–175.

[10]   Zhen Yang, Zhixiao Sun, Haiyin Piao, Yiyang Zhao, Deyun Zhou, Weiren Kong, and Kai Zhang, "An autonomous attack guidance method with high aiming precision for UCAV based on adaptive fuzzy control under model predictive control framework," *Applied Sciences* 10, no. 16 (2020): 5677.

[11]   Geraldo Mulato de Lima Filho, André Rossi Kuroswiski, Felipe Leonardo Lôbo Medeiros, Mark Voskuijl, Herman Monsuur, and Angelo Passaro, "Optimization of unmanned air vehicle tactical formation in war games," *IEEE Access* (2022): 21727-21741.

[12]   Geraldo Mulato de Lima Filho, Felipe Leonardo Lôbo Medeiros, and Angelo Passaro, "Decision support system for unmanned combat air vehicle in beyond visual range air combat based on artificial neural networks," *Journal of Aerospace Technology and Management* 13 (2021).

[13]   Air Combat Command. "Multi-command handbook 11-F16, Flying Operations, F-16 combat aircraft fundamentals" (Vol.5), 1996.

[14]   Herman Monsuur, "Assessing situation awareness in networks of cooperating entities: A mathematical approach," *Military Operations Research* (2007): 5–15.

[15]   Guilherme Amaral, Hugo Silva, Flávio Lopes, João Pedro Ribeiro, Sara Freitas, Carlos Almeida, Alfredo Martins, José Almeida, and Eduardo Silva, "UAV cooperative perception for target detection and tracking in maritime environment," in *Proceedings of the Oceans'17 MTS/IEEE Conference, Aberdeen, 2017*, 1–6.

[16]   Sandeep Ameet Kumar and Jito Vanualailai, "A Lagrangian UAV swarm formation suitable for monitoring exclusive economic zone and for search and rescue," in *Proceedings of the 2017 IEEE Conference on Control Technology and Applications, Kohala Coast, 2017*, 1874–1879.

[17]   Muhamad Syazwan Suteris, F.A. Rahman, and Azman Ismail, "Route schedule optimization method of unmanned aerial vehicle implementation for maritime surveillance in monitoring trawler activities in Kuala Kedah, Malaysia," *International Journal of Supply Chain Management* 7, no. 5 (2018): 245–249.

[18]   Angus Brown and David Anderson, "Trajectory optimization for high-altitude long-endurance UAV maritime radar surveillance," *IEEE Transactions on Aerospace and Electronic Systems* 56, no. 3 (2019): 2406–2421.

[19]   Chase C. Murray and Amanda G. Chu, "The flying sidekick traveling salesman problem: Optimization of droneassisted parcel delivery," *Transportation Research Part C: Emerging Technologies* 54 (2015): 86–109.

[20]   John Gunnar Carlsson and Siyuan Song, "Coordinated logistics with a truck and a drone," *Management Science* 64, no. 9 (2018): 4052–4069.

[21]   Niels Agatz, Paul Bouman, and Marie Schmidt, "Optimization approaches for the traveling salesman problem with drone," *Transportation Science* 52, no. 4 (2018): 965–981.

[22]   Greg N. Frederickson, Matthew S. Hecht, and Chul E. Kim, "Approximation algorithms for some routing problems," in *Proceedings of the 17th International Symposium on Foundations of Computer Science, Houston, 1976*, 216–227.

[23]   Charles E. Noon and James C. Bean. "An efficient transformation of the generalized traveling salesman problem." *INFOR: Information Systems and Operational Research* 31, no. 1 (1993): 39–44.

[24]   Chase C. Murray and Amanda G. Chu, "The flying sidekick traveling salesman problem: Optimization of droneassisted parcel delivery," *Transportation Research Part C: Emerging Technologies* 54 (2015): 86–109.

[25]   John Gunnar Carlsson and Siyuan Song, "Coordinated logistics with a truck and a drone," *Management Science* 64, no. 9 (2018): 4052–4069.

[26]   Niels Agatz, Paul Bouman, and Marie Schmidt, "Optimization approaches for the traveling salesman problem with drone," *Transportation Science* 52, no. 4 (2018): 965–981.

[27]   Ibtissem Ben Nejma and Hedi Mhalla, "Efficient local search algorithms for GTSP: A comparative study," in *Proceedings of the 1st International Conference on Electrical, Computer and Energy Technologies, Cape Town, 2021*, 1–6.

[28]   Ovidiu Cosma, Petrică C Pop, and Laura Cosma, "An effective hybrid genetic algorithm for solving the generalized traveling salesman problem," *Proceedings of the 16th International Conference on Hybrid Artificial Intelligence Systems, Bilbao, 2021*, 113–123.

[29]   Chunguo Wu, Yanchun Liang, Heow Pueh Lee, and Chun Lu, "Generalized chromosome genetic algorithm for generalized traveling salesman problems and its applications for machining," *Physical Review E* 70 (2004): 016701.

[30]   Charles E Noon and James C Bean, "An efficient transformation of the generalized traveling salesman problem," *INFOR: Information Systems and Operational Research* 31, no. 1 (1993): 39–44.

[31]   Charles E. Noon and James C. Bean, "An efficient transformation of the generalized traveling salesman problem," *INFOR: Information Systems and Operational Research* 31, no. 1 (1993): 39–44.

[32]   Gui-Jun Duan and Peng-Fei Zhang, "Research on application of UAV for maritime supervision," *Journal of Shipping and Ocean Engineering* 4 (2014): 322–326.

References

Agatz, Niels, Paul Bouman, and Marie Schmidt. "Optimization approaches for the traveling salesman problem with drone." *Transportation Science 52, no.4*, (2018): 965-981.

Air Combat Command. "Multi-command handbook 11-F16, Flying Operations, F-16 combat aircraft fundamentals." (1996).

Amaral, Guilherme, et al. "UAV cooperative perception for target detection and tracking in maitime environment." *Proceedings of the Oceans '17 MTS/IEEE Conference*. Aberdeen: IEEE, 2017. 1-6.

Ben Nejma, Ibtissem, and Hedi Mhalla. "Efficient local search algorithms for GTSP: A comparitive study." *Proceedings of the 1st International Conference on Electrical, Computer and Energy Technologies.* Cape Town: IEEE, 2021. 1-6.

Brazilian Air Force. "Doutina básica da Força Aérea Brasileira, Vol. II." 2020.

Brown, Angus, and David Anderson. "Trajectory optimization for high-altitude long-endurance UAV maritime radar surveillance." *IEEE Transaction on Aerospace and Electronic Systems,* 2019: 2406-2421.

Carlsson, John Gunnar, and Siyuan Song. "Coordinated logistics with a truch and a drone." *Management Science 64, no. 9,* 2018: 4052-4069.

Cosma, Ovidiu, Petrică C. Pop, and Laura Cosma. "An effective hybrid genetic algorithm for solving the generalized traveling salesman problem." *Proceedings of the 16th Internation Conference of Hybrid Artificial Intelligence Systems.* Bilbao: Springer, 2021. 113-123.

Duan, Gui-Jun, and Peng-Fei Zhang. "Research on application of UAV for maritime supervision." *Journal of Shipping and Ocean Engineering 4,* 2014: 322-326.

Frederickson, Greg N., Matthew S. Hecht, and Kim. "Approximation algorithms for some routing problems." *Proceedings of the 17th International Symposiu on Foundation of Computer Science.* Houston: IEEE, 1976. 216-227.

Hoagland, P., J. Jacoby, and M.E. Schumacher. "Law of the sea." In *Encyclopedia of Ocean Sciences (Second Edition),* by John H. Steele, 432-443. Oxford: Academic Press, 2001.

Hopcroft, Robyn, Eleanore Burchat, and Julian Vince. *Unmanned aerial vehicles for maritime patrol: human factor issues.* Defence Science and Technology Organisation Edinburgh (Australia) Air Operations Division, 2006.

Kumar, Sandeep Ameet, and Jito Vanualailai. "A Lagrangian UAV swarm formation suitable for monitoring exclusive exonomic zone and for search and rescue." *Proceedings of the 1st IEEE Conference on Control Technology and Applications.* Kohala Coast: IEEE, 2017. 1874-1879.

Lima Filho, Geraldo M. de, André R. Kuroswiski, Felipe L.L. Medeiros, Mark Voskuijl, Herman Monsuur, and Angelo Passaro. "Optimization of Unmanned Air Vehicle Tactical Formation in War Games." *IEEE Access,* 2022: 21727-21741.

Lima Filho, Geraldo M. de, Felipe L.L Medeiros, and Angelo Passaro. "Decision support system for unmanned combat air verhicle in beyond visual range air combat based on artificial neural networks." *Journal of Aerospace Technology and Management,* 2021.

McKenzie, Simon. "Autonomous technology and dynamic obligations: Uncrewed maritime verhicles and the regulation of maritime military surveillance in the exclusive economic zone." *Asian Journal of International Law 11, no. 1,* 2021: 146-175.

Monsuur, Herman. "Assessing situation awareness in networks of cooperating entities: A mathematical approach." *Military Operations Research,* 2007: 5-15.

Murray, Chase C., and Amanda G. Chu. "The flying sidekick traveling salesman problem: Optimization of drone-assisted parcel delivery." *Transportation Research Part C: Emerging Technologies, 54,* 2015: 86-109.

Noon, Charles, and James C. Bean. "An efficient transformation of the generalized traveling salesman problem." *INFOR: Information Systems and Operational Research 31, no. 1*, 1993: 39-44.

O'Young, Siu, and Paul Hubbard. "Raven: A maritime surveillance project using small UAV." *Proceeding of the 12th IEEE Conference on Emerging Technologies and Factory Automation.* Patras: IEEE, 2007. 904-907.

Suteris, Muhamad Syazwan, F.A. Rahman, and Azman Ismail. "Route schedule optimization method of unmanned aerial vehicle implementation for maritime surveillance in monitoring trawler activities in Kuala Kedah, Malaysia." *Internation Journal of Supply Chain Management*, 2018: 245-249.

United Nations General Assembly. "Convention on the law of the sea." December 10, 1982.

Wu, Chunguo, Yanchun Liang, Hoew Pueh Lee, and Chun Lu. "Generalized chromosome genetic algorithm for generalized traveling salesman problems and its applications for machining." *Physical Review E 70*, 2004: 016701.

Yang, Zhen, Zhixiao Sun, Haiyin Paio, Yiyang Zhao, Weiren Zhou, and Kai Zhang. "An autonomous attack guidance method with high aiming precision for UCAV based on adaptive fuzzy control under model predictive control framework." *Applied Sciences 10, no. 16*, 2020

CHAPTER 9

From data to effective actions

Providing actionable information for detect and avoid

Erik Theunissen

Abstract

To integrate remotely operated aircraft into non-segregated airspace, a Detect- and Avoid (DAA) capability is required. To remain Well Clear of other aircraft, a DAA system provides the pilot with actionable information derived from real-time data about cooperative and non-cooperative traffic. In 2017, the first Minimum Operational Performance Standard (MOPS) for DAA systems was published by RTCA. To promote interoperability between systems, NATO Study Group 268 is working on the validation of a performance-based requirements standard. To prevent requirements dictating a particular technology, data requirements are specified as sensor-agnostic. To meet the data requirements, advanced data-processing is required for detection, tracking and fusion/integration. This chapter describes the DAA process from an OODA perspective, discusses the potential of AI/ML techniques, and identifies questions that must be answered before a safe fully autonomous remain Well Clear capability can be realised.

Keywords: Unmanned aircraft, Detect and Avoid, Autonomy, Artificial intelligence, Machine learning

9.1. Introduction

Detect and Avoid (DAA) is an essential capability for Remotely Piloted Aircraft to share the airspace with manned aircraft. In a world without disturbances, in the presence of perfect data and with perfect knowledge of the dynamics, achieving a goal can be performed by actions that do not require any corrections. For the aviation domain such a world would allow all missions to be flown exactly along the pre-planned route and without the need for any deviations. Clearly, this is not the case. The need for a closed-loop approach to achieve desired goals is recognised both implicitly and explicitly in a wide range of disciplines. The Second World War drove the development of control theory for guidance and weapon control. It was recognised that for tasks such as aiming, using the observed data was not enough.

The movement of a target needed to be compensated through a prediction of the future state. Such prediction requires a model of the system. A model encapsulates the knowledge that allows the future state to be derived from (observations of) the current state and the inputs to the system. For certain applications it may be sufficient to use a linear extrapolation of the current state. At the other end of the spectrum are models that are created using Artificial Intelligence / Machine Learning (AI/ML) approaches. In recent years, advancements in the area of AI/ML have caused many speculations regarding its potential to replace human decision making in a wide variety of roles. For the military domain, one advantage is comparable to the use of remotely operated aircraft to replace manned aircraft for dull, dirty and dangerous missions. A potentially even more decisive advantage is the possibility to significantly increase the speed with which the Observe-Orient-Decide-Act (OODA) loop can be closed due to the elimination of the time humans need to assess the situation and make a decision.[1] Koch discusses the ethical implications of replacing conscious decisions with AI for the Defence domain.[2] This chapter discusses the current state-of-the-art in DAA systems, with a focus on how information requirements drive the sensor- and data-processing requirements. The potential benefits of using AI/ML techniques are discussed in the context of the OODA framework.

9.2. Background – navigation, guidance, control and automation

Control loops form the backbone of navigation, guidance and control systems. Aircraft have seen an incremental increase in automation, starting at the inner-most control-loops. In the Seventies, the increase in bandwidth that could be achieved with digital closed-loop control of aircraft attitude enabled artificial stabilisation of unstable aircraft. Closure of the position control loop introduced the capability of automatic navigation. Today's aircraft can be operated in a variety of automated modes, but the authority that determines in which mode still resides with the operator.

A closed-loop system can comprise multiple control-loop closures. The model of a fully automated single control-loop can be divided into the following four components: data acquisition, data processing, action selection, action execution (Figure 9.1). This is similar to the OODA classification often used in the military domain.

Whereas the desire to use dynamically unstable platforms drove the full automation of the inner-loop for manned aircraft, for remotely operated aircraft the control-link latency is a determining factor[3] for automating the inner-loop. Recognising the need and impact of automation for remotely controlled platforms, the first systematic classification of task-allocation between the human operator and the automation was introduced by Sheridan and Verplank.[4] Their one-dimensional Level of Automation (LoA) scale was later extended into a two-dimensional scale by dividing a function into the steps of data acquisition, data processing, decision making and action.[5]

Fig. 9.1. Data acquisition, data processing, action selection and action execution in a feedback control-loop intended to ensure that the state of the system under control meets the defined goals.

For more than 40 years, the automatic execution of a flight plan has been performed by a Flight Management System (FMS) that provides inputs to the Automatic Flight Control System (AFCS). Similarly, remotely piloted aircraft can fly a mission without the need for a pilot controlling stick, rudder and throttle. However, in case an event requires a deviation from the plan, pilot intervention is required. The OODA depiction of this supervisory control with the possibility to intervene is illustrated in Figure 9.2.

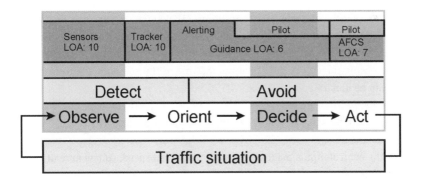

Fig. 9.2. OODA depiction of the supervisory control with the possibility to intervene. Detect and Avoid is intended to maintain a traffic situation with sufficient separation. Sensors provide the data observations that are transformed into actionable information in the Orient phase. In case of an alert, guidance is computed to remain well clear and provided to the flight control system for automatic execution. The pilot is 'on the loop' and can both alter the guidance and intervene in the execution.

During supervisory control, the role of the pilot is to compensate for the limited flexibility of automated systems in the event of an unforeseen circumstance for which the system was not designed. To benefit from the flexibility of the human operator, the system must be designed so that the pilot is able to safely and rapidly respond to unexpected events. One possible reason for the pilot to intervene is if the planned trajectory is likely to cause a loss of Well Clear with another aircraft. The pilot will have to override the automation that controls the aircraft along the planned route and command a speed, direction and/or altitude that will keep ownship Well Clear. To continue the automated closure of the navigation loop, the decision on *when and how* to manoeuvre to remain Well Clear would need to be automated. A particular challenge is associated with deciding upon the course of action in situations where the available data leaves considerable uncertainty about the potential future situation and/or the situation lacks unambiguous normative decision criteria.

This chapter will discuss the current state-of-the-art in DAA systems and identify challenges and questions that must be successfully addressed before autonomous operation through automatic closure of the DAA OODA-loop is safely possible.

9.3. Detect and avoid systems

DAA is an electronic means of staying Well Clear[6] of other traffic to meet the intent of the See-and-Avoid requirement. If remaining Well Clear fails, a situation may develop that necessitates a Collision Avoidance (CA) manoeuvre. In 2017, the first Minimum Operational Performance Standard (MOPS) for DAA systems was published by RTCA.[7] To promote interoperability between systems, NATO Study Group 268 is working on the validation of a performance-based requirements standard.

When using the OODA classification, the role of the components in the DAA system can be illustrated as follows:
1. Observe: Sensors
2. Orient: data filtering, fusion/integration, extrapolation, display
3. Decide: the pilot
4. Act: the control inputs made by the pilot that affect speed, course and altitude of the aircraft

Because the requirements for making an informed decision drive the requirements for the Observe and Orient phase, the Decide and Act will be discussed first.

9.3.1. Decide and act

Regarding *Decide*, in the Operational Services and Environment Description (OSED) of the DAA MOPS,[8] the use of a DAA system by the pilot is described as follows:

1. The remote pilot uses training, judgment, and the traffic display to assess the threat and the need to manoeuvre.
2. If the remote pilot cannot manoeuvre in response to an ATC Traffic Advisory, the remote pilot will inform ATC. During approach or departure, the manoeuvre to remain DWC must consider the procedures for making an instrument approach/departure, including flying the missed approach. When flying below the minimum safe altitude, manoeuvres are constrained to avoid obstacles and terrain.

Whereas the text in the OSED states 'traffic display', during the development of DO-365, it was concluded by Working Group 1 (WG-1) of Special Committee 228 (SC-228) that the mere depiction of traffic neither provides the pilot with sufficient information to assess the threat under all circumstances nor enables the pilot to always identify an effective manoeuvre to avoid the hazard. Currently (2022) ongoing projects that are developing requirements for DAA in European airspace have reached the same conclusion. To support the pilot with timely detection and assessment of threats, alerting functions are required. To support the pilot with manoeuvre selection, directions and altitudes that will result in an alert are colour coded yellow or red[9]. The alerting and guidance functions require a sufficiently accurate observation of the current situation and models to estimate the future situation.

9.3.2. Observe

Clearly, an informed decision to avoid a threat can only be made if the aircraft posing the hazard is detected in time and the pilot is provided with adequate information. The specification of the information requirements drives the sensor (observe) requirements. For DAA, information about the position, speed and direction of other traffic is needed. To prevent information requirements dictating a particular system and/or (sensor) technology, data requirements are specified as sensor-agnostic. Using a performance-based approach, various (combinations of) sensors have been identified that have the potential to provide this data with the required accuracy, availability, continuity and integrity. To transform the sensor measurements into state estimates, (advanced) data-processing is required for detection, tracking and fusion/integration[10].

For DAA, obvious data sources comprise Automatic Dependent Surveillance Broadcast data (ADSB-in), radar and Electro-Optical / Infrared (EO/IR). Minimum performance requirements are specified in documents such as RTCA DO-366. ADSB-in directly provides information on the position, direction and velocity of cooperative aircraft. Radar provides range and bearing to a target. Additional processing on a sequence of observations (part of the Orient phase) is required to provide directional information. EO/IR systems provide a sequence of images that require further processing to support detection of traffic and extraction of the required information about its position and velocity.

9.3.3. Orient

The required output of the Orient phase is actionable information for the Decide phase. For a DAA system, the Orient phase can be divided into three types of processes: those related to transforming the measurements into estimates of relative traffic position, direction and velocity, those related to assessing which traffic poses a threat, and those related to developing manoeuvre options to mitigate the threat.

1. From measured data to traffic state
Whereas ADSB directly provides information about the location, direction and velocity of traffic, this is only the case for cooperative traffic. Radar and EO/IR systems can detect non-cooperative targets but will also detect cooperative ones. Because of limited measurement accuracies, the same detected target will yield different estimates of position, direction and velocity. Data fusion or best-source selection is required to prevent a single target causing multiple (different) traffic entities to be displayed to the pilot.

Radar data contains range and bearing to a target but requires further processing to provide target velocity and direction. EO/IR systems require image-processing techniques for the detection of targets and the estimation of relative position, velocity and direction. AI/ML techniques have proven to provide capabilities that go well beyond today's state of the art for observed data in areas such as radar data and camera images.[11] On the other hand, to put the associated interpretation into perspective, for data on the surrounding environment that is observed by a camera, AI has matured to a level where it may detect and identify certain particular features faster, but it still falls short in terms of correct interpretation compared to humans. Research into adversarial examples illustrates the types of vulnerabilities this introduced.[12]

The technical possibility to use EO/IR for DAA was already being explored over 15 years ago.[13] Flight testing of an EO-based system revealed the need to improve declaration range and reduce false track rate. The main conclusion from the technology demonstration in 2007 was 'there is much work and technology maturation required before such a system can be deemed ready for operational use'.[14] Meanwhile, the sensor-resolution and available computing power have increased significantly, and AI/ML algorithms have further matured. Successful demonstrations of a DAA system using EO/IR with DO-365 compliant alerting and guidance is reported by Lee et al.[15]

2. From traffic state to threat assessment

Threat assessment requires the definition of what constitutes a threat. In the existing See-and-Avoid rules, the definition of Well Clear is intentionally ambiguous, thus leaving room for pilot interpretation of the specific situation. When automating traffic threat assessment, a quantitative definition of Well Clear is required. In support of RTCA SC-228, the Sense and Avoid Science and Research Panel (SARP) evaluated three UAS Well Clear candidates in four modelling and simulation environments.[16] The resulting Well Clear Volume (WCV) is described by a spatial and a temporal horizontal boundary and a spatial vertical boundary. The boundaries were quantified using the requirement that an unmitigated penetration of the volume would not exceed a 2.2% probability[17] of a Near Mid-Air Collision (NMAC). One million uncorrelated encounters were used to verify the selected values for *en-route* operations[18]. This approach, in which large, representative datasets are used to develop quantified definitions of previously subjective separation criteria, is fundamental for performance-based alerting and guidance requirements. In Europe, in the context of the SESAR URClearED project, similar analyses have been performed in 2022, using encounter models developed by EUROCONTROL. However, there are differences in the alerting concept. In the U.S., the rules of the air allow the pilot of a remotely controlled aircraft to initiate a manoeuvre to remain Well Clear in case ATC does not provide timely guidance to resolve the conflict. To support the pilot, two alerts are defined. The first alert (Caution) is issued if the system detects that a future loss of DWC will occur, but there is still time to contact ATC to resolve it. The second alert (Warning) occurs briefly before the last moment a manoeuvre is possible to prevent a loss of DWC. Appropriate manoeuvring after a Warning should prevent a loss of DWC and thus also prevent the triggering of a CA system. The current vision in the European approach does not allow the pilot to manoeuvre without ATC clearance, except in a CA situation. Therefore, in the current European concept, the Warning alert does not exist. Recognising that there will be different approaches to DAA and that different standards will exist, to ensure interoperability NATO is developing a performance-based requirements document for DAA systems.

Given that penetration of the WCV constitutes a hazard, threat assessment is performed by predicting whether any of the tracked aircraft will penetrate the ownship WCV within a specific time[19]. In the DAA MOPS, a first-order model for the prediction is deemed acceptable.[20] Besides prediction inaccuracies due to limited fidelity of the model, inaccuracies in the measurement of the ownship and traffic position, direction and velocity impact the result. In the research by Tadema et al., the impact of observation (i.e. sensor) errors on the presentation of guidance provides an example of how, in particular, limitations in accuracy of observed direction and velocity negatively impact the look-ahead time that can be used as a basis for the guidance.[21] However, even in the presence of very accurate data about the current state, the lack of information on traffic manoeuvre intent limits the look-ahead time for the prediction function. To constrain the number of unjustified / nuisance alerts, the DAA MOPS require sensor uncertainty mitigation and impose a no-earlier limit[22] on the alert time.

3. From threat assessment to decision support

If a threat is detected, decision support to select a manoeuvre must be generated. Whereas for the threat assessment a prediction was only performed using the current ownship and traffic states, for decision support the future state will be evaluated for variations to ownship current state. Those variations that are predicted to prevent the loss of Well Clear are candidate solutions. The design philosophy of the decision support determines which candidate solutions are conveyed to the pilot. Levels 2 to 4 of the LoA scale[23] classify this decision support from '*The computer offers a complete set of decision/action alternatives*' to '*suggests one alternative*'. With a level 4 (command-based) concept, the manoeuvre direction is determined by the design of the particular algorithm. Pilot judgement will only impact the manoeuvre when the pilot is convinced that the commanded manoeuvre is unsafe, but it is to be expected that in almost all other situations the pilot will manoeuvre into the commanded direction. A system that uses a set of criteria to reduce the available options to a single one takes an important part of the decision-making process away from the pilot. In such a case, one cannot claim an informed pilot decision[24]. To mitigate the loss of the pilot's ability to deal with uncertainty and exercise judgement, the design process must assure that the likelihood of a hazardous situation resulting from obeying the command is extremely low. In contrast, by presenting information about the predicted locations of hazards and have the pilot make a manoeuvre decision based on this and other available information (LoA 2), the burden on the design and certification process of the decision support is reduced.

A useful construct to classify depiction of data is based on the required mental processing needed to make an informed decision. Using the 3-level Situation Awareness (SA) classification[25], data that must be combined with other data (integrated) by the user only provides level 1 SA. If, instead of requiring the user to (mentally) combine/integrate the data, this is already achieved with the depiction, that presentation provides level 2 SA. A display can support level 3 SA by showing predicted data that answers the 'what if' question that otherwise would require the pilot to mentally extrapolate the current situation into the future.

Fig. 9.3. Simulation environment used for the evaluation of the intercept guidance concept in March 2022. The picture was taken at the moment the intercepting aircraft had reached the desired position behind the TU-160 Blackjack.

Supporting pilot decision making by providing level 3 SA support has been the focus area of several studies into the design of DAA displays. In 2008, researchers from the Netherlands Defence Academy (NLDA) presented a DAA display concept relying on the depiction of the predicted conflict space.[26] The basis for the computation of the conflict space is a process referred to as 'probing'. With such a concept, the manoeuvre decision of the pilot is based on his/her perception of the current encounter geometry, the projection of this geometry into the future for the considered manoeuvre options and information about other relevant constraints. In other words, the presentation of the predicted future conflict space supports the pilot in obtaining level 3 SA.

The underlying idea was based on concepts pursued in the nautical domain from the early Seventies, such as the Integrated Collision Avoidance System (INCAS) [Chase and Tiblin, 1971]. One of the reasons for pursuing the concept of conflict space depiction was the possibility to create an integrated hazard avoidance system that combines the hazard space formed by traffic with that of terrain and other potential hazards.[27] Together with experts from General Atomics Aeronautical Systems Inc. (GA-ASI), the idea to integrate a conflict space display into their Unmanned Aircraft System (UAS) Control Station Architecture was pursued. In 2013 and 2014 follow-on studies, various ideas for hybrid spatial temporal thresholds were pursued.[28] The GA-ASI Conflict Prediction and Display System (CPDS) became the first operational DAA system to use conflict space depiction[29]. In 2018, CPDS was used to support NASA MQ-9 Ikhana, making it the first unmanned aircraft to achieve a No Chase Certificate of Waiver or Authorisation (COA) flight without the need of a chase plane or visual observers as it operated in various classes of airspace.[30,31]

To be able to evaluate the impact of the differences in alerting philosophy between the existing DO-365 and emerging European requirements, NLDA has implemented both concepts in the DAA system used in its UAS laboratory. As a spin-off to the DAA research, underlying algorithms have been used to develop a concept to support pilots with so-called 'intercept guidance'. The guidance algorithms use the same data, but instead of using it to show predicted conflict space for all relevant nearby traffic, guidance is provided towards a selectable target, independent of range. This is a useful pilot support tool for air-policing missions. A first evaluation of these algorithms was performed in April 2022.

A first step towards automatic Remain Well Clear is to reduce the number of manoeuvre options generated in the Orient phase to a single one. The Decide phase could be implemented with a LoA between 5 to 7. Rather than selecting a manoeuvre from an understanding of the directions to be avoided, the pilot now has to decide whether to accept or reject the scheduled manoeuvre. Given the dynamics of the situation, the proposed manoeuvre may change during the time that is available for the pilot to make a decision. In the research study by Theunissen and Zijlstra an experiment is described in which pilot decisions were compared to those generated by an algorithm.[32] Pilots were also asked to comment on the manoeuvre suggested by the automation. Comments comprised:

- "The guidance is provided too soon during a manoeuvring encounter. Initially it seems beneficial to directly get a heading to fly, but it is counterproductive if a reversal follows."
- "The moment the intruders manoeuvre the pilot sees that the initial heading of the guidance command is incorrect. The pilot has to be able to overrule the guidance."
- "Due to the guidance reversals the pilot develops a sense of doubt about the system. Because of this, use of the system will be limited."

In this context it is important to note that the automation in most cases achieved sufficient separation, but the fundamental difference with the pilot manoeuvre decision was in situations with significant uncertainty regarding the intruder manoeuvre. Typically, pilots decided to wait a few seconds to reduce this uncertainty. This strategy caused pilots to need fewer reversals of manoeuvre direction, which was the main cause they outperformed the algorithm (that always decided at the required time before CPA).

The difference in perception of the best manoeuvre direction is a fundamental problem for advisory systems, which was identified 40 years ago during research into the so-called Pilot's Associate[33]: *'Especially in critical situations, pilots mistrust the advice from automation in case it differs from what they expect'* [Statler, 1993]. If this problem is not solved, LoA 5 to 7 may be technically feasible, but the mismatch between the pilot's perception and the intent of the system may result in undesirable operational issues.

9.4. Autonomy?

Autonomous[34] DAA requires a closure of the OODA loop described in the previous section without the need for pilot selection of the manoeuvre and without the need for pilot consent on the decision to execute the selected manoeuvre. Automation of such action selection and execution is a capability that already exists for CA.

During CA any manoeuvre is allowed to prevent imminent collision. The pilot does not have to consider Right of Way rules, ATC instructions or any other constraints that apply during a normal operation. The associated reduction in constraints that must be considered simplifies the decision making, and CA logic is based on maximising the miss-distance. Automation is beneficial as it reduces the time of the Orient-Decide-Act phase, but the most important justification is that it ensures availability of the CA function in case of a control link failure. Automated CA systems for remotely piloted aircraft have been designed and flight-tested. In 2016, as part of NASA DAA Flight Test 4,[35] automatic Traffic-alert and Collision Avoidance System (TCAS) manoeuvring was demonstrated in scenarios where the pilot intentionally did not manoeuvre based on DAA guidance and continued to trigger the TCAS Resolution Advisory (RA). Unless the pilot intervened, the RA was executed by the AFCS. In 2017, automatic CA manoeuvring using ACAS-Xu[36] was demonstrated.

Unlike CA, for the time-horizon that is associated with remaining Well Clear, a pilot must consider several other factors that impact the decision whether, and if so when and how to manoeuvre. This makes automation of the decision process more complex and raises fundamental legal and ethical questions.

9.4.1. Legal questions

If use of the automation is identified as a factor in an accident, it must be determined whether the manufacturer is (partly) liable. In the lawsuit following the 1995 accident of American Airlines flight 965, the aircraft operator went to court in an effort to force the companies that designed the flight management system and the navigation database to help pay monetary damages. In response, the companies claimed that these were not at fault, and that the pilots committed a series of errors.

An increased transfer of decision authority from the operator to the system will certainly impact how liability is allocated between operator and manufacturer. For example, after the automated system that was intended to counter the tendency of the aircraft to increase pitch angle during certain manoeuvres was regarded as the primary cause of two fatal accidents, the aircraft manufacturer accepted sole responsibility for the crashes.[37]

In the automotive industry, in anticipation of 'self-driving' cars, Volvo has stated it will accept "full liability" for whatever happens when one of its self-driving vehicles is operating in autonomous driving mode. It remains to be seen whether the anticipated liability in case of an accident involving an autonomous DAA system will impact the decision to develop such systems for the commercial market.

9.4.2. Ethical questions

In situations for which multiple manoeuvre options exist and rules/regulations do not provide sufficient guidance for a particular choice, remaining Well Clear relies on pilot judgment to decide. In other words, to remain Well Clear, pilot judgment is required for those situations that lack unambiguous normative decision criteria.

In exceptional situations, the result of such judgement can even be the decision to violate a certain rule. Belzer describes this as *"Exceptional violations occur more as a one-time occurrence where the person assumed the risk basically under the premise that things would work out that one time, or with the idea that their judgement was best in that single situation even though it violated the laws of the flight"*.[38] Clearly this is not necessarily always a positive attribute, but it is one of the aspects that needs to be considered/weighed in the context of 'automating' judgment.

This results in the question of whether it is ethically acceptable to replace pilot judgment by automation in the absence of unambiguous normative decision criteria. Very likely the answer is 'no'.

However, similar to the transition from a subjective use of Well Clear to an objective, quantified definition, unambiguous normative decision criteria might be developed through the definition of risk-based cost-criteria and weighing factors. This raises questions such as:

- For a multi-intruder encounter with both manned and unmanned aircraft, how will the weighing factors and cost criteria be determined?
- Does a predicted loss of Well Clear with a manned aircraft allow the right-of-way of another unmanned aircraft to be violated[39]?

Before an attempt is made to answer such questions, the following more fundamental ethical question must be answered with a 'yes':

Given the current reliance on pilot judgement, is it ethically acceptable to define and implement agreed-upon criteria in the automation for the types of questions illustrated in the example situation?

9.4.3. Other challenges

If it is concluded that the legal and ethical questions can be adequately answered, there are challenges in the area of integration and certification. As indicated earlier, the pilot must consider several other factors that impact the decision whether, and if so when and how to manoeuvre. At present, a DAA system, a Terrain Awareness & Warning System (TAWS), a weather radar and a Flight Management System (FMS) are independently designed and certified. Replacing the pilot as the integrator of information will require a considerable change to existing architectures. Although the technical feasibility is not questioned, the associated development cost needs to be justified by operational gains achieved with the increase in autonomy.

If it is concluded that conventional approaches to the design of the decision-making logic are prohibitive and an AI/ML approach for the decision-making shows promise, lack of determinism will certainly introduce issues with current certification methods. Here too, research is looking into new techniques such as Behaviour Bounded Assurance to support certification of these types of systems.[40] Also, a different approach to certification is being proposed.[41]

9.4.4.Discussion

Because of aforementioned legal, ethical, integration and certification challenges, a question that deserves particular attention concerns the expected benefits of an autonomous remain Well Clear capability. Since automatic CA is already required to mitigate for the inability of commanding a manoeuvre in case of a control-link failure, the addition of automatic remain Well Clear is not required to mitigate for this type of failure. For airspace classes in which communication with ATC is required after a Caution alert, the operational benefits are difficult to imagine. Since automatic CA is already accepted, the only gain in response-time that could be achieved with an autonomous remain Well Clear manoeuvre at the Warning alert is in the order of

20 seconds before an automatic CA manoeuvre would be executed. Given that such an aircraft operates in an airspace in which most pilots of other aircraft will resort to voice-communication after a Caution alert in order to resolve an impending loss of Well Clear, additional problems could result from the lack of pilot involvement on the side of the remotely controlled aircraft during this phase[42]. Perhaps such a mix of human decision making, and automation is just not a good idea.

9.5. Summary and conclusions

The benefits of using unmanned aircraft for military operations such as ISR, target designation and attacking targets is undisputed. The inability to meet the See-and-Avoid requirement and the resulting need to use a segregated airspace approach for operation of these aircraft is a significant constraining factor[43].

The introduction of DAA as an alternative means of compliance to the See-and-Avoid requirement necessitated a change from the subjective Well Clear requirement to a quantified definition, the so-called DWC boundary. A DAA alerting function supports the pilot with timely detection of a future loss of DWC. Directions and altitudes that will cause a DAA alert are colour coded on the DAA display. These alerting and guidance functions require a sufficiently accurate observation of the current situation and models to estimate the future situation. To prevent dictating a particular technology, data requirements are specified as sensor-agnostic.

The information requirements can only be met by combining data from various sources. In the Orient phase, the use of AI/ML increases the potential of data from radar and EO/IR. However, in the Decide phase, AI/ML at present cannot be used to replace pilot judgement for selecting the manoeuvre to remain Well Clear. If the situation develops into a Collision Hazard, automatic manoeuvre selection and execution is acceptable.

Due to different (emerging) standards, interoperability may be a constraining factor. NATO is addressing this issue with the development of a performance-based standard for DAA systems.

Autonomous DAA is only possible if it is agreed that it is ethically acceptable to define and implement criteria in the automation that replace current pilot judgement. Given the other factors that need to be taken into account, this is fundamentally different from automatic CA in which other constraints can be disregarded.

For autonomous DAA, the transfer of decision authority from the operator to the system will certainly impact how liability is allocated between operator and manufacturer. Also, if techniques such as AI/ML are used in the automation of the 'Decide' phase, new approaches to certification are needed that are able to prove compliance with the performance requirements for a system that is non-deterministic.

Notes

[1]   John Boyd. *Patterns of Conflict*. Briefing, 1986 (http://www.ausairpower.net/JRB/poc.pdf)

[2]   Wolfgang Koch. "On Digital Ethics for Artificial Intelligence and Information Fusion in the Defense Domain". *IEEE Aerospace and Electronics Systems Magazine* 36, no. 7, (2021): 94-111..

[3]   Remotely operated vehicles typically require platform-side closure of high-bandwidth control-loops.

[4]   Thomas B. Sheridan and William L. Verplank. *Human and Computer Control of Undersea Teleoperators*. Cambridge, MA: Massachusetts Institute of Technology, 1978.

[5]   Raja Parasuraman, Tom B. Sheridan and Christopher D. Wickens. "A model for types and levels of human interaction with automation". *IEEE Transactions on Systems, Man, and Cybernetics-Part A: Systems and Humans* 30, no. 3, (2000): 286–97.

[6]   The quantified definition of the Well Clear boundary is referred to as DAA Well Clear (DWC).

[7]   RTCA. *DO-365 – Minimum Operational Performance Standards (MOPS) for Detect and Avoid (DAA) Systems*. 2017.

[8]   RTCA. *DO-365 – Minimum Operational Performance Standards (MOPS) for Detect and Avoid (DAA) Systems*. 2017.

[9]   Yellow is used for directions and altitudes that will result in a Caution-level alert, red for a Warning-level alert.

[10] In ACAS-Xu, a system that combines Remain Well Clear with a CA capability, the Surveillance and Tracking Module (STM) performs both detection and tracking of aircraft [FAA, 2019].

[11] Andrea Wrabel, Roland Graef and Tobias Brosch. "A survey of artificial intelligence approaches for target surveillance with radar sensors". Aerospace and Electronics Systems Magazine 36, no. 7, (2021): 26-43.

[12] Tencent Keen Security Lab. *Experimental Security Research of Tesla Autopilot*. 2019..

[13] James Utt, John McCalmont and Mike Deschenes. Development of a Sense and Avoid System. *AIAA@ Infotech*, paper 7177, Arlington, Virginia, 2005.; Omid Shakernia, Won-Zon Chen, Scott Graham, John Zvanya, Andrew White, Norman Weingarten and Vincent M. Raska. Sense And Avoid (SAA) Flight Test and Lessons Learned. *AIAA Infotech@Aerospace Conference and Exhibit*, paper 3003, 2007.

[14] Shakernia et al., Sense And Avoid..., 2007.

[15] Jaehyun Lee, David H. Shim and Seung J. Kim. Electro-Optical Sensor Selection based on Flight Test for Detect-and-Avoid in Phase Two UAS. *Proceedings of the 38th Digital Avionics Systems Conference*, Sept. 8-12, San Diego, California, 2019; Jaehyun Lee, Hanseob Lee and David H. Shim. Vision-Based State Estimation and Collision Alerts Generation for Detect-and-Avoid. *Proceedings of the 39th Digital Avionics Systems Conference*, Oct. 11-15, San Antonio, Texas, 2020.

[16] Stephen P. Cook, Dallas Brooks, Rodney Cole, R., Davis Hackenberg and Vincent Rasks. Defining Well Clear for Unmanned Aircraft Systems. *Proceedings of the AIAA Infotech@Aerospace conference*, Jan. 5-9, Kissimmee, FL, 2015. https://doi.org/10.2514/6.2015-0481.

[17] The reference mentions a 1.5% probability, but this was later changed to 2.2% because of an increase in the vertical threshold from 450 ft to 700 ft.

[18] In a follow-on activity, the quantification of DAA Well Clear for the airspace in the terminal area was performed.

[19] The alert-time.

[20] RTCA. *DO-365 – Minimum Operational Performance Standards (MOPS) for Detect and Avoid (DAA) Systems*. 2017.

[21] Jochum Tadema, Erik Theunissen, Richard M. Rademaker and Maarten Uijt de Haag. Evaluating the impact of sensor data uncertainty and maneuver uncertainty in a conflict probe. *Proceedings of the 29th Digital Avionics Systems Conference*, Salt Lake City, Utah, 2010.

[22] 75 seconds before predicted loss of Well Clear.

[23] Parasuraman, *A Model for Types* ..., 2000.

[24] From a conceptual perspective, there is a fundamental difference between a pilot action that is the result of following a command versus a pilot decision that is based on actionable, real-time information.

[25] Level 3 Situation Awareness (SA) is indicative of the ability to correctly anticipate how the situation is likely to develop.

[26] Jochum Tadema and Erik Theunissen. A concept for UAV Operator Involvement in Airborne Conflict Detection and Resolution. *Proceedings of the 27th Digital Avionics Systems Conference*, St. Paul, Minnesota, 2008.

[27] Jochum Tadema and Erik Theunissen. Design of an Integrated Traffic, Terrain and Energy Awareness Display Concept for UAVs. *Proceedings of the AIAA Guidance, Navigation and Control Conference*, pp.1-12, August 10-13, Chicago, Illinois, 2009.

[28] Erik Theunissen, Brandon Suarez and Maarten Uijt de Haag. The Impact of a Quantitative Specification of a Well Clear Boundary on Pilot Displays for Self Separation. Proceedings of the ICNS, April 8-10, Washington, DC, 2014.

[29] The concept of operation described in the DAA MOPS is to support pilot decision making by providing the pilot with information about the directions and altitudes that are predicted to cause a loss of Well Clear. The pilot uses this information together with information from other sources about other relevant constraints to determine the best maneuver in a direction and altitude that is not indicated as hazardous. The Conflict Prediction and Display System (CPDS) complies with DO-365, but the addition of conflict space depiction goes beyond the minimum requirements specified in the DAA MOPS.

[30] In 2019 this achievement was recognized during the 62nd Annual Laureate Awards, in the category of Commercial Aviation, Unmanned Systems.

[31] NASA, https://www.nasa.gov/centers/armstrong/features/ikhana-receives-laureate-award.html

[32] Erik Theunissen and Wytze Zijlstra. Comparing Regain Well Clear Guidance. *Proceedings of the 20th ICNS Conference*, September 9-11 (Virtual Conference), 2020.

[33] An AI-based second crewmember intended for the Advanced Tactical Fighter that became the F-22.

[34] In this context autonomy is achieved with loop closure at LoA 9 or 10.

[35] Michael Marston. *Integrated Test and Evaluation Flight Test Series 4 Flight Test Plan*. Armstrong Flight Research Center, Edwards, California, 2016.

[36] ACAS-Xu combines the threat assessment and decision support in the TRM: the Threat Resolution Module (TRM), which identifies threats and provides resolution guidance.

[37] U.S. Department of Justice, https://www.justice.gov/opa/pr/boeing-charged-737-max-fraud-conspiracy-and-agrees-pay-over-25-billion

[38] Jessica A. Belzer. *Unmanned Aircraft Systems in the National Airspace System: Establishing Equivalency in Safety and Training Through a Fault Tree Analysis Approach*. M.Sc. thesis, Department of Electrical Engineering and Computer Science and the Russ College of Engineering and Technology. Ohio University, 2017.

[39] Note that in case the situation continues into a collision hazard geometry, such constraints are no longer deemed relevant.

[40] Sarathy Prakash, Sanjoy Baruah, Stephen Cook and Marilyn Wolf. Realizing the Promise of Artificial Intelligence for Unmanned Aircraft Systems through Behavior Bounded Assurance. *Proceedings of the 38th Digital Avionics Systems Conference*, Sept. 8-12, San Diego, California, 2019.

[41] Stephen Cook, Anna Dietrich, Loyd R. Hook and Andy Lacher. Promoting Autonomy Design and Operations in Aviation. *Proceedings of the 38th Digital Avionics Systems Conference*, Sept. 8-12, San Diego, California, 2019.

[42]   This problem would not exist if the remain Well Clear function of all aircraft involved was automated.
[43]   This is not just a limitation during a specific mission, but also during transit flights, joint exercises etc.

References

Belzer, Jessica A. . *Unmanned Aircraft Systems in the National Airspace System: Establishing Equivalency in Safety and Training Through a Fault Tree Analysis Approach*. M.Sc. thesis, Department of Electrical Engineering and Computer Science and the Russ College of Engineering and Technology. Ohio University, 2017.

Blasch, Erik, Tien Pham, Chee-Yee Chong, Wolfgang Koch, Henry Leung, Dave Braines and Tarek Abdelzaher. "Machine Learning/Artificial Intelligence and Information Fusion in the Defense Domain". *Aerospace and Electronics Systems Magazine* 36, no. 7, (2021): 80-93.

Boyd, John. *Patterns of Conflict*. Briefing, 1986 (http://www.ausairpower.net/JRB/poc.pdf)

Chase, K.H. and B.V. Tiblin. "INCAS Integrated Navigation and Collision Avoidance System." *Journal of the Institute of Navigation* 18, no. 2, (1971): 205-214.

Cook, Stephen P., Dallas Brooks, Rodney Cole, R., Davis Hackenberg and Vincent Rasks. Defining Well Clear for Unmanned Aircraft Systems. *Proceedings of the AIAA Infotech@Aerospace conference*, Jan. 5-9, Kissimmee, FL, 2015. https://doi.org/10.2514/6.2015-0481.

Cook, Stephen, Anna Dietrich, Loyd R. Hook and Andy Lacher. Promoting Autonomy Design and Operations in Aviation. *Proceedings of the 38th Digital Avionics Systems Conference*, Sept. 8-12, San Diego, Califiornia, 2019.

Federal Aviation Administration. *Algorithm Design Description of the Xu Airborne Collision Avoidance System X*. Document ACAS_ADU_19_001_V5R0, 2019.

Koch, Wolfgang. "On Digital Ethics for Artificial Intelligence and Information Fusion in the Defense Domain". *IEEE Aerospace and Electronics Systems Magazine* 36, no. 7, (2021): 94-111.

Lee, Jaehyun, David H. Shim and Seung J. Kim. Electro-Optical Sensor Selection based on Flight Test for Detect-and-Avoid in Phase Two UAS. *Proceedings of the 38th Digital Avionics Systems Conference*, Sept. 8-12, San Diego, California, 2019.

Lee, Jaehyun, Hanseob Lee and David H. Shim. Vision-Based State Estimation and Collision Alerts Generation for Detect-and-Avoid. *Proceedings of the 39th Digital Avionics Systems Conference*, Oct. 11-15, San Antonio, Texas, 2020.

Marston, Michael. Integrated Test and Evaluation Flight Test Series 4 Flight Test Plan, Armstrong Flight Research Center, Edwards, California, 2016.

NASA. https://www.nasa.gov/centers/armstrong/features/ikhana-receives-laureate-award.html. 2019.

Parasuraman, Raja, Tom B. Sheridan and Christopher D. Wickens. "A model for types and levels of human interaction with automation". *IEEE Transactions on Systems, Man, and Cybernetics-Part A: Systems and Humans* 30, no. 3, (2000): 286–97.

Prakash Sarathy, Sanjoy Baruah, Stephen Cook and Marilyn Wolf. Realizing the Promise of Artificial Intelligence for Unmanned Aircraft Systems through Behavior Bounded Assurance. *Proceedings of the 38th Digital Avionics Systems Conference*, Sept. 8-12, San Diego, California, 2019.

RTCA. *DO-365 – Minimum Operational Performance Standards (MOPS) for Detect and Avoid (DAA) Systems*. 2017.

RTCA. *DO-366 – Minimum Operational Performance Standards (MOPS) for Air-to-Air Radar for Traffic Surveillance*. 2017.

Shakernia, Omid, Won-Zon Chen, Scott Graham, John Zvanya, Andrew White, Norman Weingarten and Vincent M. Raska. Sense And Avoid (SAA) Flight Test and Lessons Learned. *AIAA Infotech@ Aerospace Conference and Exhibit*, paper 3003, 2007.

Sheridan, Thomas B. and William L. Verplank. *Human and Computer Control of Undersea Teleoperators*. Cambridge, MA: Massachusetts Institute of Technology, 1978.

Statler, Irving C. 'Technical Evaluation Report'. In: *Combat Automation for Airborne Weapon Systems: Man/Machine Interface Trends and Technologies*. AGARD CP-520, pp. T1-T19. 1993.

Tadema, Jochum and Erik Theunissen. A concept for UAV Operator Involvement in Airborne Conflict Detection and Resolution. *Proceedings of the 27th Digital Avionics Systems Conference*, St. Paul, Minnesota, 2008.

Tadema, Jochum. and Erik Theunissen. Design of an Integrated Traffic, Terrain and Energy Awareness Display Concept for UAVs. *Proceedings of the AIAA Guidance, Navigation and Control Conference*, pp.1-12, August 10-13, Chicago, Illinois, 2009.

Tadema, Jochum, Erik Theunissen, Richard M. Rademaker and Maarten Uijt de Haag. Evaluating the impact of sensor data uncertainty and maneuver uncertainty in a conflict probe. *Proceedings of the 29th Digital Avionics Systems Conference*, Salt Lake City, Utah, 2010.

Tencent Keen Security Lab. *Experimental Security Research of Tesla Autopilot*. 2019.

Theunissen, Erik, B. Suarez and Maarten Uijt de Haag (2013). From Spatial Conflict Probes to Spatial/Temporal Conflict Probes: Why and How. *Proceedings of the 32nd Digital Avionics Systems Conference*, Syracuse, NY.

Theunissen, E., Brandon Suarez and Maarten Uijt de Haag. The Impact of a Quantitative Specification of a Well Clear Boundary on Pilot Displays for Self Separation. *Proceedings of the ICNS*, April 8-10, Washington, DC, 2014.

Theunissen, Erik. and Wytze Zijlstra. Comparing Regain Well Clear Guidance. *Proceedings of the 20th ICNS Conference*, September 9-11 (Virtual Conference), 2020.

U.S. Department of Justice. https://www.justice.gov/opa/pr/boeing-charged-737-max-fraud-conspiracy-and-agrees-pay-over-25-billion, 2021.

Utt, James, John McCalmont and Mike Deschenes. Development of a Sense and Avoid System. *AIAA@ Infotech*, paper 7177, Arlington, Virginia, 2005.

Wrabel, Andrea, Roland Graef and Tobias Brosch. "A survey of artificial intelligence approaches for target surveillance with radar sensors". *Aerospace and Electronics Systems Magazine* 36, no. 7, (2021): 26-43.

CHAPTER 10

Battling information overload in military intelligence & security organisations

Tess Horlings, Roy Lindelauf & Sebastiaan Rietjens

Abstract

Military intelligence & security organisations are confronted with information overload as a result of continuously increasing amounts of data, high levels of uncertainty of the data, as well as multiple data formats. Information overload has serious consequences hampering the effectiveness and efficiency from the individual to the organisational level. Drawing upon insights from adjacent fields such as business administration, psychology, and communication science, this chapter analyses how people, practices and tools relate to information overload in military intelligence & security organisations. Findings of the chapter include a need for reevaluation of the type of personnel and their specific skillsets; the identification of several issues that hamper the organisational practices of information overload (collection outpacing analysis, data governance, distinguishing signals from noise, and changing intelligence processes); and a need to integrate appropriate tools (e.g., software packages, custom-made algorithms and automatisation scripts).

Keywords: Information overload; Intelligence & security organisations; Data governance

10.1. Introduction

In the current information age, military intelligence & security organisations are confronted with information overload, which is generally defined as a situation in which 'decision-makers face a level of information that is greater than their information processing capacity'.[1] Such information overload is not only the result of the continuously increasing amount of data that needs to be processed (e.g. due to the introduction of new technological platforms such as the F35, the MQ-9 Reaper or the advancement of the soldier modernisation program), but also of the high levels of uncertainty of the data, the increase in data that is available in real-time, as well as multiple data formats with structured and unstructured data from text, image, multimedia content, audio, video, sensor data or noise.[2]

Information overload has serious consequences hampering the effectiveness and efficiency of military intelligence & security organisations from the individual to the organisational level.[3] Despite this consequence 'it is surprising how little academic attention has been devoted to the changes that are taking place in the technology, management and integration of the intelligence systems that will underpin any Revolution in Military Affairs.'[4]

This chapter provides an understanding of the main components and processes leading to information overload in military intelligence & security organisations. In section 10.2 a conceptual framework of information overload is presented based upon literature from business administration, psychology, and communication science.[5] Section 10.2 concludes that people, practices and tools are the main components that influence information overload.

In section 10.3 this conceptual framework is used to analyse the concept of information overload within military intelligence & security organisations. Finally, section 10.4 discusses how a thorough understanding of the concept of information overload may benefit military intelligence & security organisations.

10.2. Information overload

10.2.1. Defining information overload

The attention for information overload in academic debate was at its highest level in the 1980s and 1990s, mainly focusing on the effects of wasted time, decreasing efficiency and health concerns, after which attention decreased in the 2000s.[6] Nearly thirty years after its initial peak, information overload research is regaining attention as the information age is exponentially increasing the amount of available data.[7] For organisations, information overload is regarded as one of the main challenges of the digital age.[8]

A broad spectrum of academic debates has been theorising on the issue of information overload. Especially management disciplines, such as accounting, organisational science, and marketing, but also studies in health care and psychology have been discussing the topic.[9] Regardless – or, as a result – of the broad spectrum of literature, recent studies indicate that a standardised definition of information overload is still lacking.[10] An often-used definition is that 'overload is regarded as that situation which arises when there is so much relevant and potentially useful information available that it becomes a hindrance rather than a help.'[11] A more practical representation of information overload is the short formula: *information processing requirements > information processing capacities.*[12]

This chapter uses the term *information overload* rather than *data overload*, and in the context of information components leading to knowledge. Following Meadow and Yuan (1997), data refers to 'a set of symbols that does not have meaning or significance to their recipient', whereas *information* does meet that requirement. The 'accumulation and integration of information received and processed by a recipient' is what constitutes *knowledge*.[13] Data can be seen as *potential information*; thus, an overload of data can be an important factor in the perceived information load.

Finally, multiple studies provide a schematic representation referring to the issue of information (over)load as an inverted U-curve, which is shown in Figure 10.1.[14] Here, the focus is on the relation between the performance (in this case the decision-making accuracy) of an individual and the information load he or she is exposed to. According to Eppler and Mengis (2004), 'the performance [...] of an individual correlates positively with the amount of information he or she receives – up to a certain point. If further information is provided beyond this point, the performance of the individual will rapidly decline.'[15]

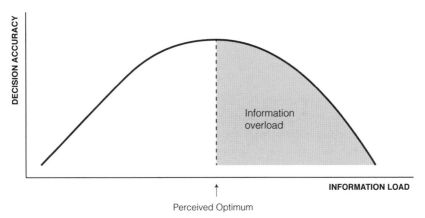

Fig. 10.1. Information overload as an inverted U-curve.[16]

Some approaches not only take into account the amount of information when discussing information load, but also the complexity of the information.[17] Indeed, as can be seen in Figure 10.1, the x-axis does not portray the *amount* of information. Rather, it portrays the *perception* of that amount, or *information load*. In order to improve decision-making accuracy, organisations need to find ways to process more information without increasing the experienced information load. Multiple facets are of influence to the experienced information load: people, practices and tools. The next section will discuss a framework for understanding these facets.

10.2.2. People, practices, and tools

The academic debate on information overload tends to make a distinction between an individual cognitive perspective and an organisational *equipment-related* perspective such as a limitation in time or budget.[18] Allen and Wilson (2003) describe these different perspectives as follows:

> At the personal level, we can define information overload as a perception on the part of the individual (or observers of that person) that the flows of information associated with work tasks is greater than can be managed effectively, and a perception that overload in this sense creates a degree of stress for which his or her coping strategies are ineffective. Similarly, at the organisational level, information overload is a situation in which the extent of perceived individual information overload is sufficiently widespread within the organisation as to reduce the overall effectiveness of management operations.[19]

Thus Figure 10.1, which presents information overload with respect to the individual cognitive dimension, can also be used to conceptually present information overload at the organisational level. In his research on information overload, Pijpers (2010) also recognises these different perspectives. His perspective proposes a triangular relationship between people, practices and tools.

> For some scholars, information overload is concerned with the individual. Others contend that it is related to the organizations where people work. Some researchers, however, believe that information overload is caused by the limitations of computer hardware and software, and therefore the problem may be solved as specific technology becomes available. What is at least clear is that the ability to deal with information depends on the relationship between tools, people, and practices within organizations, rather than on an isolated outcome variable such as information overload.[20]

According to Pijpers, people are an integral part of the information overload issue. He even argues that 'information only exists by the grace of human perception.'[21] The individual level of information overload has traditionally been the core component of this topic. In 1970, the book *Future Shock* by Alvin Toffler, introduced the phenomenon to a broad audience, describing it as 'causing both physical and physiological distress due to overloading of perception, cognition, and decision-making process.'[22] Even though individual and equipment barriers to information processing are two sides of the same coin, academic literature on information overload has primarily been focusing on cognitive (individual) issues.[23]

Despite the core focus on the individual perspective, cognitive overload does not occur in isolation. The practices within organisations, the second element in

Pijpers' trinity, shape the environment in which the individual encounters the information load. Think for instance of the organisational (information) culture or the way the organisation arranges its data governance practices.

Especially in knowledge-intensive organisations such as military intelligence organisations, there is a strong link between the information culture and the perception of information overload.[24] A positive information culture is one in which 'the value and utility of information in achieving operational and strategic success is recognised, where information forms the basis of organisational decision making and Information Technology is readily exploited as an enabler for effective Information Systems.'[25]

Finally, people within the organisation – to a greater or lesser extent – use tools to process the information they encounter. In this paper we use the term *tools* loosely to encompass a wide array of data processing technologies. Think of software packages like i2 Analyst notebook that aid in analysing networks, custom-made complex algorithms implemented using programming languages like Python and R but also scripts used to automate standard processes. The way these tools are integrated into organisational practices and the extent to which they meet the level of skill and demand of the individual using them, is the third major component of information overload.

Although the importance of information practices and technology in knowledge-intensive organisations is widely recognised, information technology is still often interpreted as a main *cause* of information overload, rather than an integral part of the solution.[26] Following Koltay (2017): 'The problem is not that there is too much information. In an information-driven society there cannot be too much information. Information overload represents the challenge to make effective use of information, thus instead of blocking or limiting, there is a need for finding appropriate tools for discovery.'[27]

As can be explained through the inverted U-curve graph in Figure 10.1, in order to counter information overload – shifting the perceived optimum to the right – processing capacities, among others, need to be improved. By doing so, the entity that processes information (being an organisation or an individual) can benefit from increasing *amounts* of information, without necessarily increasing the perception of the *load* of information, hence improving decision accuracy.[28] Algorithms from fields like data science and artificial intelligence can be a major asset in increasing processing capacities:

> One major aspect of information processing is not in the focus of researchers yet: how information can be processed and evaluated by "intelligent" information systems. While decision support systems or decision aid are widely investigated, the wide field of machine learning, deep learning and artificial intelligence, which is one of the most important drivers of digitalization, is not linked with information overload literature yet.[29]

The broad spectrum of information overload literature provides the necessary understanding of its main components in an organisation: the individual cognitive aspect (people), the organisational practices, and the tools used.

10.3. Information overload in military intelligence and security organisations

Making use of the trinity of people, practices and tools, this section reflects on the way military intelligence and security organisations can cope with information overload.

10.3.1. People

The observation that information overload is not only caused by the volume of data, but also by its complexity and the way that is experienced by an individual (or organisation), especially holds true for intelligence and security organisations.[30] Nicander (2011) points out that today's intelligence targets are 'smaller, more agile and mobile, and time-sensitive, [which] must be reflected in the management and requirement systems.'[31] Indeed, the type of personnel and specific skillsets need to be reevaluated on the basis of this expanding information load, to create the necessary conditions for creating a data-centric culture. According to Hare & Coghill (2016), this should eventually lead to the situation in which analysts will become a 'curator' or 'librarian' for information relevant to decision making. Data savviness must permeate the intelligence organisation if it is to become data driven, i.e., a minimum level of data literacy should be expected of every employee within such an organisation. Furthermore, in order to instigate these developments, management should become data literate to optimally integrate data methodologies within the intelligence process. Vogel et al. (2021) even expand the desired skillset for managers, quoting one of their interviewees who stated that a manager lacking 'the mathematical background necessary to understand the statistical meaning behind the probabilities of analytic results can be a serious detriment to achieving a reasonable analytic conclusion.'[32]

10.3.2. Practices

With regard to practices of information overload in military intelligence & security organisations, four main issues are identified. These are (1) collection outpacing analysis, (2) data governance, (3) distinguishing signals from noise, and (4) changing intelligence processes.

Collection outpacing analysis. This issue reflects on the role of data as something that may in fact hinder analysis. The cause of information overload is considered to be found within the intelligence process: the combination of 'mass collection of data by technological means' with analysts not being able to match the pace to transform it into timely and relevant products, vividly called *analysis paralysis*.[33] Lieutenant General Deptula stated the problem as: "We're swimming in sensors and we need to be careful we don't drown in the data [...] the unavoidable truths are that data velocities are accelerating and the current way we handle data is really overwhelmed by this tsunami."[34] *Analysis paralysis* may result from a combination of factors of the functional chain (e.g. information sources, complexity of the effort, context, time pressure), organisational tools and practices dealing with these factors, and individual characteristics.

Data governance. Organisational practices are one of the main components influencing the problem of information overload. Similar to large, complex, private sector organisations, the intelligence community has to 'govern, manage, secure and use data' to enhance its organisational efficiency and results.[35] Approaching the importance of data governance through the lens of the conceptual framework, a number of elements appear to be related. Well-integrated data governance will decrease the perceived load of the information by assisting the analyst in efficiently searching the information that is already stored, assessing the source reputation, and merging data from different collection disciplines before presenting it to analysts (multi-INT fusion). Data governance, albeit in a limited scope, has surfaced as an essential theme in understanding information overload in an intelligence context. As Palfy (2015) explains: 'Regardless of the subject, a key to timely, relevant, and accurate intelligence output is an organization's intelligence processing and integration capacity, which in-turn depends heavily, though not exclusively, on its approach to data governance.'[36] He argues that organisations will experience information overload when they do not establish powerful processing capacities combined with the implementation of a data governance framework. In line with the presented conceptual framework, Palfy asserts that implementing IT tools to counter information overload would be inadequate without solid data governance.[37]

Distinguishing signals from noise. The third issue is the principle of distinguishing 'signals' from 'noise'. The challenge to only concentrate on the *right* pieces of information is not unique to the current information age. The difficulty of this issue was, for example, already found to be a fundamental obstacle leading to the 'intelligence failure of Pearl Harbor in 1941. Since then, however, the amount of both signals and noise has been ever-increasing.[38] According to Winston (2007), as both secret and open-source intelligence collection has been increasing, the role of the analyst has changed from being a *gatherer* to becoming a *hunter*, indicating the search for signals in the masses of the already collected data.[39] Following Pedersen and Jansen (2019), 'instead of trying to collect sparse information from a limited supply, one now needs to locate the right pieces of information in the increasingly vast reservoir containing also large volumes of non-relevant information.'[40] Most of the aforementioned points refer to the large *amount* of signals and noise. Adding to the complexity of the effort, is the relevance of collected data that is judged by the analyst by using their understanding of the intelligence problem. Without a thorough understanding of the problem, it is impossible to correctly establish 'which pieces of collected intelligence are relevant (signals) and which are irrelevant (noise).[41]

Changing intelligence processes. In order to be beneficial to its customers, products of military intelligence and security organisations should adhere to certain principles. Well-known characteristics are that they should be timely, relevant, accurate and actionable.[42] In order to produce intelligence products that meet these requirements, intelligence organisations traditionally employ a process defined as the intelligence cycle, in which five sequential steps take place: planning and direction, collection, processing, analysis and dissemination. Recently, a considerable amount of literature has discussed to what extent the concept of the intelligence cycle is (still) accurate in the current information age, and suggestions have been made to adjust or 'move beyond' the intelligence cycle, or to change the way we conceptualise intelligence processes as a whole.[43] A data-driven approach to the intelligence process can provide intelligence and security organisations with the opportunity to both optimally use the large volumes of data *and* decrease their experienced information load. To 'discern general trends and anomalies in very large datasets – through anomaly detection and association algorithms—can help to identify potential intelligence targets, thus driving intelligence requirements.'[44]

10.3.3. Tools

There is a widely acknowledged need to integrate appropriate tools (software packages, custom-made algorithms, automatisation scripts, etc.) to deal with the problem of information overload. The unsurpassed amounts of information analysts are confronted with, require an immediate implementation of 'advanced analytic techniques.'[45] According to Eldridge et al., 'without the support of automated processes, analysts would simply be unable to cope with the deluge of information swirling around the ether.'[46] The intelligence literature suggests that the proposed tools should primarily be of assistance to the intelligence analyst, exemplifying the important connection between tools and people in an organisation. Along these lines, these tools should have the net consequence of narrowing down the information feed of the analyst.[47] Adding to this extraction of useful information, effective tools will 'present information to analysts in a way that minimizes their cognitive load [...], this primarily means visualizing information where possible, through mapping, timelining, charting and other methods.'[48] Artificial Intelligence

(AI) is one of the suggested technologies to solve the information overload problem, as it is able to, for instance, 'pre-process raw data to offload human analysts.'[49] The integration of data science and artificial intelligence tools could have a major impact on the perception of information load to intelligence analysts. Tools may be used to assess the sources of information, extract relevant information, provide alternative hypotheses based on the available data, or visualise data to temper the information load as experienced by the analyst. An important consideration here is that these uses may be mentioned in the relevant literature, but most of them are not yet operationalised.

10.4. Discussion and recommendations

The objective of this chapter is to provide understanding of the main components and processes leading to information overload in military intelligence & security organisations.

To meet this objective, we first explored the literature on information overload in several adjacent literatures. We provided several definitions and representations of the concept of information overload and explored the main components underlying information overload in an organisation. These components are: the individual cognitive aspect (people), the organisational practices, and the tools used. Applying each of these components to military intelligence & security organisations provided understanding on the type of personnel needed, the challenges in operational practices (collection outpacing analysis, data governance, distinguishing 'signals' from 'noise', and changing intelligence processes) and what tools are appropriate to use.

While these components individually contribute to our understanding of information overload in military intelligence & security organisations, it is important to stress the mutual relationships between people, practices and tools as well. Two of these relationships stand out.

First, for using tools to decrease the experienced information load, these tools need to be effectively integrated into intelligence practice. Multiple challenges have been identified in this domain. In some cases a problem was misdiagnosed as being rooted in IT tool inadequacy, while in other cases tools treated the wrong symptoms without fully understanding the root causes.[50] Furthermore, according to Treverton, the development of tools in intelligence and security organisations has mostly been bottom-up, with little push from the top.[51] Here, problems for adoption arise 'when these (newly produced) bottom-up tools clash with older knowledge regimes, organizational structures, and insufficient support mechanisms.'[52] An insufficient analysis of how people experience their information load may leave intelligence and security organisations with 'a renewed perception that their tools are both inadequate in meeting business needs and insufficient in helping analysts deal with exponential increases in data holdings relative to a given theme, network, or target set.'[53]

A second relevant issue regarding the relationship between people, practices, and tools is the extent to which people trust tools and how organisations learn to cope with information overload. The development of trust across users of the tools needs to be central and must be reflected in the policies and practices of the organisation.[54] Paramount to establishing people's trust in these tools is learning, consisting of multiple components. Laney, Chief Data Officer at Gartner, asserts that 'within organizations, as information starts to become recognized as an actual asset, we need a new language for businesspeople, for IT people—for information people—to be able to communicate effectively.'[55] The integration of tools will eventually be a core asset in decreasing the experienced information load, but in the phase prior to that integration 'the tension between exploitation and innovation can be a constant source of stress for managers. Do we focus on what we know, or do we create new products and services?'[56] Finally, Nicander (2011) stresses the importance of learning at both the organisational and individual level, as the memory of an organisation is 'derived from employees who act to adapt the *theory-in-use* that reflects an organization's past experience. [...] No organizational learning can exist without individual learning, although individual learning as a condition for organizational learning is by itself insufficient.'[57]

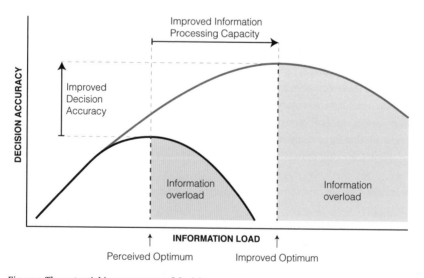

Fig. 10.2. The potential improvement of decision accuracy.

Finally, as we have seen, the individual components of information overload (people, practices, tools) as well as their mutual relationships provide several future avenues to address information overload. When being implemented, these avenues should result in an increase of information processing, leading to the ultimate goal of improving decision accuracy. In the intelligence domain this would mean to reach an improved optimum of providing timely, relevant, and actionable intelligence to gain decision advantage. Figure 10.2 presents the potential gain in decision accuracy when an organisation is able to adequately deal with information overload. The black curve in the figure portrays the initial situation, which can also be found in Figure 10.1. The blue curve shows the potential benefit when information load increases due to improved information processing capacities, leading to an improved optimum of decision accuracy. Indeed, more information can be processed before information overload sets in. This clearly visualises how intelligence & security organisations can benefit from the large volumes of information that is available to them, by dealing with the causes and effects of information overload on their people, practices, and tools.

Notes

[1]   Peter Gordon Roetzel, "Information overload in the Information Age: a review of the literature from business administration, business psychology, and related disciplines with a bibliometric approach and framework development", *Business Research* 12, no. 2 (December 2019): 480, https://doi.org/10.1007/s40685-018-0069-z.

[2]   Christopher Eldridge, Christopher Hobbs, and Matthew Moran, "Fusing algorithms and analysts: open-source intelligence in the age of "Big Data"", *Intelligence and National Security* 33, no. 3 (16 April 2018): 392, https://doi.org/10.1080/02684527.2017.1406677.

[3]   Anthony J. Cotton, 'Information Technology – Information Overload for Strategic Leaders' (U.S. Army War College, 18 March 2005), 1, https://doi.org/10.1037/e457182006-001.

[4]   Alan Dupont, "Intelligence for the twenty-first century", *Intelligence and National Security* 18, no. 4 (December 2003): 15, https://doi.org/10.1080/02684520310001688862.

[5]   Roetzel, "Information Overload in the Information Age", 481; Anne-Françoise Rutkowski and Carol S Saunders, *Emotional and Cognitive Overload: The Dark Side of Information Technology*, 2019, https://search.ebscohost.com/login.aspx?direct=true&scope=site&db=nlebk&db=nlabk&AN=1776962.

[6]   Roetzel, 'Information Overload in the Information Age', 480.

[7]   Roetzel, 480–82; David Bawden and Lyn Robinson, "Information Overload: An Introduction", in *Oxford Research Encyclopedia of Politics*, by David Bawden and Lyn Robinson (Oxford University Press, 2020), 4, https://doi.org/10.1093/acrefore/9780190228637.013.1360.

[8]   Liia Lauri, Sirje Virkus, and Mati Heidmets, "Information cultures and strategies for coping with information overload: case of Estonian higher education institutions", *Journal of Documentation* 77, no. 2 (1 January 2020): 520, https://doi.org/10.1108/JD-08-2020-0143.

[9]   Martin J. Eppler and Jeanne Mengis, "The concept of information overload: a review of literature from organization science, accounting, marketing, MIS, and related disciplines", *The Information*

Society 20, no. 5 (November 2004): 325, https://doi.org/10.1080/01972240490507974; Roetzel, 'Information Overload in the Information Age', 481.

10   Bawden and Robinson, 'Information Overload', 3; Roetzel, 'Information Overload in the Information Age', 481.

11   Bawden and Robinson, 'Information Overload', 3.

12   Eppler and Mengis, 'The Concept of Information Overload', 326.

13   Charles T. Meadow and Weijing Yuan, "Measuring the impact of information: defining the concepts", *Information Processing & Management* 33, no. 6 (November 1997): 701, https://doi.org/10.1016/S0306-4573(97)00042-3.

14   Roetzel, 'Information Overload in the Information Age', 483; Eppler and Mengis, 'The Concept of Information Overload', 326; originally from: Harold M. Schroder, Michael J. Driver, and Siegfried Streufert, *Human Information Processing: Individuals and Groups Functioning in Complex Social Situations* (Holt, Rinehart and Winston, 1967).

15   Eppler and Mengis, 326.

16   Figure adjusted from: Eppler and Mengis, 'The Concept of Information Overload', 326.

17   Roetzel, 'Information Overload in the Information Age', 482–83.

18   Roetzel, 483.

19   David Allen and T. D. Wilson, "Information overload: context and causes", *The New Review of Information Behaviour Research* 4, no. 1 (December 2003): 34, https://doi.org/10.1080/1471631031000 1631426.

20   Guus Pijpers, *Information Overload: A System for Better Managing Everyday Data*, Microsoft Executive Leadership Series (Hoboken, N.J: Wiley, 2010), 23.

21   Pijpers, 8.

22   Bawden and Robinson, 'Information Overload', 9.

23   Roetzel, 'Information Overload in the Information Age', 483.

24   Lauri, Virkus, and Heidmets, 'Information Cultures and Strategies for Coping with Information Overload', 519.

25   Adrienne Curry and Caroline Moore, "Assessing information culture—an exploratory model", *International Journal of Information Management* 23, no. 2 (1 April 2003): 94, https://doi.org/10.1016/S0268-4012(02)00102-0.

26   See, for example: Bawden and Robinson, 'Information Overload', 12; Roetzel, 'Information Overload in the Information Age', 506; Eppler and Mengis, 'The Concept of Information Overload', 331.

27   Tibor Koltay, 'Information Overload in a Data-Intensive World', in *Understanding Information: From the Big Bang to Big Data*, ed. Alfons Josef Schuster, Advanced Information and Knowledge Processing. Springer International Publishing, 2017, 207.

28   Decision accuracy metrics deal with finding 'optimal solutions' and are widely analysed and discussed in data science literature, see for instance Mohammad et al., 2015.

29   Roetzel, 'Information Overload in the Information Age', 507.

30   This paragraph draws upon: Tess Horlings, "Dealing with data: coming to grips with the Information Age in intelligence studies journals", *Intelligence and National Security*, 3 August 2022, 1–23, https://doi.org/10.1080/02684527.2022.2104932.

31   Nicander, 'Understanding Intelligence Community Innovation in the Post-9/11 World', 547.

32   Vogel et al., 'The Impact of AI on Intelligence Analysis', 836.

33   Robert D Folker, "Arming the intelligence analyst for information warfare", *American Intelligence Journal* 30, no. 2 (2012): 13.

34   James L. Lawrence II, "Activity-based intelligence: coping with the "unknown unknowns" in complex and chaotic environments", *American Intelligence Journal* 33, no. 1 (2016): 17.

35 Jeffrey Drezner et al., 'Benchmarking Data Use and Analytics in Large, Complex Private-Sector Organizations: Implications for Department of Defense Acquisition' (RAND Corporation, 2020), 1, https://doi.org/10.7249/RRA225-1.

36 Arpad Palfy, "Bridging the gap between collection and analysis: intelligence information processing and data governance", *International Journal of Intelligence and CounterIntelligence* 28, no. 2 (3 April 2015): 366, https://doi.org/10.1080/08850607.2015.992761.

37 Palfy, 370.

38 Wesley K Wark, "Introduction: "learning to live with intelligence"", *Intelligence and National Security* 18, no. 4 (December 2003): 3, https://doi.org/10.1080/02684520310001688853.

39 Thomas Winston, "Intelligence challenges in tracking terrorist internet fund transfer activities", *International Journal of Intelligence and Counter Intelligence* 20, no. 2 (19 February 2007): 333, https://doi.org/10.1080/08850600600829833.

40 Tore Pedersen and Pia Therese Jansen, "Seduced by secrecy – perplexed by complexity: effects of secret vs open-source on intelligence credibility and analytic confidence", *Intelligence and National Security* 34, no. 6 (19 September 2019): 882, https://doi.org/10.1080/02684527.2019.1628453.

41 Christiaan Menkveld, "Understanding the complexity of intelligence problems", *Intelligence and National Security*, 8 February 2021, 11, https://doi.org/10.1080/02684527.2021.1881865.

42 Loch K. Johnson, ed., *The Oxford Handbook of National Security Intelligence* (Oxford, New York: Oxford University Press, 2012).

43 See, for example: Peter Gill and Mark Phythian, 'From Intelligence Cycle to Web of Intelligence: Complexity and the Conceptualisation of Intelligence', in *Understanding the Intelligence Cycle*, ed. Mark Phythian, 2014, 21–42; Patrick Biltgen and Stephen Ryan, *Activity-Based Intelligence: Principles and Applications*, Artech House Electronic Warfare Library (Boston: Artech House, 2016); Nelson J.M. Rêgo, 'Intelligence in NATO – Contextualising a Doctrinal and Structural Clash', *Revista de Ciências Militares* VI, no. 1 (May 2018): 135–61.

44 Damien Van Puyvelde, Stephen Coulthart, and M. Shahriar Hossain, "Beyond the buzzword: big data and national security decision-making", *International Affairs* 93, no. 6 (1 November 2017): 1407–8, https://doi.org/10.1093/ia/iix184.

45 Benjamin Bell and Michael Marks, "Composite Signatures Analyst Learning Tool (CSALT): Supporting the analyst with scenario-based methodology training", *The International Journal of Intelligence, Security, and Public Affairs* 18, no. 2 (2016): 158.

46 Eldridge, Hobbs, and Moran, 'Fusing Algorithms and Analysts', 401.

47 Nick Hare and Peter Coghill, "The future of the intelligence analysis task", *Intelligence and National Security* 31, no. 6 (18 September 2016): 867, https://doi.org/10.1080/02684527.2015.1115238.

48 Hare and Coghill, 868.

49 Hare and Coghill, 863.

50 Palfy, 'Bridging the Gap between Collection and Analysis', 369.

51 Gregory F. Treverton, 'New Tools for Collaboration: The Experience of the U.S. Intelligence Community' (Washington, D.C: Center for Strategic and International Studies, January 2016), 7, https://www.businessofgovernment.org/sites/default/files/New%20Tools%20for%20Collaboration.pdf.

52 Kathleen M. Vogel et al., "The impact of AI on intelligence analysis: tackling issues of collaboration, algorithmic transparency, accountability, and management", *Intelligence and National Security* 36, no. 6 (19 September 2021): 841, https://doi.org/10.1080/02684527.2021.1946952.

53 Palfy, 'Bridging the Gap between Collection and Analysis', 369.

[54]   See, for example: Treverton, 'New Tools for Collaboration: The Experience of the U.S. Intelligence Community', 3; Frank Strickland and Chris Whitlock, "Understanding and creating colocated, cross-functional teams", *Studies in Intelligence* 60, no. 1 (2016): 54.

[55]   Doug Laney, "Private-sector applications of data science", *Studies in Intelligence* 61, no. 1 (March 2017): 9.

[56]   Adrian Wolfberg, "How information overload and equivocality affect law enforcement intelligence analysts: implications for learning and knowledge production", *Journal of Intelligence and Analysis* 23, no. 1 (Winter 2017): 32.

[57]   Lars D. Nicander, "Understanding intelligence community innovation in the post-9/11 world", *International Journal of Intelligence and CounterIntelligence* 24, no. 3 (1 September 2011): 541, https://doi.org/10.1080/08850607.2011.568295.

Bibliography

Allen, David, and T. D. Wilson. "Information overload: context and causes". *The New Review of Information Behaviour Research* 4, no. 1 (December 2003): 31–44. https://doi.org/10.1080/14716310310001631426.

Bawden, David, and Lyn Robinson. 'Information Overload: An Introduction'. In *Oxford Research Encyclopedia of Politics*, by David Bawden and Lyn Robinson. Oxford University Press, 2020. https://doi.org/10.1093/acrefore/9780190228637.013.1360.

Bell, Benjamin, and Michael Marks. "Composite Signatures Analyst Learning Tool (CSALT): supporting the analyst with scenario-based methodology training". *The International Journal of Intelligence, Security, and Public Affairs* 18, no. 2 (2016): 157–72.

Biltgen, Patrick, and Stephen Ryan. *Activity-Based Intelligence: Principles and Applications*. Artech House Electronic Warfare Library. Boston: Artech House, 2016.

Cotton, Anthony J. 'Information Technology – Information Overload for Strategic Leaders'. U.S. Army War College, 18 March 2005. https://doi.org/10.1037/e457182006-001.

Curry, Adrienne, and Caroline Moore. "Assessing information culture—an exploratory model". *International Journal of Information Management* 23, no. 2 (1 April 2003): 91–110. https://doi.org/10.1016/S0268-4012(02)00102-0.

Drezner, Jeffrey, Jon Schmid, Justin Grana, Megan McKernan, and Mark Ashby. 'Benchmarking Data Use and Analytics in Large, Complex Private-Sector Organizations: Implications for Department of Defense Acquisition'. RAND Corporation, 2020. https://doi.org/10.7249/RRA225-1.

Dupont, Alan. "Intelligence for the Twenty-First Century". *Intelligence and National Security* 18, no. 4 (December 2003): 15–39. https://doi.org/10.1080/02684520310001688862.

Eldridge, Christopher, Christopher Hobbs, and Matthew Moran. "Fusing algorithms and analysts: open-source intelligence in the age of 'Big Data'". *Intelligence and National Security* 33, no. 3 (16 April 2018): 391–406. https://doi.org/10.1080/02684527.2017.1406677.

Eppler, Martin J., and Jeanne Mengis. "The concept of information overload: a review of literature from organization science, accounting, marketing, MIS, and related disciplines". *The Information Society* 20, no. 5 (November 2004): 325–44. https://doi.org/10.1080/01972240490507974.

Folker, Robert D. "Arming the intelligence analyst for information warfare". *American Intelligence Journal* 30, no. 2 (2012): 5.

Gill, Peter, and Mark Phythian. 'From Intelligence Cycle to Web of Intelligence: Complexity and the Conceptualisation of Intelligence'. In: *Understanding the Intelligence Cycle*, edited by Mark Phythian, 21–42, 2014.

Hare, Nick, and Peter Coghill. "The future of the intelligence analysis task". *Intelligence and National Security* 31, no. 6 (18 September 2016): 858–70. https://doi.org/10.1080/02684527.2015.1115238.

Johnson, Loch K., ed. *The Oxford Handbook of National Security Intelligence*. Oxford, New York: Oxford University Press, 2012.

Koltay, Tibor. 'Information Overload in a Data-Intensive World'. In: *Understanding Information: From the Big Bang to Big Data*, edited by Alfons Josef Schuster, 197–217. Advanced Information and Knowledge Processing. Springer International Publishing, 2017.

Laney, Doug. "Private-sector applications of data science". *Studies in Intelligence* 61, no. 1 (March 2017): 1–24.

Lauri, Liia, Sirje Virkus, and Mati Heidmets. "Information cultures and strategies for coping with information overload: case of Estonian higher education institutions". *Journal of Documentation* 77, no. 2 (1 January 2020): 518–41. https://doi.org/10.1108/JD-08-2020-0143.

Lawrence II, James L. "Activity-based intelligence: coping with the "unknown unknowns" in complex and chaotic environments". *American Intelligence Journal* 33, no. 1 (2016): 17–25.

Menkveld, Christiaan. "Understanding the complexity of intelligence problems". *Intelligence and National Security*, 8 February 2021, 1–21. https://doi.org/10.1080/02684527.2021.1881865.

Nicander, Lars D. "Understanding intelligence community innovation in the post-9/11 world". *International Journal of Intelligence and CounterIntelligence* 24, no. 3 (1 September 2011): 534–68. https://doi.org/10.1080/08850607.2011.568295.

Palfy, Arpad. "Bridging the gap between collection and analysis: intelligence information processing and data governance". *International Journal of Intelligence and CounterIntelligence* 28, no. 2 (3 April 2015): 365–76. https://doi.org/10.1080/08850607.2015.992761.

Pedersen, Tore, and Pia Therese Jansen. "Seduced by secrecy – perplexed by complexity: effects of secret vs open-source on intelligence credibility and analytic confidence". *Intelligence and National Security* 34, no. 6 (19 September 2019): 881–98. https://doi.org/10.1080/02684527.2019.1628453.

Pijpers, Guus. *Information Overload: A System for Better Managing Everyday Data*. Microsoft Executive Leadership Series. Hoboken, N.J: Wiley, 2010.

Rêgo, Nelson J.M. "Intelligence in NATO – contextualising a doctrinal and structural clash". *Revista de Ciências Militares* VI, no. 1 (May 2018): 135–61.

Roetzel, Peter Gordon. "Information overload in the information age: a review of the literature from business administration, business psychology, and related disciplines with a bibliometric approach and framework development". *Business Research* 12, no. 2 (December 2019): 479–522. https://doi.org/10.1007/s40685-018-0069-z.

Rutkowski, Anne-Françoise, and Carol S Saunders. Emotional and Cognitive Overload: The Dark Side of Information Technology, 2019. https://search.ebscohost.com/login.aspx?direct=true&scope=site&db=nlebk&db=nlabk&AN=1776962.

Schroder, Harold M., Michael J. Driver, and Siegfried Streufert. *Human Information Processing: Individuals and Groups Functioning in Complex Social Situations.* Holt, Rinehart and Winston, 1967.

Strickland, Frank, and Chris Whitlock. "Understanding and creating colocated, cross-functional teams". *Studies in Intelligence* 60, no. 1 (2016): 53–58.

Treverton, Gregory F. 'New Tools for Collaboration: The Experience of the U.S. Intelligence Community'. Washington, D.C: Center for Strategic and International Studies, January 2016. https://www.businessofgovernment.org/sites/default/files/New%20Tools%20for%20Collaboration.pdf.

Van Puyvelde, Damien, Stephen Coulthart, and M. Shahriar Hossain. "Beyond the buzzword: big data and national security decision-making". *International Affairs* 93, no. 6 (1 November 2017): 1397–1416. https://doi.org/10.1093/ia/iix184.

Vogel, Kathleen M., Gwendolynne Reid, Christopher Kampe, and Paul Jones. "The impact of AI on intelligence analysis: tackling issues of collaboration, algorithmic transparency, accountability, and management". *Intelligence and National Security* 36, no. 6 (19 September 2021): 827–48. https://doi.org/10.1080/02684527.2021.1946952.

Wark, Wesley K. "Introduction: 'learning to live with intelligence'". *Intelligence and National Security* 18, no. 4 (December 2003): 1–14. https://doi.org/10.1080/02684520310001688853.

Winston, Thomas. "Intelligence challenges in tracking terrorist internet fund transfer activities". *International Journal of Intelligence and CounterIntelligence* 20, no. 2 (19 February 2007): 327–43. https://doi.org/10.1080/08850600600829833.

Wolfberg, Adrian. "How information overload and equivocality affect law enforcement intelligence analysts: implications for learning and knowledge production". *Journal of Intelligence and Analysis* 23, no. 1 (Winter 2017): 19.

PART III

Data Driven Operations

CHAPTER 11

A conceptual investigation of the trade-off between privacy and algorithmic performance

Job Timmermans & Roy Lindelauf

Abstract

Data-driven methodologies come with a huge benefit of optimising solutions for decision problems, but also potentially introduce new ethical risks, both in a general societal context and for defence and military organisations in particular. This chapter therefore conceptually investigates the trade-off between privacy on the one hand and algorithmic performance on the other, concerning the use of Ministry of Defence (MoD) relevant (bulk) datasets from a technical, moral and socio-political view. Because the MoD operates in a global context, the possible consequences of choices by the MoD are also considered in comparison with the choices made by other countries with respect to this trade-off.

Keywords: Privacy, Algorithmic performance, Machine learning, Security, Decision-making

11.1. Introduction

The last decade has seen an exponential development and integration of big data and machine learning algorithms within almost every domain thinkable, varying from informing bail and parole decisions[1], which patient is seen[2], college admissions[3] to filtering applicants for financial loans[4]. Within many decision-making challenges relevant to the Ministry of Defence (MoD), and defence and security in general, machine learning is gaining prominence too. Solutions exist for instance in automating tasking of technical collection platforms, in predicting the likelihood of terror attacks[5], in modelling terrorist and insurgent group behaviour[6], in deciding which surveillance image footage to focus on[7], fake news detection[8] and in operating (semi-) autonomous weapon systems such as tanks or drones[9], to name just a few.

This current exponential increase in data gathering and analysis capabilities confronts governments and organisations worldwide with ever-increasing *bulk* datasets to mine for information. Bulk datasets are sets of information about a large number of individuals, the majority of whom are not of interest to intelligence agencies. These bulk datasets, such as, for example, large-scale communications data or names of passengers on airline flights are gathered by defence and intelligence agencies with the goal of providing both tactical and strategic foresight and warning[10]. These data-driven methodologies come with the huge benefit of optimising solutions for decision problems, but also potentially introduce new ethical risks, both in the societal context in general, and for defence and military organisations in particular. There is a fierce debate on several aspects related to mass surveillance using bulk datasets, such as the effectiveness of such surveillance programs in preventing terrorist attacks and serious crime[11], privacy concerns relating to surveillance and technology[12], fairness and transparency of the algorithms used to analyse such datasets[13], and concerning their judicial regulation[14]. In light of these developments, in 2017, the Dutch Parliament approved a revision of the 2002 Dutch Intelligence and Security Services Act, where the new version explicitly incorporates the use of modern technology and data gathering and analysis capabilities. An evaluation of this law concluded that the Intelligence and Security Services Act 2017 (WIV)[15] is not optimally synced with the technological complexity and operational practice of Dutch intelligence agencies and specifically stated that a new bulk data regime is required[16].

Incorporating rules and regulations on data and algorithms for their analysis, implementing process frameworks for non-discrimination by design and using explicit ethical algorithm design are all important ways to deal with the challenges of bulk data and should therefore be pursued. However, one trade-off that should not be overlooked upon implementing such measures, and which is especially important within the defence and security domain, is the trade-off between military advantages in the use of algorithms for decision-making and democratic values. The Dutch Defence Strategy on data science explicitly states that the use of complex algorithms should enhance military strength but also should be aligned with core democratic values. In this chapter, therefore we conceptually investigate the trade-off between one core value, namely privacy, on the one hand and algorithmic performance on the other, with respect to the use of MoD relevant (bulk) datasets from a technical, moral and socio-political view.[17] In addition, because the MoD operates in a global context, we consider the possible consequences of choices by the MoD in comparison with the choices made by other countries concerning this trade-off.

We start by introducing a coordinate system that aids in the analysis of this trade-off. In section 11.2 we use this system to present a conceptual mapping of the trade-off between algorithmic performance within a defence and security context and the value of privacy. Next, in section 11.3 we sketch the technical dimensions of and some solutions to this trade-off followed by the socio-political dimensions in section 11.4. We conclude the chapter by briefly looking ahead to ways to further tackle this multi-dimensional challenge.

11.2. Privacy vs. security trade-off

An insightful way to depict challenges that are faced when designing and deploying ethical algorithms is offered by van den Hoven et al.[18] based on earlier work by Kuran[19]. In their article, the authors propose a two-dimensional coordinate system with on each axis a different (moral) value that simultaneously has to be met by the design (see Figure 11.1).

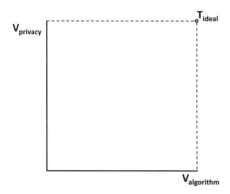

Fig. 11.1. Mapping of two moral values.[20]

In our case, the horizontal axis represents one or more moral values associated with the performance of the algorithm ($V_{algorithm}$), i.e. the benefits of deploying the algorithm, for example, its ability to discover potential targets in a large dataset. And, the vertical axis represents a moral value that may be negatively affected by deploying the algorithm, in our current example, the value of privacy ($V_{privacy}$).

The origin of the coordinate system then denotes the theoretical position in which none of the values is met. Moving upwards along the horizontal axis signifies an increase in the performance of the algorithm and hence a higher level of meeting the associated value commitment. Likewise, moving upwards along the vertical axis signifies a gradual increase in the level the value of privacy has met.

In an ideal world, both values $V_{algorithm}$ and $V_{privacy}$ can be fully met simultaneously (T_{ideal}). In that particular case, the performance of the algorithm and its associated values would not be impaired by meeting the requirement set by the opposing moral value such as privacy. In reality, however, in designing and implementing algorithms we will face situations in which different value commitments cannot (fully) be lived up to simultaneously.

It should be noted that the depiction of points and curves in our figure is a conceptual approximation of reality. In fact, a particular value may not be readily quantifiable or be mapped onto an ordinal or a rational linear scale. It, however, does allow us to display the relative positions and trends of the possible solutions in dealing with the current challenges.

In this way the coordinate system allows us to map all possible ways a trade-off, conceptually at least, can be reached between these two values. In turn, this offers a canvas to display the technical and associated socio-political challenges and their interrelationships that the MoD faces when implementing algorithmic analyses on its data. Before turning to these challenges, first we will explicate two (partly) conflicting values and what is at stake for the MoD in more detail.

11.2.1. Algorithmic performance, security and safety

As already pointed out in the introduction, the use of algorithms for data analyses can be understood to contribute to the core values of the MoD.

In its mission statement, the Dutch MoD states that its main objective is to 'protect all that we as a nation cherish.'[21] To that end, it 'fight[s] for a world of freedom and security'[22]. To be able to meet these two core values in the near future the MoD finds that it is 'vital' that it becomes a 'technologically advanced organisation'[23]. Because the use of innovative technology lies at the heart of modern warfare, meeting the core values is thus tightly connected with the use of innovative technology. The MoD, therefore, acknowledges that it relies heavily on having access to advanced technologies such as 'Artificial Intelligence (AI) and Big Data' to meet its security interests[24].

A good example to illustrate this point is the focus on information-based military decision- making ('Informatiegestuurd optreden') (IGO), which is regarded as the foundation of the future defence organisation[25]. IGO can be defined as the use of information to reduce uncertainty in (military) decision-making. Increasingly, this information comes from the (semi) automatic analysis of data using algorithmic techniques. By analysing such large quantities of data, IGO enhances the capacity to obtain insight, understanding and foresight of complex situations and operations[26]. This enables the military to detect and identify threats early on and take preventive measures when necessary.

The effective use of technology such as data analytics not only affords to meet the overarching value of security of the MoD, but also carries over to underlying values such as safety. Working with the 'most modern equipment' such as AI and big data ensures that all MoD's personnel 'can do their work as safely as possible.'[27] Similarly, by enabling more precise targeting, access to large datasets may help to reduce the risk of avoidable (civilian) causalities[28]. Additionally, there are a plethora of applications of AI and data science useful for the MoD: think of helicopter mission optimisation techniques (as described in chapter 7), predicting cache locations of IED's based on historical attack data and cultural variables[29], predicting terrorist group behaviour and optimising policy options for their destabilisation[30], predicting border incursions[31] and understanding the cyber vulnerability equity process[32] to name just a few.

As an illustration, we briefly discuss another recent example of the use of data analytics relevant to defence and security, in this case concerning the proliferation of nuclear material. There exist clandestine supply markets consisting of individuals, businesses, research institutions and governmental organisations that participate in the illicit proliferation of nuclear material. Starting from an existing list of sanctioned entities, and based upon open-source data (LinkedIn and Bloomberg) alone, Andrews et al.[33] build a machine learning model (SPINN: Suspiciousness Prediction in Nuclear Networks) to analyse a proliferation network consisting of more than 74k nodes and over 1M edges and to predict the suspiciousness of those nodes. Based upon novel network-centric features they were able to predict who is suspicious with high accuracy[34]. They were able to identify suspicious entities that were not known before, like a company in the UK that specialises in mining zinc and other base metals. Models like SPINN, provide defence intelligence organisations with decision support systems to optimise their fight against nuclear proliferation by discovering unknown targets, which enables them to rank them according to their suspiciousness and as such contribute to increasing safety and security. In the next section, we discuss the competing value of privacy in the trade-off dilemma.

11.2.2. Privacy

Besides offering the benefit of enhanced security and safety, the use of algorithms to analyse data also has several potential drawbacks, not least the risk of privacy infringement, which gives rise to challenges ranging from legal constraints to ethical information use[35]. Although there is consensus that the value of privacy is important, the concept remains hard to explicate or pin down[36]. In general, it aims to constrain access to certain types of personal data and prevent persons to acquire and use information about other persons[37]. This is necessary because contemporary information technologies threaten privacy by reducing the amount of control over personal data thereby opening up the possibility of a range of negative consequences that result from access to personal data.'[38]

The use of intelligent systems for data analytic purposes makes the concern for privacy even more acute as they offer powerful mechanisms that could increase real and anticipated assaults on individual privacy[39]. There are many examples of privacy violations that could occur when large datasets containing information about individuals are analysed. Note that a privacy violation in our sense does not entail unauthorised access to and breaches of a raw database that contains sensitive data (this is a security violation). With privacy in general the challenge is that only a surprisingly small number of seemingly unimportant facts about an individual in a large dataset is enough to identify him or her among the millions of individuals appearing in the dataset. In addition, we are concerned with the fact that specific details of an individual can be determined from the release of the neural network that is trained on the data. Consider for instance the well-known 2006 case of Netflix. The company publicly released the Netflix Prize dataset containing more than 100 million movie recommendations of almost half a million of its users as part of a competition. The goal of the prize competition was for researchers to come up with better algorithms that Netflix could use to improve their movie recommendation system. Cognisant of the Video Privacy Protection Act, Netflix replaced all user identifiers with unique but randomly selected IDs. Within a short period however, researchers from the University of Texas presented a de-anonymisation methodology and showed how to de-anonymise movie-viewing records in Netflix's dataset[40].

Even if an individual cannot be *uniquely* identified from a small set of seemingly trivial facts in a large dataset, this does not mean that no privacy violation occurs. For instance, consider the case where knowing some trivial facts about an individual leads a researcher to conclude that the specific individual is a member of a (possibly large) subset of the original dataset that all share some unique property (a certain disease for instance). Even though the individual's unique record cannot be pulled from the data, knowing that he/she has a certain disease is clearly a privacy violation. It has been shown that this can even be done if the data under consideration only contain aggregate data, for instance, when it does not contain *any* individual data at all[41].

In the case of bulk datasets used by defence organisations, similar privacy considerations can arise. It is claimed that large-scale private personal data collection efforts for counterterrorism (such as those revealed by Snowden in 2013) led to privacy violations[42]. Others suggest that bulk dataset collection is legitimate *on balance* because of its operational utility[43]. It is argued that the extent to which privacy matters for a given bulk communications dataset depends on whether the individuals in the dataset are foreigners or not and whether the data consists of mere meta-data only or also includes content (for instance mail messages). European law and international human rights treaties state that these are privacy intrusions, which might be overridden only in the interest of national security and serious crime[44]. European law states that if someone loses *control* of his/her data without their consent this is a privacy violation[45]. Both in the US and the UK, the Snowden revelations led to a reconsideration of bulk collection by the intelligence services and the adoption of a more stringent approach.

Overall, when collecting and mining bulk data, the MoD is faced with a trade-off between the desired or required operational utility based on algorithmic performance, which contributes to meeting its core values such as security and safety, and living up to moral and legal obligations to protect personal privacy. To be able to make informed decisions on this matter, a clear perspective is required on the performance of algorithms applied to those datasets (and the value they provide) as a function of the level of privacy that they guarantee. To gain some insight into this trade-off we conceptually present several privacy algorithmic designs in the next section and their corresponding plots in the value mapping (Figure 11.1).

11.3. Technical dimension of the trade-off

One way to tackle the trade-off between privacy and algorithmic performance is by technical innovation, i.e., designing algorithms that allow for privacy protection. However, although certain technical solutions may be successful in addressing the privacy issue, this comes at a price, namely a reduced performance of the algorithm (and hence its corresponding values). To illustrate this point, this subsection discusses non-exhaustively several solutions that are currently available. By plotting the technical solutions in the value mapping, we compare the trade-offs between the values that are offered by the different solutions.

In the extreme case, technical solutions to the trade-off would entail that we give up our concern for one of the rivalling values completely. This implies that we would either give up all considerations for privacy to reap maximum algorithmic performance (T_{max} in Figure 11.2) or refrain from using the technology altogether and attain maximum privacy (T_{none}). In practice, however, these solutions are not desirable or acceptable and a more nuanced position in which both values can be attained simultaneously is sought after.

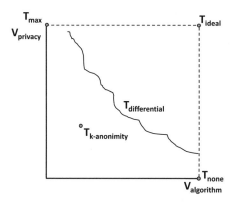

Fig. 11.2. A conceptual plot of different technical solutions to the value dilemma.

A generic solution to deal with privacy issues in big data is offered by the Privacy by Design (PBD)[46]. PBD consists of seven principles that ensure privacy (mitigation) measures are embedded throughout the design process[47], for example, requiring the design to be proactive and not reactive, privacy as the default setting and end-to-end security. An important principle of PBD that we will focus on here is the demand that privacy is to be embedded into the design itself and thus becomes an integral part of the technology.

A common way to meet this principle is by using anonymisation. This mitigates the risk of disclosing personally identifiable information (PII), for example, by removing all or a selection of such information, aggregation of data or masking PII in data[48].

One such approach that was first introduced by Samarati and Sweeney[49] is called k-anonymity. The idea behind it is that the information of each person in the data cannot be distinguished from k-1 others in the data. However, later critiques of k-anonymity showed that neither k-anonymity nor enhancements of it are entirely successful in preventing privacy leakage while keeping a reasonable data utility level[50]. This can easily be seen in the case where the data consists of medical records. Even if it satisfies k-anonymity, it might still be possible that all k individuals that a certain individual of interest belongs to, have the same disease, i.e., one can still infer highly private information of this individual even though their data is privacy protected by k-anonymity ($T_{k\text{-anonimity}}$).

Another innovative method that is gaining attention is a probabilistic anonymising technique called differential privacy[51]. This technology ($T_{differential}$) is especially fit for the current discussion because it allows for a gradual increase or decrease of the level of privacy[52], which helps us to illustrate how technical choices affect the level of realisation of moral values.

In the 2000s, the mathematical concept of differential privacy was developed by a group of computer scientists to solve the privacy challenges sketched in the second part of this chapter. The basic idea of differential privacy is that the deletion of a specific individual's data record from a large dataset does not influence the outcome of algorithmic analysis of the data very much. It is interesting to note that Google and Apple started using differential privacy on some of their applications after the first decade of the 2000s.[53] Differential privacy has an exact mathematical definition, and the amount of privacy can be 'tuned' in the sense that the stricter the privacy settings the less likely that *any* deviating analytical outcome will occur when a person's records are not included in the data compared to when they are included[54]. Thus, differential privacy guarantees for instance that the probability of getting annoying telemarketing calls does not increase when you agree to include your data in a study. It promises that the probability of getting picked out of line at an airport does not increase by very much if you allowed your past travel records to be stored. Additionally, differential privacy ensures that *for every* individual in a dataset the beliefs on anything that an analyst, who runs an experiment on the data, holds are very close to when the same analyst runs the same experiment on the data with the records of *any* individual removed. Put differently: differential privacy ensures that if someone queries the output of a differentially private algorithm to search if individual X is in the dataset, he/she will get the same result notwithstanding if individual X is included in the dataset or not. Because differential privacy provides a 'tuning knob' for the amount of privacy that is required in a given setting (based on exact mathematical arguments), $T_{differential}$ is plotted in Figure 11.2 as a curve instead of as a point. Note that the exact form of this curve is not known and depends on the particulars of the situation at hand (type of algorithm, dataset structure, etc.). However, it is clear that $T_{differential}$ is expected to be monotonically decreasing in every setting, as an increase in privacy will never yield better algorithmic performance. Hence, ethical restrictions on algorithms do not come for free.

As a consequence, the level of privacy protection implemented in an algorithm is not solely a technical challenge but above all a socio-political decision. In the next section, therefore, we will take a conceptual look at the influence of non-technical privacy demands and the corresponding implications as illustrated by the value map.

11.4. Socio-political dimension of the trade-off

As discussed in the previous section, opting for a particular data analytical solution entails a particular trade-off between the values associated with algorithmic performance such as security and safety, and the value of privacy. However, even if a particular combination of these values is practically attainable it may not be socially desirable or acceptable from a legal or moral perspective.

On the one hand, national and transnational regulation and legislation set a 'hard' limit that may not be crossed by the value of privacy. For instance, the Dutch Data Protection Authority (DPA) supervises the processing of personal data to ensure compliance with laws that regulate the use of personal data. The tasks and powers of the DPA are described in the General Data Protection Regulation (GDPR), supplemented by the Dutch Implementation Act of the GDPR[55].

On the other hand, national and organisational culture, sometimes voiced by public opinion or civil society organisations (CSOs) reflects (partly overlapping) moral standards that set further demands on the level of privacy that is considered (socially) acceptable. For instance, a survey of public opinion in the US by RAND indicates that citizens are concerned about the ethical risks of military use of AI[56]. Likewise, the Dutch digital rights organisation Bits of Freedom, for example, aims to safeguard online freedom by voicing privacy considerations in the media[57]. Additionally, on an organisational level, nowadays, the code of conduct of many large companies typically includes a statement on the handling of personal data.

The choice for a technical solution that 'solves' the trade-off dilemma thus not only depends on the technological feasibility of meeting the values but is also limited by socio-political factors such as public opinion, (organisational) culture and legislative regimes. The socio-political factors can therefore be interpreted to set a threshold value (V) that represents the minimum level of the value to meet the legal requirement and be morally and socially acceptable.

Technical solutions that are located in the coordinate system below the threshold may be technically feasible but are considered unlawful and/or not considered acceptable socially and morally. In the fictitious case plotted in Figure 11.3, for example, $T_{k-anonimity}$ is plotted below the level of privacy that generally would be considered acceptable in the Netherlands ($V_{Netherlands}$). The use of this solution therefore would be ruled out in the Netherlands because it does not guarantee the level of privacy as required by $V_{Netherlands}$, despite its algorithmic performance.

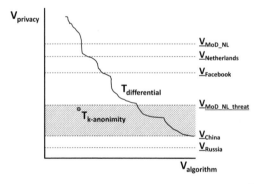

Fig. 11.3. A conceptual plot of different value thresholds.[58]

The conceptual privacy threshold value of the Netherlands differs from that of other countries due to cultural differences that also carry over to legislation and policy. Although all EU member states adhere to the GDPR, supervision and accompanying interpretations of the GDPR are the responsibility of national authorities.

In a similar vein, private and public organisations have a degree of freedom in interpreting and complying with legal and regulatory demands[59]. Organisational culture, therefore, affects the way companies address normative issues such as privacy. Alphabet Inc. and Facebook ($V_{Facebook}$), for instance, intently push the boundaries of what is permissible in terms of privacy and sometimes even cross them based on the hefty fines recently imposed by France[60].

In contrast, due to the nature of the environment it operates in, the Dutch MoD may opt for a more cautious approach toward the value of privacy. Apart from being a public organisation, the MoD is tasked with national security and has a monopoly on the use of force[61]. As a consequence, it is under close public and political scrutiny[62]. What is more, due to this special position, the culture and tradition of the MoD aspire to abide by a higher moral standard than is customary in society in general[63]. Based on this, the argument can be made that the MoD's threshold ($\underline{V}_{MoD\_NL}$) lies above that of $\underline{V}_{Netherlands}$.

At the same time, however, because of its tasks and operational requirements the MoD functions across a unique spectrum of operations – e.g. using violence on behalf of the state- in comparison with any other public or private organisation[64]. Because, as argued above, MoD's use of advanced technology is increasingly considered necessary to meet the values of security and safety[65], this also may carry over to the deployment of complex algorithms. This does not imply, however, that deployment is not bound by legal and moral restrictions. The legitimacy of taking extreme measures still depends on the general public at home and the international community, and endorsement by law and human rights thinking[66].

Nevertheless, in extraordinary circumstances, such as situations that represent a threat to national security, the threshold of the MoD ($\underline{V}_{MoD\_NL\_threat}$) may be set lower than in 'normal' – i.e. peacetime and routine conditions[67]- circumstances ($\underline{V}_{MoD\_NL}$), to allow an increase in the algorithmic performance. This is conceptually illustrated by the fictitious situation displayed in Figure 11.3, where a lower threshold allows for a higher algorithmic performance of differentially private algorithms as presented by the monotonically decreasing curve $T_{Differential}$.

Finding the right solution to the trade-off dilemma given certain circumstances can be precarious. A recent experiment by the Dutch land forces with bulk data exemplifies this. At the start of the COVID-19 pandemic, the Land Information Manoeuvre Centre (LIMC) experimented with analysing action groups by collecting and analysing data in times of crisis[68]. After attention by the media about its alleged illegitimacy, followed by political debate, the Minister of Defence decided to decommission this experiment[69].

This resonates with the broader debate on the mandate of civil services operating in the intelligence domain that is also centred around the privacy versus security argument[70]. A public consultation by the Ministry of Internal Affairs about the proposed intelligence law, for example, raised sensitivities that are felt about the necessity and proportionality of technological deployment, and the technical impossibility of compliance[71].

In the military domain, finding a solution to the trade-off dilemma is further exacerbated as increased algorithmic performance is expected to positively correlate with military power[72]. In meeting the values of safety and security, the MoD competes with adversaries that have a different culture and political system, and hence abide by different standards concerning moral values such as privacy. This lowers their privacy threshold with regards to the MoD's and hence allows their algorithms to potentially perform better. To further substantiate this point we will look into the position of China and Russia based on investigations by the RAND Corporation and (professional) online media organisations.

Both China and Russia invest heavily in the use of AI in the military domain. The Chinese People's Liberation Army (PLA) is aggressively developing AI applications such as bulk data analytics to improve its position to win military conflicts in the future[73]. The main areas of interest of bulk data use by the PLA include command & control, surveillance & reconnaissance and cybersecurity[74], but also enabling rapid operational decision-making, akin to IGO in the Netherlands[75]. Equally, Russia regards AI as essential for the future of the military and is devoting much energy to catching up with the US and other competitors[76]. Like China, Russia has taken initiatives on both the tactical and strategical levels, for example, aimed at improving command & control, decision-making, training and procurement[77].

In addition, Russia and China share a lack of sensitivity to ethical issues such as privacy compared to countries in the West[78]. Acting as a precedent, both countries have a bad track record with regards to privacy and their own citizens, using technology to censor behaviour and speech, and curtail counter-regime activities[79].

However, similar to Western countries such as the Netherlands, privacy concerns are being voiced by the Chinese public[80]. In fact, China has recently issued stringent privacy protection standards that are akin to those of Western countries[81]. However, these restrictions on data use are aimed at private enterprises and do not affect the PLA[82]. Moreover, at the same time, this new legislation has increased the power of the government to access and control private data both in China and overseas via Chinese firms[83].

In contrast, in Russia evidence for these mitigating factors is less present. Conversely, to compensate for lagging behind in the AI arms race, Russia may 'feel that ethics is a luxury it cannot afford' and be more inclined to use its technology more aggressively[84]. In addition, the weakening position of the Putin faction may increase the risk tolerance for unlawful and immoral applications[85].

Summarising, it appears that adversaries of the Dutch MoD are more willing to sacrifice privacy to attain higher algorithmic performance than Western democracies. It can be expected that the corresponding thresholds set by China ($\underline{V}_{China}$) and Russia ($\underline{V}_{Russia}$) are significantly lower than those of the Netherlands and other Western countries (see Figure 11.3). Moreover, although the societal threshold of China evidently is higher than that of Russia, this does not seem to apply to the PLA.

The area between $\underline{V}_{MoD}$ and the thresholds of the adversaries represents the opportunity cost of the Dutch MoD in terms of algorithmic performance due to its legal and moral standards. Depending on what level the algorithms are being used, these costs have to be paid out strategically, tactically and/or operationally. This raises the question of to what extent and under which circumstances Western countries are willing and can afford this potential and conceptual loss of algorithmic performance and hence of meeting the core values of security and safety?

Obviously, there are no simple answers to the trade-off dilemma between privacy and the values presented by algorithmic performance. Solutions will require due consideration and a deeper insight into all of the factors involved: technological opportunities, (moral) values, laws and regulation, public opinion, etc. Consequently, several disciplines need to be consulted such as law, engineering, ethics and the social sciences.

11.5. Conclusion

In this chapter, we explored the trade-off between privacy and security/ safety that the MoD faces when designing and implementing complex algorithms that aid in the analysis of large and complex datasets. Using a conceptual two-dimensional coordinate system with the value of privacy on the vertical axis and the value associated with algorithmic performance on the horizontal allowed us to clarify the technical, moral and socio-political dimensions of this challenge and the interwovenness of these dimensions.

First, we plotted and discussed several trade-off solutions and discussed some of their consequences. Trivially, for instance, maximum privacy yields zero algorithmic performance and hence reduces the value of safety and security to a large degree. Additionally, we showed that algorithmic design solutions to the trade-off dilemma exist – such as differential privacy – and can provide a 'tuning knob' for the amount of privacy that is desired whilst at the same time providing a certain amount of algorithmic performance. However, when one takes the socio-political context into account, only certain subsets of the possible solutions are allowed. In turn, these possible subsets of solutions are country-dependent as moral values vary around the globe. To illustrate this, we conceptually presented the cases of western countries versus China and Russia. Which solution to the trade-off is picked depends on a complex of social, political and societal factors and can have serious consequences for military power and effectiveness with respect to potential adversaries. To get agreement on the political and social mandate, it is not only recommended to involve experts such as lawyers, engineers, data scientists and foreign policy experts but also to reach out and engage with relevant stakeholders such as the general public, CSOs, military staff & troops and policy-makers. As a consequence, further investigation and eventually tackling this challenge in practice requires an interdisciplinary approach.

Notes

1 Alexandra Chouldechova and Aaron Roth, "A snapshot of the frontiers of fairness in machine learning", *Communications of the ACM* 63, no. 5 (2020): 82–89.

2 Jonathon Stewart, Peter Sprivulis, and Girish Dwivedi, "Artificial intelligence and machine learning in emergency medicine", *Emergency Medicine Australasia* 30, no. 6 (2018): 870–74.

3 Kanadpriya Basu et al., "Predictive models of student college commitment decisions using machine learning", *Data* 4, no. 2 (2019): 1–18.

4 Siddharth Bhatore, Lalit Mohan, and Y. Raghu Reddy, "Machine learning techniques for credit risk evaluation: a systematic literature review", *Journal of Banking and Financial Technology* 4, no. 1 (2020): 111–38.

5 Botambu Collins et al., 'A Survey on Forecasting Models for Preventing Terrorism', in *Advanced Computational Methods for Knowledge Engineering: Proceedings of the 6th International Conference on Computer Science, Applied Mathematics and Applications, ICCSAMA 2019*, vol. 1121 (Springer Nature, 2019), 323.

6 V. S. Subrahmanian et al., *Computational Analysis of Terrorist Groups: Lashkar-e-Taiba: Lashkar-e-Taiba* (New York: Springer, 2013); V. S. Subrahmanian et al., *A Machine Learning Based Model of Boko Haram* (Cham: Springer, 2021).

7 C. V. Amrutha, C. Jyotsna, and J. Amudha, "Deep Learning Approach for Suspicious Activity Detection from Surveillance Video", in *2020 2nd International Conference on Innovative Mechanisms for Industry Applications (ICIMIA)* (IEEE, 2020), 335–39.

8 Marina Danchovsky Ibrishimova and Kin Fun Li, "A machine learning approach to fake news detection using knowledge verification and natural language processing", *Advances in Intelligent Networking and Collaborative Systems.*, ed. L. Barolli, H. Nishino, and H. Miwa, vol. 1035 (International Conference on Intelligent Networking and Collaborative Systems, Cham: Springer, 2020), 223–34.

9 Stanislav Abaimov and Maurizio Martellini, 'Artificial Intelligence in Autonomous Weapon Systems', in *21st Century Prometheus*, ed. M. Martinelli and R. Trapp (Cham: Springer, 2020), 141–77.

10 Michelle Cayford and Wolter Pieters, "The effectiveness of surveillance technology: what intelligence officials are saying", *The Information Society* 34, no. 2 (2018): 88–103.

11 Hugo M. Verhelst, A. W. Stannat, and Giulio Mecacci, "Machine learning against terrorism: how big data collection and analysis influences the privacy-security dilemma", *Science and Engineering Ethics* 26, no. 6 (2020): 2975–84.

12 Glenn Greenwald, *No Place to Hide: Edward Snowden, the NSA, and the US Surveillance State* (New York, N.Y.: Metropolitan Books, 2014); Samuel A. Morgan, 'Security vs. Liberty: How to Measure Privacy Costs in Domestic Surveillance Programs' (Monterey, CA, Naval Postgraduate School, 2014).

13 Kiana Alikhademi et al., "A review of predictive policing from the perspective of fairness", *Artificial Intelligence and Law*, 2022, 1–17.

14 Plixavra Vogiatzoglou, "Mass surveillance, predictive policing and the implementation of the CJEU and ECtHR requirement of objectivity", *European Journal of Law and Technology* 10, no. 1 (2019): 1–18.

15 Ministry of Internal Affairs, 'Wet op de inlichtingen- en veiligheidsdiensten 2017' (2017), https://wetten.overheid.nl/BWBR0039896/2022-05-01.

16 R. Jones-Bos et al., 'Wet op inlichtingen- en veiligheidsdiensten 2017' (Den Haag: Rijksoverheid, 2021).

17 There are many ways for evaluating the performance of an algorithm. With respect to machine learning classifiers for instance common metrics are: accuracy, recall, precision, F1 score and the Matthews Correlation Coefficient, to name just a few.

[18] M.J. van den Hoven, G.J. Lokhorst, and I. van de Poel, "Engineering and the problem of moral overload", *Science and Engineering Ethics* 18, no. 1 (1 May 2011): 143–55, https://doi.org/10.1007/s11948-011-9277-z.

[19] T. Kuran, 'Moral Overload and Its Alleviation.', in *Economics, Values, Organization*, ed. A. Ben-Ner and L. Putterman (Cambridge: Cambridge University Press, 1998), 231–66.

[20] Based on van den Hoven, Lokhorst, and van de Poel, 'Engineering and the Problem of Moral Overload'.

[21] Dutch Ministry of Defence, 'The Dutch Ministry of Defence: Protecting What We Value – Tasks and Future – Defensie.Nl', onderwerp (Ministerie van Defensie, 2 September 2016), https://english.defensie.nl/topics/tasks-and-future/corporate-story.

[22] Dutch Ministry of Defence.

[23] Dutch Ministry of Defence, 'Defensievisie 2035. Vechten Voor Een Veilige Toekomst' (The Hague: Ministry of Defence, 2020), 3, https://open.overheid.nl/repository/ronl-cf4bd18b-15e0-4eff-9803-ca88a86e1122/1/pdf/defensievisie-2035-vechten-voor-een-veilige-toekomst.pdf.

[24] Dutch Ministry of Defence, 29.

[25] Dutch Ministry of Defence, 23.

[26] Dutch Ministry of Defence, 40.

[27] Dutch Ministry of Defence, 23.

[28] Dutch Ministry of Defence, 23.

[29] Paulo Shakarian, V. Subrahmanian, and Maria Luisa Spaino, 'SCARE: A Case Study with Baghdad', in *Proceedings of the Third International Conference on Computational Cultural Dynamics. AAAI*, 2009, 1–9.

[30] Subrahmanian et al., *Computational Analysis of Terrorist Groups: Lashkar-e-Taiba: Lashkar-e-Taiba*.

[31] Kevin T. Greene et al., "Understanding the timing of Chinese border incursions into India", *Humanities and Social Sciences Communications* 8, no. 1 (2021): 1–8.

[32] Haipeng Chen et al., "Disclose or exploit? A game-theoretic approach to strategic decision making in cyber-warfare", *IEEE Systems Journal* 14, no. 3 (2020): 3779–90.

[33] I.A. Andrews, S. Kumar, and F. Spezzano, 'SPINN: Suspicion Prediction in Nuclear Networks.' *IEEE International Conference on Intelligence and Security Informatics* (ISI). (2015), 19–24.

[34] Standard classifiers as Random Forest, Gaussian Naïve Bayes and support vector machines were used and analysed using the Matthews Correlation Coefficient.

[35] Christopher B. Davison et al., "Data privacy in the age of big data analytics", *Issues in Information Systems* 22, no. 2 (2021): 177–86.

[36] J.F.C. Timmermans, R. Heersmink, and M.J. van den Hoven, 'Normative Issues Report', Deliverable (FP-7 ETICA project, 2010), https://www.etica-project.eu/deliverable-files.

[37] M.J. Van Den Hoven, 'Information Technology, Privacy and the Protection of Personal Data', in *Information Technology and Moral Philosophy*, ed. M.J. van Den Hoven and J. Weckert (Cambridge ; New York, NY: Cambridge University Press, 2008), 301–21.

[38] Jeroen van den Hoven et al., 'Privacy and Information Technology', in *The Stanford Encyclopedia of Philosophy*, ed. Edward N. Zalta, Summer 2020 (Metaphysics Research Lab, Stanford University, 2020), https://plato.stanford.edu/archives/sum2020/entries/it-privacy/.

[39] R.S. Rosenberg, "The social impact of intelligent artefacts", *AI and Society* 22, no. 3 (2008): 367–83, https://doi.org/10.1007/s00146-007-0148-8.

[40] Arvind Narayanan and Vitaly Shmatikov, 'Robust de-anonymization of large sparse datasets', in *2008 IEEE Symposium on Security and Privacy (Sp 2008)* (2008 IEEE Symposium on Security and Privacy (sp 2008), IEEE, 2008), 111–25, https://doi.org/doi:10.1109/SP.2008.33.

41   Nils Homer et al., "Resolving individuals contributing trace amounts of DNA to highly complex mixtures using high-density SNP genotyping microarrays", *PLoS Genetics* 4, no. 8 (2008): e1000167.

42   David Lyon, "Surveillance, Snowden, and big data: capacities, consequences, critique", *Big Data & Society* 1, no. 2 (2014): 1–13.

43   David WK Anderson, *Report of the Bulk Powers Review* (London: HMSO, 2016).

44   Anderson.

45   Paul De Hert, "Identity management of E-ID, privacy and security in Europe. a human rights view.", *Nformation Security Technical Report* 13, no. 2 (2008): 71–75.

46   Giuseppe D'Acquisto et al., "Privacy by design in big data: an overview of privacy enhancing technologies in the era of big data analytics", *ArXiv Preprint ArXiv:1512.06000*, 2015.

47   Ann Cavoukian, 'Privacy by Design The 7 Foundational Principles' (Canada, 2009).

48   Christopher Graham, "Anonymisation: Managing Data Protection Risk Code of Practice", *Information Commissioner's Office*, 2012.

49   Pierangela Samarati and Latanya Sweeney, 'Protecting Privacy When Disclosing Information: K-Anonymity and Its Enforcement through Generalization and Suppression', 1998.

50   Josep Domingo-Ferrer and Vicenç Torra, "A Critique of K-Anonymity and Some of Its Enhancements", in *2008 Third International Conference on Availability, Reliability and Security* (IEEE, 2008), 990–93.

51   Cynthia Dwork and Aaron Roth, "The algorithmic foundations of differential privacy.", *Found. Trends Theor. Comput. Sci.* 9, no. 3–4 (2014): 211–407.

52   Cynthia Dwork, "Differential Privacy: A Survey of Results", *International Conference on Theory and Applications of Models of Computation* (Springer, 2008), 1–19.

53   Jun Tang et al., "Privacy loss in Apple's implementation of differential privacy on Macos 10.12", *ArXiv Preprint ArXiv:1709.02753*, 2016, 1–12.

54   Dwork, 'Differential Privacy: A Survey of Results'.

55   DPA, 'Tasks and Powers of the Dutch DPA', Autoriteit Persoonsgegevens, 2022, https://www.autoriteitpersoonsgegevens.nl/en/about-dutch-dpa/tasks-and-powers-dutch-dpa.

56   Forrest E. Morgan et al., 'Military Applications of Artificial Intelligence: Ethical Concerns in an Uncertain World' (RAND Corporation, 28 April 2020), xiv, https://www.rand.org/pubs/research_reports/RR3139-1.html.

57   Bits of Freedom, 'Bits of Freedom: Voor jouw internetvrijheid', Bits of Freedom, 2022, https://www.bitsoffreedom.nl/.

58   Note that the given thresholds represent an estimation of their relative rather than absolute positions for the sake of the argument.

59   Job Timmermans, 'Exploring the Multifaceted Relationship of Compliance and Integrity—The Case of the Defence Industry', in *NL ARMS Netherlands Annual Review of Military Studies 2021*, ed. Robert Beeres et al., NL ARMS (The Hague: T.M.C. Asser Press, 2022), 95–113, https://doi.org/10.1007/978-94-6265-471-6_6.

60   Siladitya Ray, 'Google And Facebook Hit With $238 Million Fines In France Over Privacy Violations', Forbes, accessed 19 May 2022, https://www.forbes.com/sites/siladityaray/2022/01/06/google-and-facebook-hit-with-238-million-fines-in-france-over-privacy-violations/.

61   Joseph Soeters, Paul C. van Fenema, and Robert Beeres, 'Introducing Military Organizations', in *Managing Military Organizations: Theory and Practice*, ed. Joseph Soeters, Paul C. van Fenema, and R.J.M. Beeres (New York, N.Y.: Routledge, 2010), 1–16.

62   Militaire Spectator, "De grenzen van veiligheid", *Militaire Spectator* 183, no. 3 (8 March 2014): 102–3; Peter Olsthoorn, Marten Meijer, and Desiree Verweij, 'Managing Moral Professionalism in Military Operations', in *Managing Militant Operations. Theory and Practice*, ed. Joseph Soeters and Paul C. van Fenema (New York, N.Y.: Routledge, 2010), 138–49.

[63] J.F.C. Timmermans et al., "Dertien Jaar Integriteitsonderzoek: Een Zoektocht Naar Defensie-Identiteit. Van Incident Naar Onderzoek Naar Incident", *Militaire Spectator* 2 (2021): 72–83.

[64] Soeters, van Fenema, and Beeres, 'Introducing Military Organizations'.

[65] Dutch Ministry of Defence, 'Defensievisie 2035. Vechten Voor Een Veilige Toekomst'.

[66] Soeters, van Fenema, and Beeres, 'Introducing Military Organizations'.

[67] Soeters, van Fenema, and Beeres.

[68] Dutch Ministry of Defence, 'Land Information Manoeuvre Centre helpt Defensie anticiperen – Nieuwsbericht – Defensie.nl', nieuwsbericht (Ministerie van Defensie, 16 November 2020), https://www.defensie.nl/actueel/nieuws/2020/11/16/land-information-manoeuvre-centre-helpt-defensie-anticiperen; Esther Rosenberg and Karel Berkhout, 'Alle alarmbellen gingen af, maar waarom greep dan niemand in bij Defensie?', *NRC*, 1 November 2021, https://www.nrc.nl/nieuws/2021/11/01/waarom-greep-niemand-in-bij-defensie-a4063866.

[69] Dutch Ministry of Defence, 'Kamerbrief over uitvoeren moties Land Information Manoeuvre Centre – Kamerstuk', kamerstuk (The Hague: Ministry of Defence, 26 October 2021), https://www.rijksoverheid.nl/documenten/kamerstukken/2021/10/26/kamerbrief-over-uitvoeren-moties-land-information-manoeuvre-centre; Rosenberg and Berkhout, 'Alle alarmbellen gingen af, maar waarom greep dan niemand in bij Defensie?'

[70] A. Claver, "The big data paradox: juggling data flows, transparency and secrets", *Militaire Spectator* 187, no. 6 (2018): 309–23.

[71] Claver.

[72] Morgan et al., 'Military Applications of Artificial Intelligence', xiv.

[73] Morgan et al., 81; Derek Grossman et al., 'Chinese Views of Big Data Analytics' (RAND Corporation, 1 September 2020), 27, https://www.rand.org/pubs/research_reports/RRA176-1.html.

[74] Grossman et al., 'Chinese Views of Big Data Analytics', viii–ix.

[75] Morgan et al., 'Military Applications of Artificial Intelligence', 82.

[76] Morgan et al., 83.

[77] Andrew Eversden, 'A Warning to DoD: Russia Advances Quicker than Expected on AI, Battlefield Tech', C4ISRNet, 24 May 2021, https://www.c4isrnet.com/artificial-intelligence/2021/05/24/a-warning-to-dod-russia-advances-quicker-than-expected-on-ai-battlefield-tech/.

[78] Morgan et al., 'Military Applications of Artificial Intelligence', xiv.

[79] Alina Polyakova, 'Weapons of the Weak: Russia and AI-Driven Asymmetric Warfare', *Brookings* (blog), 15 November 2018, https://www.brookings.edu/research/weapons-of-the-weak-russia-and-ai-driven-asymmetric-warfare/; Grossman et al., 'Chinese Views of Big Data Analytics', 49.

[80] Morgan et al., 'Military Applications of Artificial Intelligence', 79.

[81] Grossman et al., 'Chinese Views of Big Data Analytics', 11, 81.

[82] Morgan et al., 'Military Applications of Artificial Intelligence', 81.

[83] Matt Pottinger and David Feith, 'Opinion | The Most Powerful Data Broker in the World Is Winning the War Against the U.S.', *The New York Times*, 30 November 2021, sec. Opinion, https://www.nytimes.com/2021/11/30/opinion/xi-jinping-china-us-data-war.html.

[84] Morgan et al., 'Military Applications of Artificial Intelligence', 98–99.

[85] Morgan et al., 98.

References

Abaimov, Stanislav, and Maurizio Martellini. 'Artificial Intelligence in Autonomous Weapon Systems'. In *21st Century Prometheus*, edited by M. Martinelli and R. Trapp, 141–77. Cham: Springer, 2020.

Alikhademi, Kiana, Emma Drobina, Diandra Prioleau, Brianna Richardson, Duncan Purves, and Juan E. Gilbert. "A review of predictive policing from the perspective of fairness". *Artificial Intelligence and Law*, 2022, 1–17.

Amrutha, C. V., C. Jyotsna, and J. Amudha. 'Deep Learning Approach for Suspicious Activity Detection from Surveillance Video'. In 2020 2nd International Conference on Innovative Mechanisms for Industry Applications (ICIMIA), 335–39. IEEE, 2020.

Anderson, David WK. Report of the Bulk Powers Review. London: HMSO, 2016.

Andrews, I.A., S. Kumar, and F. Spezzano. "SPINN: Suspicion Prediction in Nuclear Networks." *IEEE International Conference on Intelligence and Security Informatics* (ISI)., 19–24. IEEE, 2015.

Basu, Kanadpriya, Treena Basu, Ron Buckmire, and Nishu Lal. "Predictive models of student college commitment decisions using machine learning". *Data* 4, no. 2 (2019): 1–18.

Bhatore, Siddharth, Lalit Mohan, and Y. Raghu Reddy. "Machine learning techniques for credit risk evaluation: a systematic literature review". *Journal of Banking and Financial Technology* 4, no. 1 (2020): 111–38.

Bits of Freedom. 'Bits of Freedom: Voor jouw internetvrijheid'. Bits of Freedom, 2022. https://www.bitsoffreedom.nl/.

Cavoukian, Ann. 'Privacy by Design The 7 Foundational Principles'. Canada, 2009.

Cayford, Michelle, and Wolter Pieters. "The effectiveness of surveillance technology: what intelligence officials are saying". *The Information Society* 34, no. 2 (2018): 88–103.

Chen, Haipeng, Qian Han, Sushil Jajodia, Roy Lindelauf, V. S. Subrahmanian, and Yanhai Xiong. "Disclose or exploit? A game-theoretic approach to strategic decision making in cyber-warfare". *IEEE Systems Journal* 14, no. 3 (2020): 3779–90.

Chouldechova, Alexandra, and Aaron Roth. "A Snapshot of the Frontiers of Fairness in Machine Learning". *Communications of the ACM* 63, no. 5 (2020): 82–89.

Claver, A. "The big data paradox: juggling data flows, transparency and secrets". *Militaire Spectator* 187, no. 6 (2018): 309–23.

Collins, Botambu, Dinh Tuyen Hoang, HyoJeon Yoon, and Ngoc Thanh Nguyen. 'A Survey on Forecasting Models for Preventing Terrorism'. In: *Advanced Computational Methods for Knowledge Engineering: Proceedings of the 6th International Conference on Computer Science, Applied Mathematics and Applications*, ICCSAMA 2019, 1121:323. Springer Nature, 2019.

D'Acquisto, Giuseppe, Josep Domingo-Ferrer, Panayiotis Kikiras, Vicenç Torra, Yves-Alexandre de Montjoye, and Athena Bourka. "Privacy by Design in Big Data: An Overview of Privacy Enhancing Technologies in the Era of Big Data Analytics". ArXiv Preprint ArXiv:1512.06000, 2015.

Davison, Christopher B., Edward J. Lazaros, Jensen J. Zhao, Allen D. Truell, and Brianna Bowles. "Data privacy in the age of big data analytics". *Issues in Information Systems* 22, no. 2 (2021): 177–86.

De Hert, Paul. "Identity Management of E-ID, Privacy and Security in Europe. A Human Rights View." Nformation Security Technical Report 13, no. 2 (2008): 71–75.

Domingo-Ferrer, Josep, and Vicenç Torra. "A Critique of K-Anonymity and Some of Its Enhancements". In *2008 Third International Conference on Availability, Reliability and Security*, 990–93. IEEE, 2008.

DPA. 'Tasks and Powers of the Dutch DPA'. Autoriteit Persoonsgegevens, 2022. https://www.autoriteit-persoonsgegevens.nl/en/about-dutch-dpa/tasks-and-powers-dutch-dpa.

Dutch Ministry of Defence. 'Defensievisie 2035. Vechten Voor Een Veilige Toekomst'. The Hague: Ministry of Defence, 2020. https://open.overheid.nl/repository/ronl-cf4bd18b-15e0-4eff-9803-ca88a86e1122/1/pdf/defensievisie-2035-vechten-voor-een-veilige-toekomst.pdf.

———. 'Kamerbrief over uitvoeren moties Land Information Manoeuvre Centre – Kamerstuk'. Kamerstuk. The Hague: Ministry of Defence, 26 October 2021. https://www.rijksoverheid.nl/documenten/kamerstukken/2021/10/26/kamerbrief-over-uitvoeren-moties-land-information-manoeuvre-centre.

———. 'Land Information Manoeuvre Centre helpt Defensie anticiperen – Nieuwsbericht – Defensie.nl'. Nieuwsbericht. Ministerie van Defensie, 16 November 2020. https://www.defensie.nl/actueel/nieuws/2020/11/16/land-information-manoeuvre-centre-helpt-defensie-anticiperen.

———. 'The Dutch Ministry of Defence: Protecting What We Value – Tasks and Future – Defensie.Nl'. Onderwerp. Ministerie van Defensie, 2 September 2016. https://english.defensie.nl/topics/tasks-and-future/corporate-story.

Dwork, Cynthia. 'Differential Privacy: A Survey of Results'. In: International Conference on Theory and Applications of Models of Computation, 1–19. Springer, 2008.

Dwork, Cynthia, and Aaron Roth. "The algorithmic foundations of differential privacy." *Found. Trends Theor. Comput. Sci.* 9, no. 3–4 (2014): 211–407.

Eversden, Andrew. 'A Warning to DoD: Russia Advances Quičker than Expected on AI, Battlefield Tech'. C4ISRNet, 24 May 2021. https://www.c4isrnet.com/artificial-intelligence/2021/05/24/a-warning-to-dod-russia-advances-quicker-than-expected-on-ai-battlefield-tech/.

Graham, Christopher. 'Anonymisation: Managing Data Protection Risk Code of Practice'. Information Commissioner's Office, 2012.

Greene, Kevin T., Caroline Tornquist, Robbert Fokkink, Roy Lindelauf, and V. S. Subrahmanian. "Understanding the timing of Chinese border incursions into India". *Humanities and Social Sciences Communications* 8, no. 1 (2021): 1–8.

Greenwald, Glenn. *No Place to Hide: Edward Snowden, the NSA, and the US Surveillance* State. New York, N.Y.: Metropolitan Books, 2014.

Grossman, Derek, Christian Curriden, Logan Ma, Lindsey Polley, J. D. Williams, and Cortez A. Cooper III. "Chinese Views of Big Data Analytics". RAND Corporation, 1 September 2020. https://www.rand.org/pubs/research_reports/RRA176-1.html.

Homer, Nils, Szabolcs Szelinger, Margot Redman, David Duggan, Waibhav Tembe, Jill Muehling, John V. Pearson, Dietrich A. Stephan, Stanley F. Nelson, and David W. Craig. "Resolving individuals contributing trace amounts of DNA to highly complex mixtures using high-density SNP genotyping microarrays". *PLoS Genetics* 4, no. 8 (2008): e1000167.

Hoven, Jeroen van den, Martijn Blaauw, Wolter Pieters, and Martijn Warnier. 'Privacy and Information Technology'. In *The Stanford Encyclopedia of Philosophy*, edited by Edward N. Zalta, Summer 2020. Metaphysics Research Lab, Stanford University, 2020. https://plato.stanford.edu/archives/sum2020/entries/it-privacy/.

Hoven, M.J. van den, G.J. Lokhorst, and I. van de Poel. "Engineering and the problem of moral overload". *Science and Engineering Ethics* 18, no. 1 (1 May 2011): 143–55. https://doi.org/10.1007/s11948-011-9277-z.

Ibrishimova, Marina Danchovsky, and Kin Fun Li. 'A Machine Learning Approach to Fake News Detection Using Knowledge Verification and Natural Language Processing'. In *Advances in Intelligent Networking and Collaborative Systems.*, edited by L. Barolli, H. Nishino, and H. Miwa, 1035:223–34. Cham: Springer, 2020.

Jones-Bos, R., L. van den Herik, B. Jacobs, W. Nagtegaal, and S.E. Zijlstra. 'Wet op inlichtingen- en veiligheidsdiensten 2017'. Den Haag: Rijksoverheid, 2021.

Kuran, T. 'Moral Overload and Its Alleviation.' In *Economics, Values, Organization*, edited by A. Ben-Ner and L. Putterman, 231–66. Cambridge: Cambridge University Press, 1998.

Lyon, David. "Surveillance, Snowden, and big data: capacities, consequences, critique". *Big Data & Society* 1, no. 2 (2014): 1–13.

Militaire Spectator. 'De grenzen van veiligheid'. Militaire Spectator 183, no. 3 (8 March 2014): 102–3.

Ministry of Internal Affairs. Wet op de inlichtingen- en veiligheidsdiensten 2017 (2017). https://wetten.overheid.nl/BWBR0039896/2022-05-01.

Morgan, Forrest E., Benjamin Boudreaux, Andrew J. Lohn, Mark Ashby, Christian Curriden, Kelly Klima, and Derek Grossman. "Military Applications of Artificial Intelligence: Ethical Concerns in an Uncertain World". RAND Corporation, 28 April 2020. https://www.rand.org/pubs/research_reports/RR3139-1.html.

Morgan, Samuel A. 'Security vs. Liberty: How to Measure Privacy Costs in Domestic Surveillance Programs'. Naval Postgraduate School, 2014.

Narayanan, Arvind, and Vitaly Shmatikov. 'Robust De-Anonymization of Large Sparse Datasets'. In: 2008 IEEE Symposium on Security and Privacy (Sp 2008), 111–25. IEEE, 2008. https://doi.org/doi:10.1109/SP.2008.33.

Olsthoorn, Peter, Marten Meijer, and Desiree Verweij. 'Managing Moral Professionalism in Military Operations'. In: *Managing Militant Operations. Theory and Practice*, edited by Joseph Soeters and Paul C. van Fenema, 138–49. New York, N.Y.: Routledge, 2010.

Polyakova, Alina. 'Weapons of the Weak: Russia and AI-Driven Asymmetric Warfare'. Brookings (blog), 15 November 2018. https://www.brookings.edu/research/weapons-of-the-weak-russia-and-ai-driven-asymmetric-warfare/.

Pottinger, Matt, and David Feith. 'Opinion | The Most Powerful Data Broker in the World Is Winning the War Against the U.S.' The New York Times, 30 November 2021, sec. Opinion. https://www.nytimes.com/2021/11/30/opinion/xi-jinping-china-us-data-war.html.

Ray, Siladitya. 'Google And Facebook Hit With $238 Million Fines In France Over Privacy Violations'. Forbes. Accessed 19 May 2022. https://www.forbes.com/sites/siladityaray/2022/01/06/google-and-facebook-hit-with-238-million-fines-in-france-over-privacy-violations/.

Rosenberg, Esther, and Karel Berkhout. 'Alle alarmbellen gingen af, maar waarom greep dan niemand in bij Defensie?' NRC. 1 November 2021. https://www.nrc.nl/nieuws/2021/11/01/waarom-greep-nie-mand-in-bij-defensie-a4063866.

Rosenberg, R.S. "The social impact of intelligent artefacts". *AI and Society* 22, no. 3 (2008): 367–83. https://doi.org/10.1007/s00146-007-0148-8.

Samarati, Pierangela, and Latanya Sweeney. 'Protecting Privacy When Disclosing Information: K-Ano-nymity and Its Enforcement through Generalization and Suppression', 1998.

Shakarian, Paulo, V. Subrahmanian, and Maria Luisa Spaino. 'SCARE: A Case Study with Baghdad'. In *Proceedings of the Third International Conference on Computational Cultural Dynamics*. AAAI, 1–9, 2009.

Soeters, Joseph, Paul C. van Fenema, and Robert Beeres. 'Introducing Military Organizations'. In *Managing Military Organizations: Theory and Practice*, edited by Joseph Soeters, Paul C. van Fenema, and R.J.M. Beeres, 1–16. New York, N.Y.: Routledge, 2010.

Stewart, Jonathon, Peter Sprivulis, and Girish Dwivedi. 'Artificial intelligence and machine learning in emergency medicine'. *Emergency Medicine Australasia* 30, no. 6 (2018): 870–74.

Subrahmanian, V. S., Aaron Mannes, Amy Sliva, Jana Shakarian, and John P. Dickerson. *Computational Analysis of Terrorist Groups*: Lashkar-e-Taiba: Lashkar-e-Taiba. New York: Springer, 2013.

Subrahmanian, V. S., Chiara Pulice, James F. Brown, and Jacob Bonen-Clark. *A Machine Learning Based Model of Boko Haram*. Cham: Springer, 2021.

Tang, Jun, Aleksandra Korolova, Xiaolong Bai, Xueqiang Wang, and Xiaofeng Wang. "Privacy loss in Apple's implementation of differential privacy on Macos 10.12". *ArXiv Preprint ArXiv:1709.02753*, 2016, 1–12.

Timmermans, J.F.C., M.P. Bogers, R.M.M. Bertrand, and R.J.M. Beeres. 'Dertien Jaar Integriteitsonder-zoek: Een Zoektocht Naar Defensie-Identiteit. Van Incident Naar Onderzoek Naar Incident'. Militaire Spectator 2 (2021): 72–83.

Timmermans, J.F.C., R. Heersmink, and M.J. van den Hoven. 'Normative Issues Report'. Deliverable. FP-7 ETICA project, 2010. https://www.etica-project.eu/deliverable-files.

Timmermans, Job. 'Exploring the Multifaceted Relationship of Compliance and Integrity—The Case of the Defence Industry'. In: *NL ARMS Netherlands Annual Review of Military Studies 2021*, edited by Robert Beeres, Robert Bertrand, Jeroen Klomp, Job Timmermans, and Joop Voetelink, 95–113. NL ARMS. The Hague: T.M.C. Asser Press, 2022. https://doi.org/10.1007/978-94-6265-471-6_6.

Van Den Hoven, M.J. 'Information Technology, Privacy and the Protection of Personal Data'. In: *Information Technology and Moral Philosophy*, edited by M.J. van Den Hoven and J. Weckert, 301–21. Cambridge ; New York, NY: Cambridge University Press, 2008.

Verhelst, Hugo M., A. W. Stannat, and Giulio Mecacci. "Machine learning against terrorism: how big data collection and analysis influences the privacy-security dilemma". *Science and Engineering Ethics* 26, no. 6 (2020): 2975–84.

Vogiatzoglou, Plixavra. "Mass surveillance, predictive policing and the implementation of the CJEU and ECtHR requirement of objectivity". *European Journal of Law and Technology* 10, no. 1 (2019): 1–18.

CHAPTER 12

Information manoeuvre and the Netherlands armed forces

Legal challenges ahead

Paul A.L. Ducheine, Peter B.M.J. Pijpers & Eric H. Pouw

Abstract

With the Defence White Paper "Defence Vision 2035", the Netherlands have articulated that its armed forces need to be capable to execute 'information-driven operations'. This intent reflects the threats and opportunities emerging from the inception of cyberspace, with the Russia-Ukraine war a case in point. Cyberspace has unlocked the information environment, raising obvious concerns about the use of data and potential infringements of privacy since it simultaneously gives new impetus to use data to improve military intelligence and understanding, enhance decision making, but moreover to use information as a 'weapon' of influence. However, while (nascent) capacity and will to employ the armed forces in the information environment are present, parts of the conceptual component cause friction. The principal cause for this is the current legal framework applicable to information manoeuvre, that seriously hampers training and preparing for operations. The 'lacuna' must be dealt with, for it would be hypocritical to demand security without empowering the agencies with the tools that ensure their readiness for deployment.

Keywords: Information Manoeuvre, Data Protection, Armed Forces, the Netherlands, GDPR.

'War is ninety percent information'[1]

12.1. Introduction

The *raison d'existence* of the Netherlands' armed forces is to defend and protect the Kingdom's interests, and to maintain and promote the international legal order.[2] Though these objectives could stand the test of time, the dynamics of the (geopolitical) security context within which the State's interests need to be protected have changed, partially as a result of the dawn of cyberspace.[3]

Whilst organisational readjustment to the evolving security landscape is required, the domain of engagement – land, sea, air, space or cyberspace – is indifferent to the protection of interests; a view that is echoed in the Netherlands' Defence Vision 2035 (DV35).[4] This Defence White Paper articulates the defence organisation's ambition to be an effective agent in the information environment and to make better use of available data and information in the military realm.[5]

One of the baselines is to be able to execute so-called 'information-driven operations', also (and from here on) called information manoeuvre,[6] in 2035. The defence organisation should not only be capable of obtaining an authoritative information position, which is needed for integrated command and control and 'information-driven operations', it must also use information as a 'weapon', i.e. as a means or instrument of influence.[7]

From a legal point of view, manoeuvring in the information environment affects the armed forces in at least two ways. Firstly, before armed forces are deployed, they need to achieve the appropriate level of readiness. Adopting the concept of information manoeuvre demands the expertise and familiarisation of personnel with new concepts, doctrine, procedures, standards and equipment, by means of education, experimentation, exercises and training (hereinafter referred to as 3ET) with data and information: requirements that need to be achieved prior to and not whilst deployed. To the extent 3ET involves real world data and information available in the current information-environment, this unavoidably implies the processing of personal data which could infringe privacy.

Secondly, an authoritative information position needs to be obtained prior to a planned or envisioned deployment. The collection, processing and dissemination of data and information must start before the commencement of a mission mandate. If this includes a role for the armed forces, this raises questions regarding (the scope of) the purpose, tasks and legal authorities of the armed forces outside deployments and their relationship with those of others, such as the intelligence and security services.

The aim of this contribution is two-fold. First, it articulates the legal framework applicable to armed forces manoeuvring in the information environment; with a particular focus on the processing of personal data. Second, it aims to raise awareness on the implications of the legal framework for activities of the armed forces for 3ET and obtain an authoritative information position in the readiness phase and provide some suggestions to overcome these implications.

The main question in this chapter is whether the current legal framework offers adequate room for the armed forces to manoeuvre in the information environment prior to deployment.

The structure of the chapter is as follows. In the next section the authors describe the information manoeuvre concept. Section 12.3 outlines relevant aspects of the general legal framework applicable to activities of the armed forces in this concept. In section 12.4, based on a fictitious scenario in the near future, the implications of this framework for the armed forces to conduct 3ET with real world data and information, as well as for its role in obtaining an authoritative information-position, will be assessed. Finally, the chapter will reflect on the main research question, provide suggestions on how to deal with the implications and return to the secondary aim of the chapter – raising awareness of the implications of the current legal lacunae for the conceptual component of military power.

As a point of departure, unless otherwise indicated, the chapter focuses solely on activities of the armed forces in the information environment[8] in the readiness phase[9] for a (possible or planned) deployment abroad.[10] It is in the readiness phase that most legal challenges can be found.

12.2. Information manoeuvre

The information environment is the environment we live in and from which we gather data via our senses; and after storing, fusing and processing the data to information and knowledge, we can communicate in the information environment. This information environment (see figure 12.1) entails the physical, the virtual and the cognitive dimension.

The information environment is not new; however, with the emergence of cyberspace new layers were introduced,[11] most prominently the virtual layers of data and software (logical layer) and of virtual personae – our reflections in cyberspace using (inter alia) social media accounts.

Cyberspace has substantially enlarged the information environment. From a security and military perspective, the expanded access to and usability of the information environment has created (re)new(ed) opportunities and threats, including digital espionage, influence operations or subversion via cyberspace,[12] not least due to the low costs of entry, high speed of dissemination and high degree of penetration into societies.

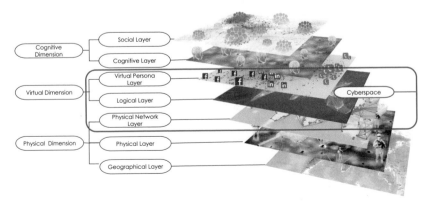

Fig. 12.1. Information Environment and Cyberspace.[13] Permission granted by Jelle van Haaster.

Previously unheard-of threats, such as DDoS-attacks,[14] have materialised,[15] and it is therefore not the question *whether* but *how* the Netherlands needs to respond to malign use of the information environment. The concept of information manoeuvre is (being) developed to facilitate this response.

Information manoeuvre can be defined as the use of information of a cognitive, virtual (digital) or physical nature by armed forces 'in the operational environment [...]to achieve a position of advantage in respect to others [...] in order to accomplish the mission'.[16] Information manoeuvre is, therefore, a way of exerting power and achieving effects by using information in any cognitive, virtual of physical form to shape the operational environment in an advantageous manner,[17] but moreover to use information as a weapon, i.e. a means of influence.

To apply this concept, obtaining an authoritative information position is crucial. This position is to be acquired through various mechanisms and sensors ('observe' in Figure 12.2), which enables adequate understanding (or sensemaking) of situations (insight) and attaining foresight for future development.[18] Based on understanding, effective decisions can be made, and action in any of the dimensions of the information environment can be taken.[19]

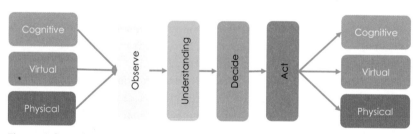

Fig. 12.2. Information Manoeuvre Conceptualised.[20]

The purpose is to be faster and better in decision making and act more effectively than others. This process can result in several activities. One is kinetic action with effects in the physical dimension to influence audiences in an indirect manner. The acme of information manoeuvre, however, is to ultimately use information (of any nature) as a means to achieve (offensive or defensive) effects.[21] The DV35 envisions 'armed forces that also use information as a weapon in its own right and that are permitted to use this weapon at an early stage and offensively where necessary.'[22] This may take place in three forms (or a combination thereof). First, the use of military and kinetic force generating effects in the physical dimension. Second, as so-called hard-cyber operations, i.e. the targeting in cyberspace itself through digital subversion or sabotage operations to achieve effects in the virtual dimension (virtual objects such as data and personae including social media accounts) and the physical network layer (computers, routers). And finally, via so-called soft-cyber operations (digital influence operations), meaning that information can be used as a weapon to influence the cognitive dimension of targeted audiences using cyberspace as a vector.[23] These influence operations aim to change the attitude of (opposing) actors by persuading them in a constructive manner or 'the deliberate use of information [...] to confuse, mislead and ultimately influence the actions that the targeted population makes.'.

12.3. The legal framework for activities in the information environment

This paragraph outlines the general legal framework for activities of the Netherlands' armed forces in the information environment.[24] The legal framework comprises the legal bases for action and the legal regimes that apply to the action itself.

12.3.1. Task, order and legal authority

The armed forces are one of the so-called 'sword powers' of the government.[25] Based on the principle of legality – one of the principles of the Netherlands' constitutional law – activities of the armed forces require a legal basis if and when their actions impair the rights and privileges of citizens and organisations. This implies that such armed forces' actions typically may only take place (1) if they can be based on, and are carried out within the boundaries of, a formally assigned task, (2) an (implied or explicit) order by the government to execute this task and, (3) a legal authority to carry out the activity in the execution of the task. The notion 'legal authority' refers to the requirement to have a national or international legal basis for the execution of a legitimate task which may constitute an infringement upon the (human) rights and freedoms of citizens – both in the Netherlands and abroad.[26]

The next sections explain where tasks, orders and legal authorities derive from in the context of an international deployment, and how they relate to the readiness phase.

12.3.2. International deployment

Article 97 of the Netherlands' Constitution is the principal provision setting out the position of the armed forces in the constitutional order:

"1. There shall be armed forces for the defence and protection of the interests of the Kingdom, and in order to maintain and promote the international legal order.
2. The Government shall have supreme authority over the armed forces."[27]

Paragraph 1 sets out the triple purposes for the Kingdom's armed forces: (1) to defend the Kingdom and allies, (2) to maintain and promote the international legal order, and (3) to protect the (other vital) interests of the Kingdom.[28]

The decision to make use of the armed forces for these purposes, i.e. to deploy them abroad, lies solely with the Government of the Netherlands and is, besides political and military strategic considerations, based on the legal framework (legal bases and legal regimes) applicable to a mission or operation.[29] Such a decision forms the mandate, and thus constitutes the basis for the task and legal authorities, for deployment.[30]

In the event of an international deployment, legal authority for activities required to execute the mandate follows from the legal basis in international law for the mission or operation and the applicable legal regime(s).[31] The following legal bases in international law are internationally recognised:

1. The consent of a host State, which in the first place governs the presence of foreign armed forces on its territory, territorial sea and airspace above it,[32] and secondly may set out which specific activities may be executed, and under which conditions they may be carried out.[33] In so far as activities in the information environment affect the sovereignty of the host State, it must be ensured that the consent of the host State governs such activities.

2. A UN Security Council Resolution (UNSCR) adopted under Chapter VII of the UN Charter, in which case the (interpretation of the) text of the resolution determines the nature and scope of the measures that may be taken.[34]

3. (Collective) self-defence in case of an armed attack, as recognised under Article 51 of the UN Charter.[35]

Deployment of the armed forces in an international context, however, may also take place without a specific legal basis in case of (ad hoc or standing) cooperation, for example in the case of the Dutch contribution to the Standing NATO Maritime Group (SNMG).[36]

In sum, when deployed abroad,[37] activities in the information environment may only take place within the confines of a mandate, which reflects the legal basis for the deployment, and takes into account the legal regimes applicable in the context of the deployment.

12.3.3. Readiness

In order to effectively execute military power in fulfilment of a mandate during deployment abroad, appropriate levels of readiness of the armed forces must be achieved, *inter alia* by means of 3ET. In the Netherlands, a distinction is made between, on the one hand, 'general readiness' (OG),[38] and, on the other hand, deployment-specific readiness (IG)[39] following a governmental decision to deploy.[40]

Deployment and readiness constitute the core elements of the Ministry of Defence's business process. A ministerial decree – the General Organisation Decree for Defence (AOD) – sets out the responsibilities for the chief authorities within the ministry, such as the Chief of Defence (CHOD) and his subordinate commanders.[41] Article 3 AOD, for example, stipulates that the CHOD is responsible for the direction of the preparation, execution and evaluation of all military operations, subject to the instructions of the Minister of Defence.[42] It is important to emphasise that while the AOD tasks the CHOD and his subordinate commanders (of the army, navy etc.) with responsibilities concerning the readiness of their forces, it is not a document providing legal authority.

General readiness activities must comply with national legislation without exemption. For the purposes of achieving general readiness, there is no specific Dutch law and therefore no explicit legal basis permitting the armed forces to carry out activities that infringe upon the rights and freedoms of citizens. Neither is there an international legal basis to fall back on since OG takes place prior to actual deployment and therefore outside the context of the international legal bases and regimes set out above.[43]

Table 12.1. Readiness & Deployment matrix for International Deployment

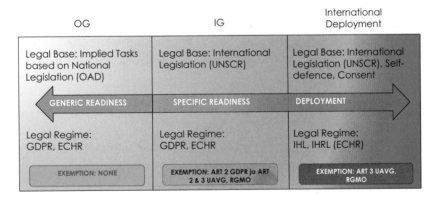

As for IG, the situation is different (see Table 12.1). In this case, legal authority derived from the Government's decision to deploy and the international legal basis for that deployment can also be used for activities to prepare for the specific deployment, including activities carried out prior to actual deployment. The IG period is understandably limited in time and subject to specific guidance by the CHOD who directs all military operations. However, in the event of a deployment without a specific legal basis, as explained above, no legal authorities will become available for the armed forces for deployment-specific readiness activities. In that case, the activity requires a basis somewhere in Dutch law.

To sum up:

- activities may only be carried out based on an order that falls within the scope of an assigned task and require a legal authority to the extent it constitutes an infringement upon (human) rights, *regardless of* deployment or readiness;
- readiness activities may only be carried out within the scope of the readiness-tasks assigned to the CHOD and his subordinate commanders. To the extent that these activities infringe on (human) rights, a legal authority must follow from Dutch law (OG and IG) or international law (IG).

The following section will make clear that the legal latitude for activities that amount to the processing of personal data or otherwise infringe with the right to privacy differs significantly between deployment and IG on the one hand, and OG on the other hand.

12.3.4. Some aspects concerning the processing of personal data

Activities of the armed forces in the information environment may infringe upon the right to privacy, irrespective of location. This is for example the case when collecting, processing and sharing personal data from open sources such as the internet or social media for purposes of Open Source Intelligence (OSINT), and when using Intelligence, Surveillance and Reconnaissance (ISR) sensors that process personal data.

The right to privacy is a constitutional right protected in Article 10 of the Netherlands' constitution:

> "1. Everyone shall have the right to respect for his privacy, without prejudice to restrictions laid down by or pursuant to Act of Parliament.
> 2. Rules to protect privacy shall be laid down by Act of Parliament in connection with the recording and dissemination of personal data.
> 3. [...]."

The Netherlands is also party to human rights treaties that protect the privacy of persons, such as the International Convention on the Protection of Civil and Political Rights (ICCPR) and the European Convention on Human Rights (ECHR).[44] Article 8 of the ECHR stipulates:

> "1. Everyone has the right to respect for his private and family life, his home and his correspondence.

2. There shall be no interference by a public authority with the exercise of this right except such as is in accordance with the law and is necessary in a democratic society in the interests of national security, public safety (...)."

Being a member of the European Union (EU), the Netherlands is also bound by the EU's General Data Protection Regulation (GDPR),[45] which has direct effect in Dutch law. The GDPR contains rules on a specific element of privacy, namely the (automated) processing[46] of personal data.[47] In order to further regulate certain aspects of the GDPR, the Dutch legislator adopted the GDPR Implementing Law (UAVG).[48]

The GDPR does not apply to the processing of personal data 'in the course of an activity which falls outside the scope of Union law',[49] nor to activities which fall 'within the scope of Chapter 2 of Title V of the TEU' which relates to the Common Foreign and Security Policy, including defence-related matters.[50]

The processing of personal data in the interests of national security, for example those carried out by a Member State's intelligence and security services,[51] but also the processing of personal data by its armed forces,[52] falls outside the scope of the GDPR.[53]

This inapplicability of the GDPR, however, does not mean that the processing of personal data by the armed forces is left unregulated. After all, both the Dutch constitution and international legal obligations such as those emanating from the ECHR demand that State interference with the right to privacy requires a basis in law.[54]

As it would be unconstitutional and contrary to international legal obligations to leave the processing of personal data by the armed forces unregulated, the legislator used the UAVG to declare the GDPR applicable to any processing of personal data by the armed forces in the execution of its own tasks.[55] This means that the GDPR and UAVG are also applicable during deployments abroad. However, realising that circumstances during deployments do not always allow for the full compliance with all provisions of the GDPR, the UAVG and a ministerial regulation – the Regulation on Data Protection Military Operations (RGMO)[56] – indicates that under certain conditions (parts of) the GDPR and the UAVG are nonetheless *not* applicable to the processing of personal data. These conditions are that:

(1) the government decides to deploy or make available[57] armed forces for the purposes set forth in Article 97 of the Netherlands' Constitution;[58] and

(2) the international legal basis for the operation for the benefit of which personal data is processed authorises such processing;[59] and

(3) the processing of personal data is necessary for the execution of the mandate or the protection of the forces;[60] and

> (4) the Minister of Defence subsequently decides to make use of the authority granted under Article 3(3)(a) UAVG to decide to exempt the armed forces from applicability of parts of the GDPR and UAVG for this particular operation or mission.[61]

While the RGMO is not explicit on this, the processing of personal data may also take place in the IG-phase, under the same conditions set out above.

Two important conclusions can be drawn from the above. First, in the IG-phase and during military deployments for the purposes of Article 97 of the Dutch Constitution, the processing of personal data is *permitted* only when the Minister of Defence decides so, based on an assessment of the necessity for the execution of the mandate or the protection of the forces. Second, without a legal basis elsewhere in national law (such as WIV 2017),[62] the processing of personal data by the armed forces is governed by the GDPR.

In view of the latter, the question then is: away from a deployment and IG period, when is the processing of personal data by the armed forces lawful under the GDPR?

12.3.5. Processing of personal data under GDPR for achieving readiness

It follows from both definitions of 'processing' and 'personal data'[63] that the threshold for processing personal data by the armed forces is easily crossed, particularly when using the Internet or employing means and methods of ISR.

If so, the processing of personal data must comply with the principles set forth in Article 5 GDPR,[64] *provided that* one of the six bases, addressed in Article 6 GDPR, that authorise the processing of personal data is applicable.[65] While some of these bases are not relevant for achieving the levels of readiness of the armed forces, others appear better suited, but are still troublesome.

This concerns, to begin with, 'consent', but the processing of personal data on the basis of consent is subject to very strict conditions.[66] In the context of 3ET it can therefore only take place in a controlled setting, which makes that 'consent' fit 3ET in very limited circumstances.[67] In the context of activities carried out to obtain an authoritative information position, this basis appears not suitable at all.

A second processing base, 'processing [that is] is necessary for the performance of a task carried out in the public interest or in the exercise of official authority vested in the controller' requires a legal base for the processing of personal data in the law of the EU or national law, setting out, *inter alia*, who is the public authority tasked to do so, a sufficient description of the task and the purpose for the processing.[68] There is no such legal basis for the armed forces for 3ET, nor for activities in support of obtaining an authoritative information position.

The third basis – processing that is necessary for the purposes of the legiti-mate interests pursued by the controller or a third party[69] – whilst applicable for deployment, is not available for public authorities that process personal data in the execution of their task.[70] The reason is that it is up to the legislator to create the legal basis for the processing of personal data by public authorities. As noted, no legal bases for the processing of personal data by the armed forces for the purposes of its (general or mission-specific) preparation for deployment are in place at this moment in either EU or Dutch law.

In sum, with the exception of (the highly cumbersome) 'consent' as mentioned in the GDPR, the armed forces appear to have *no* legal basis to process personal data when carrying out activities for the purposes of OG.

There would be a legal basis once the Government has decided to deploy or make available Dutch armed forces and:

– the CHOD has ordered the mission or operation specific preparation (IG), and;
– there is an international legal basis, and;
– a legal authority to process personal data can be derived from the international legal basis, and;
– the processing of personal data is necessary for the execution of the mandate or the protection of the forces.

Based on that, the Minister of Defence may subsequently decide to make use of the authority granted under Article 3(3)(a) UAVG to decide to exempt the armed forces from applicability of parts of the GDPR and UAVG for this particular operation or mission.

In the next part, we will apply these conclusions set out above for information manoeuvring activities as envisioned by the DV2035.

12.4. Implications for information manoeuvring activities

We will use a hypothetical scenario in the near future in which the concept of infor-mation manoeuvre as outlined in the DV35 has been implemented, while the legal framework as set out above (in paragraph 12.3) has not changed. Would that legal framework permit the armed forces to (1) educate, exercise and experiment and train (3ET) to achieve an acceptable level of readiness and (2) to collect, process and disseminate information to contribute to the authoritative information position?

12.4.1. Setting the stage

In the fictitious scenario one State has invaded another State outside the EU and NATO-alliance – similar to the Russia-Ukraine war. The Netherlands is not a belligerent party, but the vicinity of the war impedes our national interests. As part of NATO's strengthening of deterrence strategy and the defence of the eastern flank of NATO's European territory,[71] it can be imagined that the Netherlands will deploy troops to front-line States including Lithuania, Romania or Slovakia.[72] In addition, the Netherlands, together with other States, have supported the invaded victim State with weapons and other military equipment.[73]

The Dutch armed forces have the capacity to roam the information environment with a wide variety of sensors to collect, process, analyse, interpret and disseminate data and information needed to provide the CHOD and his subordinate commanders with insight and foresight required to understand the situation and to make decisions based thereon.

At this stage in the conflict an international legal basis from which a legal authority to process personal data by Dutch armed forces can be derived, is still lacking. The conflict does not amount to a case of collective self-defence involving the Netherlands,[74] a UNSCR is not at hand, and – in absence of additional enabling SOFAs – the consent of the host States Lithuania, Romania and Slovakia is insufficient to serve as a legal basis to process personal data.

The following example will be examined to illustrate if today's legal framework fits the concept of information manoeuvre. While receiving strategic intelligence from the MIVD, the CHOD orders the – then established – joint information manoeuvre unit of the armed forces (JIMU) to provide a daily update of the tactical situation on the ground, to complete the intelligence picture needed to obtain an authoritative information position in support of Article 3 AOD.[75] In addition, their products can be disseminated to the army, navy and air force commands, as well as to the commanders of the units stationed in Lithuania, Romania and Slovakia. In the execution of this order, the JIMU makes extensive use of Internet-based open sources.

The question is: in view of the legal aspects discussed in the previous paragraph(s), would the abovementioned activity be permissible in terms of task, legal basis, and legal authorities?

12.4.2. Task, order and legal authority

In this scenario, the JIMU is not deployed yet, nor is there a solid international legal basis to prepare for a specific deployment. Does the current legal framework offer a basis to task the armed forces, in this case the JIMU, to roam the information environment for data and information to enhance the information position of the CHOD *cum suis?* A search for an explicit task in law stipulating such a responsibility for the armed forces will be without result, for there is none. It can nonetheless be argued that the task is implied in so far as it falls within the scope of the responsibilities of the CHOD and the armed forces' subordinate commanders as set forth in the AOD.

This can be problematic once the implied tasks of the armed forces infringe upon human rights of citizens. Notwithstanding the threat described in the scenario, the legal situation for the Netherlands and its armed forces is one of peace, and any State action is governed by the international law, IHRL in particular. As noted, when using real world data and information, it is rather impossible to avoid personal data from being processed.

Unless the Government decides to deploy forces or make them available and the Minister of Defence makes use of the Article 3(3)(a) UAVG authority, in light of the fact that the international legal basis provides the legal authority to process personal data, the GDPR and UAVG apply in full; hence personal data cannot be processed in the course of information manoeuvring activities by the armed forces. Applying the above, JIMU is not permitted to process personal data since JIMU is not deployed or preparing to be deployed (IG) nor did the Minister of Defence invoke Article 3(3)(a) UAVG.[76]

12.4.3. The legal challenge

It is precisely here that a major vulnerability with respect to the role of the armed forces manoeuvring in the information environment surfaces. While the DV35 articulates that the Defence organisation must be able to manoeuvre in the information environment, the armed forces lack the legal framework to do so. Currently only the MIVD has an explicit task and legal authority to roam the information environment, including the processing of personal data, outside the context of a deployment. However, the MIVD is tasked for other purposes, and therefore is deliberately restricted in tasks and authority by the legislator.

12.5. Conclusion and reflection

The inception of the GDPR has provided the EU with a solid legal regime for the protection of personal data. The GDPR provides several exemptions. First, the GDPR is not applicable in cases that fall outside the scope of Union law (EU jurisdiction). Second, the GDPR does not apply to the processing of personal data in the exercise of activities that fall within the scope of the Common Foreign and Security Policy. These exemptions in the GDPR have been revoked by the Netherlands via the UAVG, meaning that the GDPR and the UAVG apply to the Netherlands' armed forces *in full*. Based on national legislation, the Minister of Defence can decide to lift application of (parts of) the GDPR and the UAVG ex art 3(3)(a) if the armed forces are deployed or made available for deployment under the auspices of Article 97 of the Constitution, subject to certain conditions.

12.5.1. Conclusion

The Netherlands and its citizens rely on its armed forces to protect the vital interests in all domains of the information environment. But, does the current legal framework offer adequate room for the armed forces to manoeuvre in the information environment? The answer is no. As illustrated, if the current legal framework would still apply in the hypothetical case in the near future, the role for the armed forces as envisioned in the DV35 would be seriously hampered.

It can be assessed that for deployments,[77] there is a sufficient – or rather a 'workable' – legal framework for executing information manoeuvre operations by the armed forces provided that the international legal basis and the political mandate provide the required legal authorities and tasks.

However, if the conditions related to task and authority are not in place, or if the activity is not an actual deployment, the GDPR applies in full and the armed forces are in effect disabled from roaming the information environment as they are not authorised to process personal data. It can furthermore be concluded that the current legal framework affects the day-to-day activities of the armed forces related to generic readiness tasks, but also to activities that might include the processing of personal data, related to the preparation of an actual deployment.

In general terms, the capability to educate, exercise and experiment and train (3ET) to achieve an acceptable level of readiness in order to manoeuvre effectively in the information environment would be flawed as a result of the failing legal part of the conceptual framework to support the existing and desired capacities. The same flaw hampers the ambition to have the capability to collect, process and disseminate information in order to contribute to the authoritative information position that is necessary for deployment.

12.5.2. Reflection

Safeguarding privacy and the prior legal authority for infringements of privacy is not the bone of contention in this contribution. However, cognisant of the fact that the GDPR existed before the concept of information manoeuvre came in vogue, it is questionable whether the legislator could have foreseen that the decision to apply the GDPR in full to the armed forces would have caused a near-unworkable situation for the armed forces' information manoeuvre.

This predicament endangers the ambition of the DV35. However, we are not there yet. Aside from the deployment of armed forces for information manoeuvre activities, it is required – to achieve the DV35 ambition – that military personnel should be able to adequately train and exercise (3ET) in order to achieve desired levels of readiness (OG and IG).

The question is how to do so? Three suggestions are offered. First, take the given legal framework, comply with it and accept the restrictions. In this context armed forces could work with fictitious data sets for training and exercise purposes or use virtual training 'grounds' that offer virtual and synthetic realistic environments, such as the existing 'second life' society. Though basic skills can be trained, the Internet evolves too fast to make these virtual ranges realistic.

Secondly, one can also find legitimate loopholes or workarounds. Efforts can be made to optimise the legal possibilities offered via the national police or the MIVD. Though the Police Act or the WIV 2017 are not meant to facilitate information manoeuvre activities of the armed forces, organisational constructions including secondments or placing units under command of the Police and MIVD during specific activities might make the gaps smaller for the readiness phase.

Or – thirdly – pursue a path advocating a change in legislation. The challenge then is to strike the right balance between the protection of privacy and the provision of security, a challenge not solely for the government but society as a whole. Citizens and companies are entitled to enjoy their rights and privileges, for which a certain level of security is required. It is understandable that striking this balance can be difficult since these attempts are likely to be met with opposition, as was witnessed with the adoption of COVID related regulations, the WIV 2017 referendum and the attempts by other public authorities to be granted with similar legal authorities.[78] But security comes at a price and given the conflict in Ukraine security proves to be a poor fit with complacency.

Ultimately, the question is whether the society is willing to accept that the armed forces are expected to master the ins and outs of information manoeuvre whilst being deployed for one of its constitutional aims, while acceptable readiness levels prior to deployment cannot be achieved because the necessary preconditions are absent. The same applies to a role for the armed forces as a contributor to obtaining an authoritative information position. Perhaps a more confronting way of putting this is: is society ready to accept casualties that clearly could have been prevented had the armed forces been fully equipped to do what is necessary and expected of them in the information environment? If the armed forces are to fulfil their task as first responder and the last line of defence in protecting the security interests of the Kingdom, the legal predicaments need to be dealt with. For it would be hypocritical to demand security without providing the armed forces with the appropriate tools and powers to do so.

Notes

1   Attributed to Napoleon Bonaparte.

2   Art. 97(1) Constitution of the Kingdom of the Netherlands.

3   Marcus Willett, "Assessing cyber power," *Survival* 61, no. 1 (2019): 85–90. pp. 85-87.

4   Netherlands Ministry of Defence, "Defence Vision 2035: Fighting for a Safer Future," 2020. Accessible here: https://english.defensie.nl/downloads/publications/2020/10/15/defence-vision-2035

5   Though similar trends are visible in other states, see e.g.: Christine E. Wormuth, "Message from the Secretary of the Army to the Force," *US Army*, 2022., in this chapter, 'defence organisation' refers to the defence organisation of the Kingdom of the Netherlands. The same principle applies to 'Ministry of Defence', 'armed forces' and 'intelligence and security services'. MoD and defence organisation encompass the armed forces (army, navy, airforce, marechaussee) as well as the Military Intelligence and Security Service (MIVD).

6   British Army, "Force Troops Command Handbook," 2019.; Paul A.L. Ducheine, Jelle van Haaster, and Richard van Harskamp, "Manoeuvring and Generating Effects in the Information Environment," in *Winning Without Killing: The Strategic and Operational Utility of Non-Kinetic Capabilities in Crisis – NL ARMS 2017*, ed. Paul A.L. Ducheine and Frans P.B. Osinga, 2017.

7   Netherlands Ministry of Defence, "Defence Vision 2035: Fighting for a Safer Future." Annex p. XII

8   Wherever this (implicitly or explicitly) involves activities of military intelligence entities (such as reconnaissance platoons of infantry battalions, or the units belonging to Joint ISTAR-Command), this concerns units of the armed forces, not of the (defence) intelligence and security service(s). In the Netherlands, the Military Intelligence and Security Service (MIVD) is an entity within the Ministry of Defence, under the authority of the Secretary-General of the Ministry of Defence. It is not part of the armed forces, acting under the authority of the Chief of Defence (CHoD).

9   This phase entails activities related to 3ET prior to deployment. In Dutch: "gereedstelling".

10   A 'deployment' is a military operation sanctioned by the Netherlands' government. In Dutch: "inzet".

11   Ducheine, Haaster, and Harskamp, "Manoeuvring and Generating Effects in the Information Environment." § 9.2

12   AIVD, MIVD, and NCTV, "Dreigingsbeeld Statelijke Actoren," 2021. pp. 32-34.

13   Ducheine, Haaster, and Harskamp, "Manoeuvring and Generating Effects in the Information Environment." § 9.2.2.; see also: Jelle van Haaster, "On Cyber: The Utility of Military Cyber Operations During Armed Conflict" (2018). p. 173 (note 898).

14   DDoS means a distributed denial of service wherein an overwhelming number of computers let a website crash, rendering it (temporarily) unavailable, see e.g.: United Kingdom Government, "UK Assess Russian Involvement in Cyber Attacks on Ukraine," 2022.

15   For example, the cyber-attacks on the Iranian nuclear facility in 2010 or the attack on the Ukrainian Electricity Grip in 2015. Kim Zetter, *Countdown to Zero: Stuxnet and the Launch of the World's First Digital Weapon*, 2015.; Kim Zetter, "Inside the Cunning, Unprecedented Hack of Ukraine's Power Grid," *Wired*, 2016.

16   Derived from NATO's (AAP-6) definition of 'to manoeuvre': "The Employment of forces on the battlefield through movement in combination with fire, or fire potential, to achieve a position of advantage in respect to the enemy in order to accomplish the mission."

17   Army, "Force Troops Command Handbook."

18   DCDC, "JDP 04 2nd Edition – Understanding and Decision-Making," 2016.

19   See e.g. Henk Warnar, "Warships as tools for international diplomacy," *Atlantisch Perspectief*, no. 2 (2022): 9–13.

20 Peter B.M.J. Pijpers and Paul A.L. Ducheine, "'In You Have A Hammer': Reshaping the Armed Forces' Discourse on Information Maneuver," *ACIL Research Paper 2021-34*, 2021. p. 8, accessible here: https://papers.ssrn.com/sol3/papers.cfm?abstract_id=3954218

21 Pijpers and Ducheine. pp. 11-13.

22 Netherlands Ministry of Defence, "Defence Vision 2035: Fighting for a Safer Future." Annex p. XII

23 For the difference between these cyberoperations see also: Peter B.M.J. Pijpers and Kraesten L. Arnold, "Conquering the invisible battleground," *Atlantisch Perspectief* 44, no. 4 (2020). pp. 12-14; Sean Cordey, "Cyber Influence Operations: An Overview and Comparative Analysis," *Center for Security Studies (CSS), ETH Zurich*, 2019. pp. 15-19.

24 The framework for the intelligence & security services, MIVD and the General Intelligence and Security Service (AIVD), is quite distinct and set out in the Intelligence and Security Act 2017 (WIV 2017).

25 The two 'sword powers' are the armed forces and the police.

26 For international deployment, these three steps are combined in the so-called Article 100 procedure referring to Article 100 of the Dutch Constitution. See also: Government of the Netherlands, "Nederlandse Deelname Aan Vredesmissies (Participation in Peace-Keeping Missions) 2013-2014/ 29 521 – 226," *Parliamentary Papers II*, 2014.; Paul A.L. Ducheine, Kraesten L. Arnold, and Peter B.M.J. Pijpers, "Decision-Making and Parliamentary Control for International Miiltary Cyber Operations by The Netherlands Armed Forces," in *Liber Amicorum*, ed. Rogier Bartels et al., 2020, 59–81.

27 Official English translation of the Constitution of the Kingdom of the Netherlands, available at https://www.denederlandsegrondwet.nl/id/vkwrf6d92rnm/de_tekst_van_de_grondwet_met_toelichting.

28 Ducheine, Arnold, and Pijpers, "Decision-Making and Parliamentary Control for International Military Cyber Operations by The Netherlands Armed Forces." Available at SSRN: https://ssrn.com/abstract=3540732.

29 See e.g. Paul A.L. Ducheine and Eric H. Pouw, "Legitimizing the Use of Force Legal Bases for Operations Enduring Freedom and ISAf," in *Mission Uruzgan: Collaborating in Multiple Coalitions for Afghanistan*, ed. Jan van der Meulen et al. Amsterdam: Amsterdam University Press, 2012, 67–80.

30 In the execution of his responsibility to direct the execution of all military operations, the CHOD, by means of an order of his Director of Operations, will 'translate' the mandate in specific orders for units assigned for deployment.

31 Such as International Human Rights Law or – if appropriate – International Humanitarian Law.

32 North Sea Continental Shelf Cases, ICJ Reports (1969). Para 59, p. 37; Gleider Hernandez, *International Law* Oxford, United Kingdom: Oxford University Press, 2019. 474-475.

33 It must be noted, however, that the consent of the host State does not serve as an independent source for authorities. This may require an international agreement or arrangement to ensure mutual understanding on the terms and conditions. This is to prevent that armed forces carry out executive powers abroad that, while allowed under the laws of the host State, are not allowed or would violate the law of the Sending nation, including obligations under international law, for example those following from international human rights treaties to which the Sending nation is a party, such as the European Convention on Human Rights (ECHR).

34 An international deployment of Dutch armed forces on the basis of a UNSCR may take place in the context a UN-led mission (as was the case in for example MINUSMA, or UNMISS), or a UN-mandated mission (e.g. SFOR, or ISAF). In addition to the legal basis of the UNSCR and depending on the context (armed conflict or peace), the legal regimes of IHL and/or IHRL determine the legal room for manoeuvre.

35 In which case the right to self-defence and most notably International Humanitarian Law (IHL or the Law of Armed Conflict LOAC) and – when applicable – International Human Rights Law (IHRL) determine the authority to act.

36 See: Hans J.F.R. Boddens Hosang, "Force Protection, Unit Self-Defence and Extended Self-Defence," in *The Handbook of the International Law of Military Operations*, 2015. pp. 476-501.

37 The decisions to deploy armed forces domestically, rest with the authorities designated by the legislator in formal law, such as the Minister of Justice and Security in case of military support to the Police. See e.g. Articles 57-59 of the Police Act 2012.

38 In Dutch: 'Operationele Gereedstelling' or OG.

39 In Dutch: 'Inzetspecifieke Gereedstelling' or IG.

40 Based on Article 97 Constitution for deployment abroad (or on specific formal acts for domestic deployment).

41 In Dutch: Algemeen Organisatiebesluit Defensie 2021, or AOD 2021.

42 Article 3(c) AOD 2021.

43 The exception being readiness-activities taking place abroad with the consent of the host State through e.g. a Status of Forces Agreement determining what is permitted and what is not. Such activities will be governed by International Human Rights Law, implying that activities amounting to any infringement of human rights requires a basis in law, in this case meaning that both the law of the host State as well as the Sending state's (e.g. Dutch) national law must provide such basis.

44 See also Protocol 223 to the 1981 Convention for the Protection of Individuals with regard to Automatic Processing of Personal Data, of the Council of Europe, to which the Netherlands is a signatory. The Protocol will enter into force as of 11 October 2023.

45 The "European Union General Data Protection Regulation (2016/679)," L119 Official Journal of the European Union § (2016). in Dutch: Algemene Verordening Gegevensbescherming (AVG). – is accessible here: https://eur-lex.europa.eu/eli/reg/2016/679/oj

46 'Processing' is defined in Article 4(2) of the GDPR as "any operation or set of operations which is performed on personal data or on sets of personal data, whether or not by automated means, such as collection, recording, organisation, structuring, storage, adaptation or alteration, retrieval, consultation, use, disclosure by transmission, dissemination or otherwise making available, alignment or combination, restriction, erasure or destruction."

47 'Personal data' is defined in Article 4(1) of the GDPR as "any information relating to an identified or identifiable natural person ('data subject'); an identifiable natural person is one who can be identified, directly or indirectly, in particular by reference to an identifier such as a name, an identification number, location data, an online identifier or to one or more factors specific to the physical, physiological, genetic, mental, economic, cultural or social identity of that natural person."

48 In Dutch: Uitvoeringswet AVG, or UAVG.

49 Article 2(2)(a) GDPR.

50 Article 2(2)(b) GDPR; Articles 23-46 of the Consolidated Version of the Treaty on European Union, C–326 Official Journal of the European Union (2012). This includes Article 42(7), the mutual assistance clause.

51 With respect to the Dutch intelligence and security services AIVD and MIVD, the Intelligence and security services Act 2017 (WIV 2017) provides such basis. For that reason, the GDPR jo UAVG Article 3(3)(b) is not applicable to the processing of personal data by the AIVD and MIVD to the extent authorised by WIV 2017. This means that the processing of personal data by both AIVD and MIVD that does not find a legal basis in the WIV 2017 is still governed by the GDPR and UAVG.

52 See e.g. Marten C. Zwanenburg, "Het Gebruik van Biometrie in Militaire Missies : Aanzet Voor Een Studie van Het Juridisch Kader," *Militair Rechtelijk Tijdschrift*, 2021, 1–18. pp. 11-15.

53 Military deployments outside of EU missions are exempt based on Article 2(2)(a) GDPR, deployments within the EU scope based on Article 2(2)(b) GDPR. See Zwanenburg. p. 13.

54 Article 6 sub 3 TEU considers the Fundamental rights as guaranteed in the ECHR are part of Union Law. Consolidated Version of the Treaty on European Union, C–326 Official Journal of the European Union.

55 *Parliamentary Papers II 2017-2018*, 34851 no 3 (2018), footnote 4 on page 9 & p 14. The Explanatory Memorandum to the law is accessible here: https://zoek.officielebekendmakingen.nl/kst-34851-3.html. See also: Article 3(1) and (2) UAVG. See also the commentary to the UAVG, available at https://zoek.officielebekendmakingen.nl/kst-34851-3.html.

56 In Dutch: Regeling Gegevensbescherming Militaire Operaties, or RGMO.

57 In Dutch: beschikbaar stellen. 'Making available' refers to the making available of units to EU or NATO, for example to NATO's VJTF.

58 This includes national deployment in support of national authorities. This type of deployment falls outside the scope of the RGMO, because in these instances any processing of personal data by the armed forces takes place in the execution of tasks and legal authorities of the national authorities. This explains why the focus in the RGMO lies on international operations only, as illustrated by the second condition ("to the extent this is authorized by the international legal basis for the operation for the benefit of which personal data is processed").

59 Article 1(1) and (2) RGMO.

60 Article 1(1) RGMO. This includes foreign coalition forces, for example NATO partners.

61 In other words: without an explicit decision by the Minister of Defence, the processing of personal data during the military operation is governed by the GDPR (with the exception of UN-led missions, such as MINUSMA. In those missions, the processing of personal data is governed by the rules of the UN). In the event of such decision, the Dutch Data Protection Authority (in Dutch: Autoriteit Persoonsgegevens, or AP) must be informed as soon as possible, Article 3(4) UAVG.

62 General Intelligence and Security Service, "The Intelligence and Security Services Act 2017," Ministry of Interior and Kingdom Relations, 2018, https://english.aivd.nl/about-aivd/the-intelligence-and-security-services-act-2017. In Dutch: Wet op de inlichtingen- en veiligheidsdiensten 2017, or WIV 2017.

63 See footnotes 46 and 47.

64 Article 5 GDPR: "1. Personal data shall be: (a) processed lawfully, fairly and in a transparent manner in relation to the data subject ('lawfulness, fairness and transparency'); (b) collected for specified, explicit and legitimate purposes and not further processed in a manner that is incompatible with those purposes; further processing for archiving purposes in the public interest, scientific or historical research purposes or statistical purposes shall, in accordance with Article 89(1), not be considered to be incompatible with the initial purposes ('purpose limitation'); (c) adequate, relevant and limited to what is necessary in relation to the purposes for which they are processed ('data minimisation'); (d) accurate and, where necessary, kept up to date; every reasonable step must be taken to ensure that personal data that are inaccurate, having regard to the purposes for which they are processed, are erased or rectified without delay ('accuracy'); (e) kept in a form which permits identification of data subjects for no longer than is necessary for the purposes for which the personal data are processed; personal data may be stored for longer periods insofar as the personal data will be processed solely for archiving purposes in the public interest, scientific or historical research purposes or statistical purposes in accordance with Article 89(1) subject to implementation of the appropriate technical and organisational measures required by this Regulation in order to safeguard the rights and freedoms of the data subject ('storage limitation'); (f) processed in a manner that ensures appropriate security of the personal data, including protection against unauthorised or unlawful processing and against accidental loss, destruction or damage, using appropriate technical or organisational measures ('integrity and confidentiality'). 2. (...)

65   Article 6 GDPR, paragraph 1: "Processing shall be lawful only if and to the extent that at least one of the following applies:(a) the data subject has given consent to the processing of his or her personal data for one or more specific purposes; (b) processing is necessary for the performance of a contract to which the data subject is party or in order to take steps at the request of the data subject prior to entering into a contract; (c) processing is necessary for compliance with a legal obligation to which the controller is subject; (d) processing is necessary in order to protect the vital interests of the data subject or of another natural person; (e) processing is necessary for the performance of a task carried out in the public interest or in the exercise of official authority vested in the controller; (f) processing is necessary for the purposes of the legitimate interests pursued by the controller or by a third party, except where such interests are overridden by the interests or fundamental rights and freedoms of the data subject which require protection of personal data, in particular where the data subject is a child."

66   Article 7 GDPR.

67   An example in case is the education, training, exercise and experimentation with biometrical data, in which it is possible to gain consent of role players, however, thereby taking notice of the fact that consent must be given entirely freely. This means, for example, that it is not possible to use role players from the same unit to conduct the exercise.

68   In addition, there must be publicly available law that informs the citizen with sufficient accuracy which personal data will be collected and processed for the purposes of a certain public task, and under which conditions these data will be adapted, kept and used.

69   E.g. for reasons of fraud prevention or network and information security.

70   Article 6(1) AVG.

71   NATO, "SACEUR Statement on the Activation of the NATO Response Force," *Shape Newsroom*, 2022.

72   Government of the Netherlands, "NAVO (NATO) 2021-2022/ 28 676 – 404," *Parliamentary Papers II*, 2022.

73   Sebastien Roblin, "The Dutch Are Sending Huge German Armored Howitzers To Ukraine," *Forbes*, 2022.

74   There has been no armed attack by Russia or by non-State agents under authority and control of Russia on a NATO-member that would trigger the right to self-defence of that member, based on Article 5 of the NATO-treaty.

75   Besides the responsibility to direct the preparation, execution and evaluation of all military operations and the responsibility to direct the readiness of the armed forces, his responsibilities also include the task of primary military advisor to the Minister of Defence, as well as directing the operational commands of the navy, air force, army and Royal Marechaussee (the latter to the extent their activities are a responsibility of the Minister of Defence; most activities of the Royal Marechaussee are carried out under authority of the Minister of Justice and Security.

76   Deployment, though, is not inconceivable, as the JIMU could, for example, function as reach back-capacity for the units deployed in Lithuania, Romania and Slovakia. But, as noted, the mere fact of deployment does not generate legal authority to process personal data; this must follow from the international legal basis underlying the deployment which in this scenario is missing.

77   See also Zwanenburg & Van de Put "The use of biometrics in military operations abroad and the right to private life", Chapter 13, Section 5 Conclusions, in this volume.

78   An example is a recent proposal for a law permitting the NCTV, the national coordinator for combatting terrorism and security, to process personal data, including special data such as religion or political preference, in the exercise of its analyses of trends and phenomena. See: an interview with Frederik Zuiderveen Borgesius, "'Wetsvoorstel Beloont NCTV Voor Iets Wat Niet Mag,'" *Radboud Recharge*, 2022. (The title reads: "Proposal for Law rewards the NCTV for something it is not allowed to do").

References

AIVD, MIVD, and NCTV. "Dreigingsbeeld Statelijke Actoren," 2021.

Army, British. "Force Troops Command Handbook," 2019.

Boddens Hosang, Hans J.F.R. "Force Protection, Unit Self-Defence and Extended Self-Defence." In *The Handbook of the International Law of Military Operations*, 2015.

Borgesius, Frederik Zuiderveen. "'Wetsvoorstel Beloont NCTV Voor Iets Wat Niet Mag.'" *Radboud Recharge*, 2022.

Consolidated Version of the Treaty on European Union, C–326 Official Journal of the European Union (2012).

Cordey, Sean. "Cyber Influence Operations: An Overview and Comparative Analysis." *Center for Security Studies (CSS), ETH Zurich*, 2019.

DCDC. "JDP 04 2nd Edition – Understanding and Decision-Making," 2016.

Ducheine, Paul A.L., Kraesten L. Arnold, and Peter B.M.J. Pijpers. "Decision-Making and Parliamentary Control for International Military Cyber Operations by The Netherlands Armed Forces." In *Liber Amicorum*, edited by Rogier Bartels, Jeroen C. van den Boogaard, Paul A.L. Ducheine, Eric Pouw, and Joop E.D. Voetelink, 59–81, 2020.

Ducheine, Paul A.L., Jelle van Haaster, and Richard van Harskamp. "Manoeuvring and Generating Effects in the Information Environment." In *Winning Without Killing: The Strategic and Operational Utility of Non-Kinetic Capabilities in Crisis – NL ARMS 2017*, edited by Paul A.L. Ducheine and Frans P.B. Osinga, 2017.

Ducheine, Paul A.L., and Eric H. Pouw. "Legitimizing the Use of Force Legal Bases for Operations Enduring Freedom and ISAf." In *Mission Uruzgan: Collaborating in Multiple Coalitions for Afghanistan*, edited by Jan van der Meulen, Robert Beeres, Joseph Soeters, and Ad Vogelaar, 67–80. Amsterdam: Amsterdam University Press, 2012.

European Union General Data Protection Regulation (2016/679), L119 Official Journal of the European Union § (2016).

General Intelligence and Security Service. "The Intelligence and Security Services Act 2017." Ministry of Interior and Kingdom Relations, 2018. https://english.aivd.nl/about-aivd/the-intelligence-and-security-services-act-2017.

Government of the Netherlands. "NAVO (NATO) 2021-2022/ 28 676 – 404." *Parliamentary Papers II*, 2022.

———. "Nederlandse Deelname Aan Vredesmissies (Participation in Peace-Keeping Missions) 2013-2014/ 29 521 – 226." *Parliamentary Papers II*, 2014.

Haaster, Jelle van. "On Cyber: The Utility of Military Cyber Operations During Armed Conflict," 2018.

Hernandez, Gleider. *International Law*. Oxford, United Kingdom: Oxford University Press, 2019.

NATO. "SACEUR Statement on the Activation of the NATO Response Force." *Shape Newsroom*, 2022.

Netherlands Ministry of Defence. "Defence Vision 2035: Fighting for a Safer Future," 2020.

North Sea Continental Shelf Cases, ICJ Reports (1969).

Pijpers, Peter B.M.J., and Kraesten L. Arnold. "Conquering the invisible battleground." *Atlantisch Perspectief* 44, no. 4 (2020).

Pijpers, Peter B.M.J., and Paul A.L. Ducheine. "'If You Have a Hammer': Reshaping the Armed Forces' Discourse on Information Maneuver." *ACIL Research Paper 2021-34*, 2021.

Roblin, Sebastien. "The Dutch Are Sending Huge German Armored Howitzers To Ukraine." *Forbes*, 2022.

United Kingdom Government. "UK Assess Russian Involvement in Cyber Attacks on Ukraine," 2022.

Warnar, Henk. "Warships as tools for international diplomacy." *Atlantisch Perspectief*, no. 2 (2022): 9–13.

Willett, Marcus. "Assessing cyber power." *Survival* 61, no. 1 (2019): 85–90.

Wormuth, Christine E. "Message from the Secretary of the Army to the Force." *US Army*, 2022.

Zetter, Kim. *Countdown to Zero: Stuxnet and the Launch of the World's First Digital Weapon*, 2015.

———. "Inside the Cunning, Unprecedented Hack of Ukraine's Power Grid." *Wired*, 2016.

Zwanenburg, Marten C. "Het Gebruik van Biometrie in Militaire Missies : Aanzet Voor Een Studie van Het Juridisch Kader." *Militair Rechtelijk Tijdschrift*, 2021, 1–18.

CHAPTER 13

The use of biometrics in military operations abroad and the right to private life

Marten Zwanenburg & Steven van de Put

Abstract

This chapter analyses the use of biometrics by military operations extraterritorially from the perspective of the right to private life in Article 8 of the European Convention on Human Rights (ECHR). Such an analysis is called for in view of the increasing use of biometrics by armed forces. The chapter concludes that it follows from the case law of the European Court of Human Rights (ECtHR) that the ECHR is applicable to certain conduct of armed forces outside of their own State's territory, and that this includes situations involving the use of biometrics. Similarly, based on this case law there are good grounds for concluding that all collection, storage and disclosure of biometric data falls within the scope of Article 8 (1) ECHR, at least where the data is systematically collected, stored and shared as is the case in military operations. This means that such use must meet the requirements set out in Article 8 (2) ECHR in order not to constitute a violation of the right to private life. The chapter discusses these requirements and concludes that although States have a certain margin of appreciation, compliance with the right to private life during extraterritorial military operations appears to be a tall order.

Keywords: Biometrics, ECHR, Right to privacy, Jurisdiction, Military operations.

13.1. Introduction

The collection and use of data is an increasingly important feature of military operations. An example of this is the use of biometric systems by armed forces. Biometric systems are systems used for the purpose of the biometric recognition of individuals based on their behavioural and biological characteristics. Such characteristics include fingerprints, face and finger topography, gait, voice and DNA. These characteristics are unique, which makes them ideal for recognising persons. This makes biometric systems a valuable tool for military operations, as they can be used to deny anonymity.

In order for a biometric system to function, it is necessary to collect, store and exchange data. The more data available, the more effective the system is in recognising persons. This has led to the collection of enormous amounts of biometric data in certain recent operations, for example in Afghanistan. This increasing use of biometric systems by military operations raises the question of what the legal parameters are for such use. Only recently has this question been started to be addressed in academic literature. The discussion has mainly focused on the application of International Humanitarian Law (IHL).[1] Thus far, there has been little academic attention for the implications of human rights, in particular the interaction between the right to privacy and the use of biometric data in military operations. Yet the use of data by military operations raises important questions concerning the applicability and application of the right to privacy.

This chapter explores the applicability and application of the right to privacy to the use of biometrics by military operations. It focuses in particular on the right as it has been included in Article 8 of the European Convention on Human Rights (ECHR). This article provides that "Everyone has the right to respect for his private and family life, his home and his correspondence." This choice is motivated by the rich case law of the European Court of Human Rights (ECtHR) concerning the right to private life, as the right to privacy is termed in Article 8 ECHR.

Particular attention will be given in this chapter to the right to private life as it applies to military operations outside of the territory of States. Such extraterritorial operations raise questions concerning the right to private life, including whether that right applies at all outside of such territory. After all, the ECHR was designed to apply primarily in the territory of States Parties. Yet currently it is widely accepted that the application of the ECHR is not limited to the territory of States. This chapter submits that the right to privacy would also be relevant during military operations abroad.[2]

Due to limitations of space, the issue of the interrelationship between the right to private life and IHL during armed conflict will not be explored.[3] It is possible however that the latter may have impact on limitations placed on the use of data by the former.

The chapter is structured as follows. After this introduction, a brief introduction will be given to biometric systems and their use in military operations (section 13.2). Having defined the object of study, section 13.3 will focus on the (extraterritorial) application of the ECHR to military operations. The following section will introduce the right to private life in Article 8 ECHR and discuss the applicability and application of the right to private life to the use of biometric data in military operations abroad. In other words, how does the right to privacy impact such use? (section 13.4). The chapter concludes with a number of final observations and recommendations for further research (section 13.5).

13.2. Biometrics and its use in military operations

An authoritative definition of "biometrics" or "biometric recognition" is that this concerns the automated recognition of individuals based on their biological and behavioural characteristics.[4] A biometric system is essentially a pattern recognition system that operates by acquiring biometric data from an individual, extracting a feature set from the acquired data, and comparing this feature set against the template set in the database.[5] It uses the physical, physiological or behavioural characteristics of individuals to recognise them.[6] Examples of such characteristics are face topography, hand topography, finger topography, iris structure, vein structure of the hand, voice, gait, and DNA.[7] These characteristics are unique, which makes them ideal for recognising persons.[8]

A biometric system can be used for verification or for identification. Verification refers to validating a person's identity by comparing the captured biometric data with his or her own biometric template(s) stored in the system database.[9] This is a one-to-one process, which answers the question of whether the person concerned is who he or she claims to be. Identification refers to recognising an individual by searching the templates of all the users in the database for a match.[10] Identification is a one-to-many comparison to establish an individual's identity, without the person concerned having to claim an identity.

To confirm the identity of individuals, biometric systems make use of various biometric characteristics of individuals that are unique to that individual. This potential, with biometrics representing a unique identifier, makes them very valuable within military operations. It has led to a great number of applications of biometrics in the military domain.[11] Examples of this are base access, identifying persons eligible for host nation training, identifying persons connected to Improvised Explosive Devices (IEDs), identifying persons involved in piracy at sea, and targeting.

The use of biometrics in military operations appears to have been first introduced by the United States (US) after the invasion of Iraq in 2003.[12] The US has taken a leading role when it comes to the use of biometrics, visible in its extensive use during operations in Afghanistan and Iraq. Although the US has remained at the forefront of military use of this technology, other States' armed forces have also started using it during military operations. This is not surprising, considering the broad variety of potential applications referred to above. Currently, NATO has recognised biometrics as an important operational capacity.[13] This highlights that biometrics will be something that is expected to be relevant for the foreseeable future.

13.3. (Extraterritorial) application of the ECHR to military operations

The right to private life in Article 8 ECHR can only be relevant to the use of biometrics by military operations abroad if it applies extraterritorially, i.e. outside of the territory of the State. A key element in this context has always been the concept of jurisdiction. Article 1 of the ECHR provides that "The High Contracting Parties shall secure to everyone within their jurisdiction the rights and freedoms" set out in the Convention. This represents 'a threshold criterion which determines whether the state incurs obligations under the treaty, and consequently whether any particular act of the state can be characterised as internationally wrongful.'[14] It is based on the conception that there needs to be a substantial relationship between the State and a potential victim of infringement, to ensure that this relationship is not arbitrary.

As the human rights field matured, courts, including the ECtHR, were increasingly confronted with alleged violations that took place outside a State's own territory. Based on the universal character of human rights, it was argued that it was to be considered arbitrary if States would be allowed to commit violations across their borders.[15] Through a developing jurisprudence it became accepted that whereas 'the jurisdiction of States is primarily territorial, it may sometimes be exercised outside the national territory.'[16] Known as extraterritorial jurisdiction, this meant that States, in certain cases, could also be legally responsible for violations outside of their territory.

Crucial within this notion has been the concept of effective control. This can manifest itself through effective control of territory, or effective control over persons through State agents.[17] Effective control over territory can be exercised either directly by the State through its armed forces, or through a subordinate administration.[18] The case law of the ECtHR also makes clear that there is jurisdiction in the sense of Article 1 ECHR when there is effective control by a State over persons through its agents, which the Court has referred to as "State agent authority and control". This includes, in certain circumstances, the use of force, in particular when an individual is taken into custody.[19] It also includes cases where a State party to the ECHR, through the consent, invitation or acquiescence of the Government of that territory, exercises some or all of the public powers normally exercised by that Government.[20] The precise contours of "State agent authority and control" are however difficult to distil from the ECtHR case law. The Court has however emphasised that the control is actually effective: in situations of chaos,[21] or in situations in which there is an insufficient link between the conduct and the victim[22] the Court has argued against the violation falling within the jurisdiction of the state.

Situations in which a State would hold effective control could thus amount to a sufficient jurisdictional link within operations. Examples relevant to the current research could include the taking of biometrical data of detainees, in which the individual would without a doubt be seen to fall within the effective control of the State. Individuals aboard a (war-)ship flying the flag of the State could,[23] based on the case law of the ECtHR, also be seen to fall within the control of a state party.

The Court has, however, in applying the notion of effective control to situations of armed conflict, been relatively conservative. In Georgia v. Russia (II), the Court held with regard to Article 2 of the Convention (concerning the right to life) that in the event of military operations carried out during an international armed conflict, it was not possible to speak of "effective control" over an area or over an individual. The very reality of armed confrontation and fighting between enemy military forces seeking to establish control over an area in a context of chaos meant that there could be no such control.[24]

Commentators have noted that this might be due to political concerns and the Court trying to maintain a fine balance between being considered a legitimate court and the compliance of states.[25] Whereas there is legal precedent for arguing that under certain conditions States Parties to the ECHR have effective control through either agents or territorial control, this cannot be presumed.

Recent cases at the Court have however highlighted two relevant factors which would entail that the gathering and also storage of private data could fall within the jurisdiction of a member State. Relevant precedent could be found when considering the Court emphasising close, physical proximity between agents of the State and the victim. A second option would be relying upon the newly established doctrine of "special features" for finding extraterritorial jurisdiction in the sense of Article 1 ECHR, recently employed by the Court when considering a procedural duty to investigate. Whereas both these notions were used by the ECtHR in the context of the right to life, the authors submit that there is no reason in principle why they could not also apply in the context of Article 8 ECHR.

Starting with the concept of proximity, the Court considered this notion in the recent *Carter v. Russia* case, which dealt with the assassination of Alexander Litvinenko by Russian agents in the United Kingdom.[26] In its seminal decision in the case of *Bankovic*, the Court had previously held that the "simple" fact that lethal force was used did not bring an individual within the personal control of a state agent. In *Georgia v. Russia (II)* however, the Court recognised that there had been an evolution in its case law in this respect.[27] It acknowledged that in a number of cases it has applied the concept of "State agent authority and control" over individuals to scenarios going beyond physical power and control exercised in the context of arrest or detention. The Court added that these cases were restricted to situations of 'isolated and specific acts involving an element of proximity'.[28]

Building on this development, in *Carter* the Court considered whether Russian State agents who poisoned Mr. Litvinenko exercised physical power and control over the life of Mr. Litvinenko. The Court held that this was indeed the case, inter alia because:

> when putting poison in the teapot from which Mr. Litvinenko poured a drink, they knew that, once ingested, the poison would kill Mr. Litvinenko. The latter was unable to escape the situation. In that sense, he was under the physical control of Mr. Lugovoy and Mr. Kovtun, who wielded power over his life.[29]

The Court thus looked at whether the Russian agents exercised physical power and control over the life of Mr. Litvinenko, rather than over Mr. Litvinenko as a person. This is an important difference because State agents can impact an individual's rights without having physical control over that individual. It follows from the judgements in *Georgia v. Russia (II)* and *Carter v. Russia* that it is vital that an element of proximity is involved. If this is the case, as was the case in *Carter v. Russia*, it could be argued that the violation 'amounted to the exercise of physical power and control over his life in a situation of proximate targeting.'[30]

Due to the very nature of the gathering of biometric data, this would quite often involve a similar element of proximity. On these grounds, it could be argued that this would fall within the jurisdiction of the State.[31] Examples of this would be the taking of DNA samples, the taking of iris scans or the registration of fingerprints. The gathering of biometric data however does not necessarily involve an element of proximity. Data can also be gathered without involving such an element. This lack of distance however does not necessarily create a fundamental issue for the application of the ECHR.

Here arguments could be presented on the grounds that in the context of the use of biometric data there may be "special features" which bring the situation within the jurisdiction of the State. In a line of case law on the duty to investigate under Article 2 ECHR, the Court has considered that "special features" can also establish a "jurisdictional link" with a State. It remains to be seen whether the Court will also apply the notion of "special features" outside of the context of the procedural obligation to investigate under Article 2 ECHR. If it does, this could provide an additional basis for bringing the use of biometric data within the jurisdiction of the State.[32]

Supporting the notion that the gathering and storage of data could fall within the jurisdiction is the fact that States have so far not contested this notion. Most notably, in the *Big Brother Watch* cases, the Court was directly asked to consider the gathering and storage of data by States. In both cases the respondent State raised no objection to the applicability of the Convention;

> the Government raised no objection under Article 1 of the Convention, nor did they suggest that the interception of communications was taking place outside the State's territorial jurisdiction. Moreover, during the hearing before the Grand Chamber the Government expressly

confirmed that they had raised no objection on this ground as at least some of the applicants were clearly within the State's territorial jurisdiction. Therefore, for the purposes of the present case, the Court will proceed on the assumption that, in so far as the applicants complain about the section 8(4) regime, the matters complained of fell within the jurisdictional competence of the United Kingdom.[33]

Relying on the *Temple of Preah Vihear* precedent, this silence can also be constructed as legally relevant. In this case, the International Court of Justice (ICJ) held that for any silence to be constructed as legally significant, or as a form of acquiescence, it must be "clear that the circumstances were such as called for some reaction, within a reasonable period."[34] As the ECtHR directly asked (and confirmed) the opinion of the respondent State in these cases, their non-contesting of the fact that the Court found the gathering of data would be within the jurisdiction can be considered legally relevant. It could be seen as a confirmation that the State considers the gathering (and subsequent storage) of personal data outside of their territory to fall within their competence.[35]

The above indicates there are legal arguments for concluding that the gathering and use of biometric data during a military operation by a State outside of its own territory may fall within the jurisdiction of that State. This would make the ECHR relevant, even when this takes place outside of the territory of the State. During military operations, it might thus be the case that the obligations from the ECHR would still apply.

13.4. The right to private life in Article 8 ECHR, its applicability and application to the use of biometric data in military operations abroad

13.4.1. Applicability of the right to private life to the use of biometric data in military operations abroad

The right to privacy, or the right to private life as it is referred to in the ECHR, is laid down in Article 8 ECHR. That article provides:

1. Everyone has the right to respect for his private and family life, his home and his correspondence.
2. There shall be no interference by a public authority with the exercise of this right except such as is in accordance with the law and is necessary in a democratic society in the interests of national security, public safety or the economic well-being of the country, for the prevention of disorder or crime, for the protection of health or morals, or for the protection of the rights and freedoms of others.

The ECtHR has made clear that the "essential object of Article 8 is to protect the individual against arbitrary interference by public authorities."[36] The Court has explained that "private life" is a broad term encompassing the sphere of personal autonomy within which everyone can freely pursue the development and fulfilment of his or her personality and establish and develop relationships with other persons and the outside world.[37] As is clear from the text of Article 8, it first needs to be established whether something constitutes an interference with the exercise of the right to private life. If this is the case, then such interference will constitute a breach of Article 8 if it does not meet the criteria in the second paragraph of the article ("except such as").

This chapter focuses on the use of biometric data in military operations. The first question that needs to be addressed in this context is whether the collection, storage and sharing of biometric data falls within the scope of application of "private life" so that Article 8 is applicable. The case law of the ECtHR makes clear that the storage of information relating to an individual's private life and the release of such information falls within the application of Article 8 (1).[38] In this context, it has underlined that the term "private life" must not be interpreted restrictively.[39]

In Amann v. Switzerland, the ECtHR linked a broad interpretation of the right to privacy in the context of data to international instruments in the field of data protection. It held that such a broad interpretation:

> corresponds with that of the Council of Europe's Convention of 28 January 1981 for the Protection of Individuals with regard to Automatic Processing of Personal Data, which came into force on 1 October 1985 and whose purpose is "to secure in the territory of each Party for every individual ... respect for his rights and fundamental freedoms, and in particular his right to privacy, with regard to automatic processing of personal data relating to him" (Article 1), such personal data being defined as "any information relating to an identified or identifiable individual" (Article 2).[40]

This is particularly relevant for biometric data because the use of a biometric system by definition constitutes "automatic processing of personal data".

This case law, and the link it established to the protection of personal data under international data protection instruments, suggest that Article 8 (1) ECHR applies to all collection and further processing of personal data.[41] It would therefore always apply to the collection and further processing of biometric data, which is a particular kind of personal data. This would be particularly so because biometric data is considered a subset of personal data that requires specific protection beyond that provided to "regular" personal data.[42]

This conclusion finds support inter alia in the Grand Chamber judgment in *S. and Marper v. United Kingdom*, in which the Grand Chamber of the ECtHR considered that as fingerprints objectively contain unique information about the individual concerned, allowing his or her identification with precision in a wide range of circumstances, the retention of this information without the consent of the individual concerned cannot be regarded as neutral or insignificant.[43] The Court continued to hold that:

> while it may be necessary to distinguish between the taking, use and storage of fingerprints, on the one hand, and samples and profiles, on the other, in determining the question of justification, the retention of fingerprints constitutes an interference per se with the right to respect for private life.[44]

Ultimately, whether the collection, storage and sharing of biometric data falls within the scope of Article 8 ECHR needs to be determined on a case-by-case basis. De Vries states that in determining whether this is the case, the ECtHR:

> Takes into account the specific context in which the information has been recorded and retained, the nature of the records, the way in which they are used and processed, the results that may be obtained and the applicant's reasonable expectations as to the private character of the information.[45]

The authors consider that there are good grounds for concluding that all collection, storage and disclosure of biometric data falls within the scope of Article 8 (1) ECHR, at least where the data is systematically collected, stored and shared as is the case in military operations. All biometric data objectively contain unique information about the individual concerned, allowing his or her identification with precision in a wide range of circumstances. This means that the rationale given by the Court in *S. and Marper v. UK* for concluding that the retention of fingerprints constitutes an interference per se with the right to respect for private life applies to all biometric data collected, stored and shared in military operations.

13.4.2. Application of the right to private life to the use of biometrics by military operations abroad

1. Requirements in Article 8 (2) ECHR

In the previous section, it was concluded that there are good grounds to conclude that all collection, storage and disclosure of biometric data by a military operation is an interference with the right to private life. Such an interference constitutes a violation of Article 8 ECHR, unless the requirements set out in the second paragraph of that article are met cumulatively. These are that the interference is

a)   in accordance with the law;

b)   necessary in a democratic society;

c) in the interests of national security, public safety or the economic well-being of the country, for the prevention of disorder or crime, for the protection of health or morals, or for the protection of the rights and freedoms of others.

These requirements are discussed in more detail in this section.

2. Lawfulness

In order not to fall foul of Article 8, the use of biometric data by a military operation must first be "in accordance with the law". This requirement has a formal and substantive sense.[46] In the formal sense, there must be an authorisation by a rule recognised in the national legal order.[47] This prevents the collection, storage and use of this biometric data from being arbitrary and demands a strong legal basis. A number of States have adopted legislation giving their armed forces the power to collect, store and share biometric data. The authors are aware of at least the Netherlands and Germany having such legislation.[48] Such domestic legislation constitutes the "law" referred to in Article 8.

Where such domestic legislation of the State using biometrics abroad is lacking, it may be asked whether domestic legislation of the State where the data is being collected or a resolution of the United Nations Security Council adopted under Chapter VII of the UN Charter can be understood as "law" in the sense of Article 8 (2) ECHR. With regard to the latter, it may be noted that although Article 8 (2) does not specify that "the law" must be domestic law, the ECtHR does refer explicitly to "domestic" law in case law on Article 8.[49] Academic commentaries have however also recognised that a right could potentially be read into some IHL clauses.[50] In a similar fashion, it can be argued that alternative authorisation could be provided by UNSC resolutions. The Court has so far however not considered any international legal or Security Council obligations directly. This can be explained by the fact that the Court has only been asked to consider domestic legislation of the State interfering with the right to privacy. In theory, a host State could also adopt domestic legislation allowing the use of biometrics on its territory by those other States. The ECtHR has not yet addressed the question of whether such legislation could qualify as the "law" under Article 8 (2) ECHR. It has however not excluded this possibility.

The substantive sense of the "in accordance with the law" criterion requires that the rule must be accessible and foreseeable. The ECtHR held in this respect that the:

expression "in accordance with the law" further refers to the quality of the law in question, requiring that it should be compatible with the rule of law and accessible to the person concerned who must, moreover, be able to foresee its consequences for him.[51]

Foreseeability implies that the law must be sufficiently foreseeable in its terms to give individuals an adequate indication as to the circumstances in which, and the conditions on which, the authorities are entitled to resort to measures affecting their rights under the Convention.[52] If the law grants discretion to public authorities, it must indicate with reasonable clarity the scope and manner of exercise of the relevant discretion so as to ensure to individuals the minimum degree of protection to which they are entitled under the rule of law in a democratic society.[53] In the context of the use of biometrics, albeit not in a military context, the ECtHR has held that:

> The level of precision required of domestic legislation – which cannot in any case provide
> for every eventuality – depends to a considerable degree on the content of the instrument in
> question, the field it is designed to cover and the number and status of those to whom it is
> addressed.[54]

With regard to the requirement that the law is adequately accessible, this means that the person concerned must be able to have an indication that is adequate in the circumstances of the legal rules applicable to a given case. The requirement of accessibility appears to be difficult to meet for the law of the State using biometrics in another State. The very fact that the law is part of another legal system than that of the host State suggests that it is less accessible to persons in the host State. This is all the more so if the law is in a different language than that of the host State.

The requirement in Article 8 ECHR that any interference with the right to private life must be "in accordance with the law" also requires that adequate safeguards be in place to ensure that an individual's Article 8 rights are respected. The law must provide adequate safeguards to offer the individual adequate protection against arbitrary interference.[55] The Court provided some indication of what such safeguards can consist of in its judgment in *S. and Marper v. UK*:

> It is as essential [...], to have clear, detailed rules governing the scope and application of
> measures, as well as minimum safeguards concerning, inter alia, duration, storage, usage,
> access of third parties, procedures for preserving the integrity and confidentiality[56]

It follows from this requirement that the law providing for the use of biometric data by armed forces will have to provide in some detail when and in respect of whom biometric data can be collected, stored and shared.[57]

3. Necessary in a democratic society

The interference must be "necessary in a democratic society". This means that a fair balance must be struck between the competing interests of the individual and society as a whole.[58] The interference must correspond to a pressing social need, and, in particular, must remain proportionate to the legitimate aim pursued.[59] This entails an assessment of the proportionality of the interference, i.e. balancing the right of the individual against the interests of the State and the society it represents.[60] Such a balancing will need to be done taking into account the specific circumstances of the case. The ECtHR has held that the requirement of "necessary in a democratic society" must be interpreted narrowly and that the need for restrictions must be convincingly argued in a given case.[61]

In principle, the State has a certain margin of appreciation in determining what it considers necessary in a democratic society.[62] The breadth of the margin varies and depends on a number of factors including the nature of the Convention right at issue, its importance for the individual, the nature of the interference and the object pursued by the interference.[63] In the context of national security, the authorities enjoy a wide margin of appreciation.[64] However, this margin is subject to the supervision of the Court, and the Court must be satisfied that there exist adequate and effective guarantees against abuse.[65] As the ECtHR has held that the need for safeguards to prevent the use of personal data that would be in violation of Article 8 "is all the greater where the protection of personal data undergoing automatic processing is concerned", it appears that the margin of appreciation is more limited in the case of biometric data.[66]

A number of relevant aspects of proportionality can be derived from the case law of the ECtHR. One is that the amount of data that is collected and stored should be as limited as possible.[67] This means that it is unlikely that the large-scale collection and storage of biometric data as was undertaken by the US in Afghanistan would be considered proportional.[68] The data should only be used for the purpose for which it was collected.[69] This means that the sharing of the data by the armed forces with other government agencies in their own State is likely to not be considered proportional. Data should not be kept for longer than is necessary for the purposes for which it has been collected.[70] This implies that normally the data should be deleted at the latest when the military operation in which the data has been collected ends. The data should also be relevant, accurate and up-to-date.[71]

There must also be a system of supervision in place. As the ECtHR held in *Roman Zakharov v. Russia*, it has to determine "whether the procedures for supervising the ordering and implementation of the restrictive measures are such as to keep the 'interference' to what is 'necessary in a democratic society'".[72] The supervision should normally be carried out by the judiciary. However, the Court has held in the context of secret surveillance that supervision by non-judicial authorities can be sufficient, provided that they are independent of the authorities carrying out the surveillance and are vested with sufficient powers and competence to exercise effective and continuous control.[73] This means that supervision by a person within the chain of command of the person ordering the use of biometrics in a military operation would not be sufficient. Although supervision by the judiciary is preferred by the ECtHR, it can be argued that at least certain uses of biometrics in military operations, such as when used for intelligence purposes, are similar to secret surveillance and that with regard to such use supervision by a non-judicial authority could meet the requirements of Article 8. Based on publicly available information, however, it appears that there is no independent supervision of the use of biometrics by the armed forces of States Parties to the ECHR.

Closely related to the requirement of supervision is that the individual whose biometric data are used be able to challenge the measure to which he or she has been subjected. A procedure for such a challenge must thus be available.[74]

A second element of the 'necessary within a democratic society' condition is that it is deemed to serve a legitimate purpose. For an interference with the right to private life not to breach Article 8, it must be for one of the purposes referred to in Article 8 (2) ECHR. The ECtHR has interpreted the terms of these purposes rather broadly.[75] Depending on the mandate of the operation, it appears that "national security", "the prevention of disorder or crime" and "the protection of the rights and freedoms of others" would be the most likely legitimate aims for the use of biometrics in a military operation. For example, the use of biometrics in an operation based on national self-defence can contribute to such self-defence, and thereby to the security of the State defending itself. The use of biometrics in an operation aimed at combating piracy, which is an international crime and criminalised as such in the domestic law of most States, can contribute to the prevention of crime. The use of biometrics in an operation supporting another State in fighting a terrorist group can contribute to the protection of the rights and freedoms of others which are threatened by that terrorist group. In such a way the use of biometric data could serve a legitimate purpose within military operations.

13.5. Conclusion

This chapter has demonstrated the relevance of the right to private life in Article 8 of the ECHR during military operations abroad, in particular to the use of biometrics in such operations. It has done so by using a two-pronged approach. In the first section, it has aimed to establish that there are situations in which the use of biometric data abroad could fall within the jurisdiction of a State party to the ECHR. The developing concept of extraterritorial jurisdiction allows for several situations in which a State would be bound to respect the human rights obligations that are relevant for the collection and further processing of biometric data. This is a result of the obligations found within Article 8 of the ECHR. In its case law the ECtHR has established that any infringement of an individual's privacy may not be arbitrary. States must ensure that any use of biometric data must have a legal basis. Likewise, this must serve a legitimate purpose and be proportionate with regard to the goals of any program. These conditions will still apply to States conducting military operations abroad. Although States have a certain margin of appreciation, compliance with the right to private life during extraterritorial military operations appears to be a tall order. In theory, States may derogate from Article 8, although there is some debate on the possibility of derogation in the context of extraterritorial application of the ECHR.[76] But in any event, derogation is subject to strict limits and therefore is not a panacea.

The appearance of biometric data within doctrines and in the practice of States highlights that its use is here to stay. This makes it necessary to keep conducting research as to the legal framework surrounding this data. This chapter has offered some first considerations on the European perspective towards the human rights obligations that are relevant in these cases. In doing so, it aims to offer a more complete picture surrounding the legal obligations of states when using these new technologies. As technologies and the uses of said technologies keep developing, it remains important to consider the broad range of legal obligations that could potentially influence these uses.

Notes

[1] Alison Mitchell, "Distinguishing Friend from Foe: Law and Policy in the Age of Battlefield Biometrics Notes and Comments," *Canadian Yearbook of International Law* 50 (2012): 289; William H. Boothby "Biometrics" in *New Technologies and the Law in War and Peace*, ed. William H. Boothby (Cambridge University Press, 2019); Rohan Talbot, "Automating occupation: international humanitarian and human rights law implications of the deployment of facial recognition technologies in the occupied Palestinian territory" *International Review of the Red Cross* 102 (2020): 823; Robin Geiß and Henning Lahmann, "Protection of data in armed conflict" *International Law Studies* 97 (2021): 555; Marten Zwanenburg, "Know thy enemy: the use of biometrics in military operations and International Humanitarian Law" *International Law Studies* 97 (2021: 1403.

[2] This application is however still dependent on several conditions which would need to be satisfied.

[3] See e.g. Asaf Lubin, "The Rights to Privacy and Data Protection under International Humanitarian Law and Human Rights Law," in *Research Handbook on Human Rights and Humanitarian Law: Further Reflections and Perspectives*, eds. Robert Kolb, Gloria Gaggioli and Pavle Kilibarda. Cheltenham: Edward Elgar Publishing, 2022), 463.

[4] ISO/IEC International Standard 2382-37, "Information Technology – Vocabulary – part 37: biometrics 2" (2012).

[5] Anil Jain, Arun Ross, Salil Prabhakar, "An introduction into biometric recognition," *IEEE Transactions on Circuits and Systems for Video Technology* 14 (2004): 4, 5.

[6] See for an extensive description of biometrics inter alia Nancy Y. Liu, *Bio-privacy: Privacy Regulations and the Challenge of Biometrics*. Milton Park, Abingdon, Oxon; New York: Routledge, 2012, 29–59.

7   For additional characteristics see Boothby, *supra* note 1, p. 192.

8   "Recognising" is used here as a term encompassing verification and identification as defined below.

9   Jain et al, "An Introduction into Biometric Recognition," 5.

10 Jain et al, "An Introduction into Biometric Recognition," 6.

11 See e.g. William Buhrow, *Biometrics in Support of Military Operations: Lessons from the Battlefield* (Boca Raton: CRC Press, 2017) 41–68.

12 Buhrow, *Biometrics in Support of Military Operations*, 3.

13 Mark Lunan, "New Doctrinal Concepts: Biometrics," *The Three Swords Magazine* 33 (2018): 37.

14 Marko Milanovic, *Extraterritorial Application of Human Rights Treaties: Law, Principles, and Policy* Oxford: Oxford University Press, 2011, 46.

15 See e.g. Issa and Others v. Turkey, no. 31821/96, § 71, 16 November 2004.

16 Legal Consequences of the Construction of a Wall in the Occupied Palestinian Territory (Advisory Opinion) [2004] ICJ Rep 883, 9 July 2004, para. 109.

17 Al-Skeini and Others v. the United Kingdom [GC], no. 55721/07, 7 July 2011, para. 130-140.

18 See e.g. M.N. and Others v. Belgium (decision) [GC], no. 3599/18, 5 May 2020, para. 103; Sandu and Others v. the Republic of Moldova and Russia, nos. 21034/05 and 7 others, 17 July 2018, paras. 36-38.

19 Al Skeini and others v. United Kingdom, para. 136.

20 Al Skeini and others v. United Kingdom, para. 135.

21 Georgia v. Russia (II) [GC], no. 38263/08, 21 January 2021.

22 See e.g. Banković and Others v. Belgium and Others (decision) [GC], no. 52207/99, 12 December 2001.

23 See e.g. Medvedyev and Others v. France, no. 3394/03, 10 July 2008.

24 Georgia v. Russia (II), paras. 126, 137. For a discussion of this judgment see Floris Tan and Marten Zwanenburg, "One step forward, two steps back? *Georgia v Russia (II)*, European Court of Human Rights, Appl No 38263/08," *Melbourne Journal of International Law* 22 (2021): 136.

25 Marko Milanovic, "Al-Skeini and Al-Jedda in Strasbourg," *European Journal of International Law* 23 (2012): 121.

26 Carter v. Russia, no. 20914/07, 21 September 2021.

27 Georgia v. Russia (II), para. 114.

28 Georgia v. Russia (II), para. 132.

29 Carter v. Russia, para. 160.

30 Carter v Russia, para. 161.

31 One of the crucial questions for the court to consider would thus be if the gathering of data took place with an element of proximity or through a more remote method. This could influence if the Court would consider this to fall within a state's jurisdiction. See; Marko Milanovic, "European Court Finds Russia Assassinated Alexander Litvinenko," EJIL:Talk!, September 23, 2021. https://www.ejiltalk.org/european-court-finds-russia-assassinated-alexander-litvinenko/.

32 See for a further explanation of special features: Eugénie Delval, "The Kunduz Airstrike before the European Court of Human Rights: a glimmer of hope to expand the Convention to UN Military Operations, or a tailored jurisdictional link?" *The Military Law and the Law of War Review* 59 (2021): 244, 256-258.

33 Big Brother Watch and Others v. the United Kingdom [GC], nos. 58170/13 and 2 others, 25 May 2021, para. 272.

34 Case concerning the Temple of Preah Vihear (Cambodia v. Thailand) (judgement) [1962] ICJ Rep 260, 15 June 1962, para 23.

35 Whereas beyond the scope of this chapter, the Court seems to consider that this would also be the case when receiving data from third parties. See; Big Brother Watch and Others v. the United Kingdom, paras. 495-497.

36 See e.g. P. and S. v. Poland, no. 57375/08, 30 October 2012, para. 94.

37 Jehova's Witnesses of Moscow v. Russia, no. 302/02, 10 June 2010, para. 117.

38 Leander v. Sweden, Series A no. 116, 26 March 1987, para. 48; Rodica Mihaela Rotaru v. Romania, no. 34325/05, 10 November 2009, para. 43.

39 Amann v. Switzerland [GC], no. 27798/95, 16 February 2000, para. 65.

40 Amann v. Switzerland, para. 65. See also Satakunnan Markkinapörssi Oy and Satamedia Oy v. Finland [GC], no. 931/13, 27 June 2017, para. 133.

41 David Harris, Michael O'Boyle, Ed Bates and Carla Buckley, *Law of the European Convention on Human Rights*. Oxford: Oxford University Press, 2018, 524. But see also Kokott and Sobotta, who suggest that the protection of Article 8 ECHR only starts to apply as an event recedes into the past. Juliane Kokott and Christoph Sobotta, "The distinction between privacy and data protection in the jurisprudence of the CJEU and the ECtHR," *International Data Privacy Law* 3 (2013): 222, 224.

42 See e.g. Article 9 of Regulation (EU) 2016/679 of the European Parliament and of the Council of 27 April 2016 on the protection of natural persons with regard to the processing of personal data and to the free movement of such data, and repealing Directive 95/46/EC (General Data Protection Regulation), OJ 2016 L 119/1.

43 S. and Marper v. the United Kingdom [GC], nos. 30562/04 and 30566/04, 4 December 2008.

44 S. and Marper v. the United Kingdom, para. 86.

45 Karin de Vries, "Right to Respect for Private and Family Life," in *Theory and Practice of the European Convention on Human Rights*, eds. Pieter van Dijk, Fried van Hoof, Arjen van Rijn and Leo Zwaak. Cambridge, Antwerp, Portland: Intersentia, 2018, 667, 673.

46 Wiliam Schabas, *The European Convention on Human Rights: A Commentary*. Oxford: Oxford University Press, 2015, 402.

47 Schabas, *The European Convention on Human Rights*, 402.

48 See for the Netherlands the Uitvoeringswet Algemene Verordening Gegevensbescherming and the Regeling Gegevensbescherming Militaire Operaties. For Germany see e.g. Sebastian Cymutta, "Biometric Data Processing by the German Armed Forces during Deployment," *NATO CCDCOE Research Paper* (2021): 9-12

49 See e.g. S. And Marper v United Kingdom, para. 95.

50 See Zwanenburg, "Know Thy Enemy".

51 Klaus Müller v. Germany, no. 24173/18, 19 November 2020, para. 49.

52 Fernández Martínez v. Spain, no. 56030/07, 15 May 2012, para. 117.

53 Piechowicz v. Poland, no. 20071/07, 17 April 2012, para. 212.

54 S and Marper v. United Kingdom, para. 96.

55 Bykov v. Russia [GC], no. 4378/02, 10 March 2009, para. 81.

56 S. and Marper v. United Kingdom, para. 99.

57 It has been argued that the domestic legislation of the Netherlands does not meet this requirement. See Jeanice Koorndijk, "De Verwerking van Biometrische Gegevens tijdens Stability-policing-operaties van de Nederlandse Krijgsmacht: Richting Lessons Learned ter Verbetering van het Juridisch Kader" (Master's thesis, University of Amsterdam, 2021), 33.

58 Keegan v. Ireland, Series A no. 290, 26 May 1994, para. 94.

59 See e.g. Vavřička and Others v. the Czech Republic [GC], nos. 47621/13 and 5 others, 8 April 2021, para. 273.

60 Schabas, *The European Convention on Human Rights: A Commentary*, 406.

61 See e.g. M.N. and Others v. San Marino, no. 28005/12, 7 July 2015, para. 73.

62 Schabas, *The European Convention on Human Rights: A Commentary*, 406. See generally on the margin of appreciation in the context of Article 8 Yutaka Arai, "The Margin of Appreciation Doctrine

in the Jurisprudence of Article 8 of the European Convention on Human Rights," *Netherlands Quarterly of Human Rights* 16 (1998): 41.

[63]   S. and Marper v. United Kingdom, para. 102.

[64]   Big Brother Watch and others v. United Kingdom, para. 338.

[65]   Klass and Others v. Germany, Series A no. 28, 6 September 1978, para. 50.

[66]   M.K. v. France, no. 19522/09, 18 April 2013, para. 35.

[67]   See e.g. Catt v. the United Kingdom, no. 43514/15, 24 January 2019, para. 123.

[68]   See for a brief description of the large-scale collection in Afghanistan e.g. Katja Linskov Jacobsen, "Biometric data flows and unintended consequences of counterterrorism," *International Review of the Red Cross* 103 (2021): 619, 626.

[69]   See e.g. Karabeyoğlu v. Turkey, no. 30083/10, 7 June 2016, para. 117.

[70]   See S. and Marper v. United Kingdom, para. 103.

[71]   See e.g. M.K. v France, para. 36; Khelili v. Switzerland, no. 16188/07, 18 October 2011, paras. 64-70.

[72]   Roman Zakharov v. Russia [GC], no. 47143/06, 4 December 2015, para. 232.

[73]   Klass and others v. Germany, para. 56.

[74]   See e.g. Vig v. Hungary, no. 59648/13, 14 January 2021, para. 57; Szabó and Vissy v. Hungary, no. 37138/14, 12 January 2016, para. 86; M.N. and others v. San Marino, para. 78.

[75]   Schabas, *The European Convention on Human Rights: A Commentary,* 405.

[76]   See e.g. Marko Milanovic, "Extraterritorial Derogations from Human Rights Treaties in Armed Conflict," in *The Frontiers of Human Rights: Extraterritoriality and its Challenges*, ed. Nehal Bhuta. Oxford: Oxford University Press, 2016; Stuart Wallace, "Derogations from the European Convention on Human Rights: The Case for Reform," *Human Rights Law Review* 20 (2020): 769.

References

Al-Skeini and Others v. the United Kingdom [GC], no. 55721/07, 7 July 2011

Amann v. Switzerland [GC], no. 27798/95, 16 February 2000

Arai, Yutaka "The Margin of Appreciation Doctrine in the Jurisprudence of Article 8 of the European Convention on Human Rights," *Netherlands Quarterly of Human Rights* 16 (1998)

Banković and Others v. Belgium and Others (decision) [GC], no. 52207/99, 12 December 2001

Big Brother Watch and Others v. the United Kingdom [GC], nos. 58170/13 and 2 others, 25 May 2021

Boothby, William H. "Biometrics" in *New Technologies and the Law in War and Peace*, ed. William H. Boothby. Cambridge University Press, 2019.

Buhrow, William, *Biometrics in Support of Military Operations: Lessons from the Battlefield* (Boca Raton: CRC Press, 2017)

Bykov v. Russia [GC], no. 4378/02, 10 March 2009

Carter v. Russia, no. 20914/07, 21 September 2021

Case concerning the Temple of Preah Vihear (Cambodia v. Thailand) (judgement) [1962] ICJ Rep 260, 15 June 1962

Catt v. the United Kingdom, no. 43514/15, 24 January 2019

Cymutta, Sebastian, "Biometric Data Processing by the German Armed Forces during Deployment," NATO CCDCOE Research Paper (2021)

Delval, Eugénie "The Kunduz Airstrike before the European Court of Human Rights: a glimmer of hope to expand the Convention to UN Military Operations, or a tailored jurisdictional link?" *The Military Law and the Law of War Review* 59 (2021)

de Vries, Karin "Right to Respect for Private and Family Life," In: *Theory and Practice of the European Convention on Human Rights*, eds. Pieter van Dijk, Fried van Hoof, Arjen van Rijn and Leo Zwaak. Cambridge, Antwerp, Portland: Intersentia, 2018.

Fernández Martínez v. Spain, no. 56030/07, 15 May 2012

Geiß, Robin and Henning Lahmann, "Protection of data in armed conflict" *International Law Studies* 97 (2021)

Georgia v. Russia (II) [GC], no. 38263/08, 21 January 2021

Harris, David, Michael O'Boyle, Ed Bates and Carla Buckley, *Law of the European Convention on Human Rights.* Oxford: Oxford University Press, 2018.

Issa and Others v. Turkey, no. 31821/96, 16 November 2004

ISO/IEC International Standard 2382-37, "Information Technology – Vocabulary – part 37: biometrics 2" (2012)

Jain, Anil, Arun Ross, Salil Prabhakar, "An introduction into biometric recognition," *IEEE Transactions on Circuits and Systems for Video Technology* 14 (2004)

Jehova's Witnesses of Moscow v. Russia, no. 302/02, 10 June 2010

Karabeyoğlu v. Turkey, no. 30083/10, 7 June 2016

Keegan v. Ireland, Series A no. 290, 26 May 1994

Khelili v. Switzerland, no. 16188/07, 18 October 2011

Klass and Others v. Germany, Series A no. 28, 6 September 1978

Klaus Müller v. Germany, no. 24173/18, 19 November 2020

Kokott, Juliane and Christoph Sobotta, "The distinction between privacy and data protection in the jurisprudence of the CJEU and the ECtHR," *International Data Privacy Law* 3 (2013)

Koorndijk, Jeanice "De Verwerking van Biometrische Gegevens tijdens Stability-policing-operaties van de Nederlandse Krijgsmacht: Richting Lessons Learned ter Verbetering van het Juridisch Kader" (Master's thesis, University of Amsterdam, 2021)

Leander v. Sweden, Series A no. 116, 26 March 1987

Legal Consequences of the Construction of a Wall in the Occupied Palestinian Territory (Advisory Opinion) [2004] ICJ Rep 883, 9 July 2004

Linskov Jacobsen, Katja "Biometric data flows and unintended consequences of counterterrorism," *International Review of the Red Cross* 103 (2021)

Liu, Nancy Y. *Bio-privacy: Privacy Regulations and the Challenge of Biometrics.* Milton Park, Abingdon, Oxon; New York: Routledge, 2012.

Lubin, Asaf "The Rights to Privacy and Data Protection under International Humanitarian Law and Human Rights Law," forthcoming in *Research Handbook on Human Rights and Humanitarian Law: Further Reflections and Perspectives*, eds. Robert Kolb, Gloria Gaggioli and Pavle. Kilibarda. Cheltenham: Edward Elgar Publishing, 2022.

Lunan, Mark, "New Doctrinal Concepts: Biometrics," *The Three Swords Magazine* 33 (2018)

Medvedyev and Others v. France, no. 3394/03, 10 July 2008

Milanovic, Marko, *Extraterritorial Application of Human Rights Treaties: Law, Principles, and Policy.* Oxford: Oxford University Press, 2011.

Milanovic, Marko "Al-Skeini and Al-Jedda in Strasbourg," *European Journal of International Law* 23 (2012)

Milanovic, Marko "Extraterritorial Derogations from Human Rights Treaties in Armed Conflict," in *The Frontiers of Human Rights: Extraterritoriality and its Challenges*, ed. Nehal Bhuta. Oxford: Oxford University Press, 2016.

Milanovic, Marko "European Court Finds Russia Assassinated Alexander Litvinenko," EJIL:Talk!, September 23, 2021. https://www.ejiltalk.org/european-court-finds-russia-assassinated-alexander-litvinenko/.

Mitchell, Alison "Distinguishing friend from foe: law and policy in the age of battlefield biometrics notes and comments," *Canadian Yearbook of International Law* 50 (2012)

M.K. v. France, no. 19522/09, 18 April 2013

M.N. and Others v. Belgium (decision) [GC], no. 3599/18, 5 May 2020

M.N. and Others v. San Marino, no. 28005/12, 7 July 2015

P. and S. v. Poland, no. 57375/08, 30 October 2012

Piechowicz v. Poland, no. 20071/07, 17 April 2012

Rodica Mihaela Rotaru v. Romania, no. 34325/05, 10 November 2009

Roman Zakharov v. Russia [GC], no. 47143/06, 4 December 2015

S. and Marper v. the United Kingdom [GC], nos. 30562/04 and 30566/04, 4 December 2008.

Sandu and Others v. the Republic of Moldova and Russia, nos. 21034/05 and 7 others, 17 July 2018

Satakunnan Markkinapörssi Oy and Satamedia Oy v. Finland [GC], no. 931/13, 27 June 2017

Schabas, William. *The European Convention on Human Rights: A Commentary.* Oxford: Oxford University Press, 2015.

Szabó and Vissy v. Hungary, no. 37138/14, 12 January 2016

Talbot, Rohan "Automating occupation: International Humanitarian and Human Rights Law implications of the deployment of facial recognition technologies in the Occupied Palestinian Territory" *International Review of the Red Cross* 102 (2020)

Tan, Floris and Marten Zwanenburg, "One step forward, two steps back? Georgia v Russia (II), European Court of Human Rights, Appl No 38263/08," *Melbourne Journal of International Law* 22 (2021)

Vavřička and Others v. the Czech Republic [GC], nos. 47621/13 and 5 others, 8 April 2021

Vig v. Hungary, no. 59648/13, 14 January 2021

Wallace, Stuart "Derogations from the European Convention on Human Rights: the case for reform," *Human Rights Law Review* 20 (2020):

Zwanenburg, Marten "Know thy enemy: the use of biometrics in military operations and International Humanitarian Law" *International Law Studies* 97 (2021)

CHAPTER 14

2020 Nagorno-Karabakh War

A case of triple D: Diplomacy, Drones and Deception

Peter de Werd, Michiel de Jong & Han Bouwmeester

Abstract

The role of information as a source of power in Russia's foreign policy and military actions has received increasing attention of Western scholars and policymakers, especially since the start of the current war in Ukraine. Their focus has been on the importance of the information domain in Russian foreign policies and military operations in Georgia and Ukraine as 'typical' case studies. This study aims to retrieve more insight in the nature of information operations by studying the atypical case of the 2020 Nagorno-Karabakh War between Armenia and Azerbaijan, in which Russia proliferated herself as a mediator. It offers an explorative analysis, based on diplomatic communication, social media and news reporting in conjunction with military operational developments between September 27 and December 31, 2020. From a comparison of various sources and angles, this study tentatively approaches the shifting perspectives in the Russian narratives' construction and their effects on the enhancement of Russian foreign political interests towards Armenia, Azerbaijan and Turkey. This study argues that the Russian 'narrative' consists of a number of subvariants, tailored to various national and international audiences. In addition, although it is difficult to technically attribute artificial social media accounts and subsequent messaging and amplification of Turkish, Azerbaijani and Armenian narratives to Russia, these activities fit with Russian interests. Therefore, further research on Russian information operations should also focus on atypical cases in order to comprehend New Type Warfare as a broader activity.

Keywords: Nagorno-Karabakh, Turkey, Russia, Information operations, Narrative, Diplomacy, War, Drones, Deception, Armenia, Azerbaijan

14.1. Introduction[1]

Contemporary debate in military strategic studies has seen a proliferation of modes of 'warfare', such as financial, legal, cyber, hybrid[2], deception or information warfare.[3] Characterised as (mostly) non-violent or non-destructive forms of influence, these modes are situated along the spectrum with on one end coercive force and compellence and on the other non-coercive presence.

While non-kinetic approaches to warfare are not new, the development of the information environment in the last decades has caused for Western military doctrine to redefine the role of information as a source of power. In this section, the ample use of 'non' reveals the challenge of defining these different forms of action in terms of 'warfare'.

Among Western scholars and policymakers, Russia's foreign policy and military actions have frequently been viewed from the perspective of broad influence, information operations and deception warfare. The Russian Ministry of Defence itself has defined information operations as "the ability to ... undermine political, economic, and social systems; carry out mass psychological campaigns against the population of a state in order to destabilize society and the government; and force a state to make decisions in the interest of their opponents."[4] As seen in the aftermath of the downing of Malaysia Airlines flight MH-17 over Ukraine in July 2014, Russia's information operations were aimed at a broader international (Western) community.[5] Its main aim was to diffuse and create ambiguity to challenge the legal investigation by the international Joint Investigation Team. In other cases, efforts are aimed more at misleading targeted audiences to adopt an alternative to their 'truth'. Of course, truth itself is always partially subjective. 'Misleading' can thus best be understood in terms of relative effect: actors change perception, beliefs, and in turn also their actions.

An in-depth account of modern Russian deception warfare – as Western researchers have analysed it – is Bouwmeester's dissertation *'Krym Nash'* (Crimea is Ours).[6] It presents an array of means and modalities that all underline the increasing importance of the information environment in Russian military operations in Georgia and Ukraine. Firstly, the Russo-Georgian armed conflict in 2008 was primarily a physical encounter only supported to a limited extent by information operations. Secondly, during the annexation of Crimea in 2014 information operations played a more central role as the perception of Russia's opponent was manipulated while a surprise effect emerged. Although not attempting to reduce or simplify the case specific contexts and complexities in any way, despite many differences it was Russia's intervention that triggered the conflict in both cases. Thirdly, although not always directly visible on the battlefield, the Russian 'special military operation' in Ukraine launched by Putin in early 2022, seems to open a new chapter in the development of information operations.

However, in order to reflect on the nature of information operations more thoroughly, studying an atypical case is relevant. In this respect, this chapter offers a preliminary analysis of the 2020 Nagorno-Karabakh war. In September 2020 war broke out between Armenia and Azerbaijan over the disputed region of

Nagorno-Karabakh. Eventually, Russia was able to mediate and facilitate a cease-fire agreement between the belligerents. Russia also deployed peacekeeping troops and set up a joint monitoring headquarters with Turkey in Azerbaijan. Russia's involvement in this conflict was characterised by some NATO officials as a way of successfully 'consolidating Moscow's status as the key player in the [Caucasus] region'. The Russian Federation had pursued a similar strategy in the 1990s, when it had initiated Joint Control Commissions (JCCs) in Transdniestria, South Ossetia and Abkhazia. The term JCC evokes the feeling of organisations with good and peaceful intentions, but the history of the JCCs in the earlier mentioned regions showed that it had become nothing more than a cover for the Russian authorities to deploy troops and exert influence in the region. In other words, Russian authorities were repeatedly able to approach conflict situations differently and conjure up perceptions that pleased them. This raises the question of what can be learned from the Russian foreign activities in the information environment during the 2020 Nagorno-Karabakh war?

Due to the recent nature of the conflict, this case study analysis is explorative and primarily focuses on diplomatic communication, social media and news reporting, in conjunction with military operational developments and events between September 27 and December 31, 2020. Apart from a short description of the military operations, a content analysis of official communiques is combined with a meta-study of relevant social and news media research. Firstly, the case is introduced, further describing the main actors in their historical and regional context. Secondly, the course of the Azerbaijani Operation Iron Fist in late-2020 and related military developments are discussed. Thirdly, relevant narratives related to the key actors in Armenia, Azerbaijan, Turkey and Russia and some external influences are considered. Fourthly, identified key events and statements are synthesised with diplomatic and military developments and are placed in light of Russian information operation doctrine and more specific Russian deception methods. An additional postscript reflects on research findings in the context of the recent developments in Ukraine.

14.2. The conflict's origins and the First Nagorno-Karabakh War

The conflict originated between the Armenian and Azerbaijani populations and the region or enclave of Nagorno-Karabakh, located in the centre of Azerbaijan, formed its epicentre. In the ensuing century (since 1918) the conflict involved Turkey, Russia and Iran as major geopolitical players, and France, the United States and Israel as minor ones.

Between May 1918 and April 1920 Armenia and Azerbaijan briefly formed independent states[7], at a time when the Russian Empire was weak and struggling with the Communist revolution that gave rise to the Soviet Union. The Armenian Republic became a refuge for survivors of the Armenian genocide under Turkish nationalists of 1915-1916.[8]

In this period, the first conflicts broke out between the states and populations of Azerbaijan and Armenia. After the expansion of the Soviet Union's sphere of influence over the areas south of the Caucasus in 1920-1923, both states were transformed into Soviet Republics. Stalin decided to turn the Nagorno-Karabakh region with an Armenian majority and Azerbaijani minority into an oblast, an autonomous area, in the Soviet Republic of Azerbaijan. It became an enclave surrounded by seven Azerbaijani districts and separated from the Soviet Republic of Armenia. For seven decades the conflict remained frozen, as it were.

With deterioration of the Soviet Union in 1988, nationalist and ethnic tensions resurfaced in the Caucasus.[9] The Armenian government had repeatedly, but in vain, proposed to President Mikhail Gorbachev of the Soviet Union to transfer the oblast Nagorno-Karabakh to Armenia. In 1988, Armenians shattered a demonstration of Azerbaijanis in Stepanakert, resulting in two deaths.[10] Subsequently, in several cities in Azerbaijan, pogroms broke out against the Armenian minorities, and afterwards also violence by Armenians against Azerbaijanis erupted in and around the enclave. President of the Soviet Union, Gorbachev proved powerless against the rapidly spreading violence in Azerbaijan and Armenia, and his officials and military commanders reacted slowly.[11] Hundreds were killed in the chaos resulting in the first waves of refugees of both populations. After the Soviet Union's dissolvement in 1991, subsequently followed by referenda among their populations, Armenia and Azerbaijan proclaimed themselves independent republics. The following year, both joined the military alliance of former Soviet Republics led by Russia, the Collective Security Treaty Organisation (CSTO). However, Azerbaijan didn't opt for the renewal of its membership in 1999, although it maintains good ties with Russia.[12]

After the results of a referendum in Nagorno-Karabakh in 1991, in which the Armenian majority voted for independence and aimed for a long-term affiliation with Armenia, war broke out between the young Republics of Armenia and Azerbaijan.[13] Between 1992 and 1994 the Armenian armed forces succeeded in capturing 14% of the Azerbaijani territory, namely the enclave of Nagorno-Karabakh and the seven surrounding districts. Due to internal political struggles the Azerbaijanis were unable to properly defend their own territory. This situation was perpetuated by ceasefire negotiations during which Russia acted as a mediator in Bishkek in 1994.[14] The Armenians converted the enclave Nagorno-Karabakh into the Republic of Artsakh, also called the Nagorno Karabakh Republic. Yet remarkably enough the Armenian Republic didn't formally recognise it. During this war, later known as the First Nagorno-Karabakh War, 350,000-700,000 Azerbaijanis fled Armenia, the enclave and the seven districts; and in turn 300,000 Armenians fled

Azerbaijan. The refugee Armenians often found shelter in villages and towns in and around the enclave. The seven districts turned into a severely depopulated area.[15]

14.2.1. Stagnated peace talks

As of 1993, the United Nations (UN) Security Council issued four resolutions urgently calling Armenia to return the conquered territories to Azerbaijan. The council also made efforts to negotiate to settle the conflict between the warring parties. The UN upheld the viewpoint that the enclave of Nagorno-Karabakh belonged to Azerbaijan according to international law, which the Azerbaijani government also repeatedly invoked. Under the auspices of the Organisation of Security and Co-Operation in Europe (OSCE) emerged the Minsk Group, led by co-chairs Russia, the United States and France, which mediated peace talks between Armenia and Azerbaijan. Those negotiations were extremely difficult between 1994 and 2020.[16] Many proposals, came to nothing, even after the adoption in 2007 of the promising 'Basic Principles for the Peaceful Settlement of the Nagorno-Karabakh Conflict'. Also known as the Madrid protocols, they encompass six principles: the return of the territories surrounding Nagorno-Karabakh to Azerbaijani control; an interim status for Nagorno-Karabakh providing guarantees for her security and self-governance; a corridor linking Armenia to Nagorno-Karabakh; the future determination of the final legal status of Nagorno-Karabakh through a legally binding expression of popular will; the right of all internally displaced persons and refugees to return to their former places of residence; and international security guarantees that would include a peacekeeping operation.[17]

During the failed negotiations a factor of importance was the way in which the presidents of Armenia and Azerbaijan conducted their talks behind closed doors, without any accountability to their parliaments. The two presidents also feared the impact of premature concessions on their parties and those of the opposition. For the Armenian government, the influence from Artsakh also played a role, where many of the political elite and presidents in the capital of Armenia, Yerevan, came from. Of importance was also the influence of the large Armenian communities as electorates in France and the United States, which regularly were suspected of dampening the negotiations' results.[18]

14.2.2. Turkey, alienation and hostility in a simmering conflict

In the region, Turkey forged good ties with Azerbaijan, relying on the Turkish origin of the Azerbaijanis. President Hayder Aliyev (father of the current president) expressed this warm relationship as 'two states, one nation'. For Turkey, Azerbaijan became of great strategic importance in a political, economic and social sense. After

Georgia, it offers Turkey access to brother peoples in Central Asia, such as those in Kazakhstan, Turkmenistan, Uzbekistan and Kyrgyzstan, via road and railways in a 60 km Azerbaijani land strip between Iran and Russia and the Caspian Sea.[19] Close ties with Azerbaijan also fit Turkey's geopolitical aspirations in Central Asia, where it competes for influence with Iran and Russia, but also seeks affiliation with China, a trend that has accelerated since Recep Erdogan's presidency.[20]

Parallel to this trend, relations between Turkey and Armenia broke down, with serious consequences for the economies of Armenia and Nagorno-Karabakh. Although both economies became enclosed by Azerbaijan and Turkey, and became more isolated, important passenger and goods traffic was still possible via Georgia and especially Iran, with which Armenia maintained good relations (much Russian freight traffic runs over the Caspian Sea and via Iran to Armenia).[21] The increasing cooperation between Turkey and Azerbaijan aroused strong antipathy and existential feelings of insecurity among the Armenian population in Armenia and Artsakh, and the Armenian communities worldwide, because of the denials of the presidents of those states in public and the media of the Armenian genocide in 1915-1916. The Armenian state and the Armenian armed forces are therefore seen as important pillars for the survival of the Armenian population.[22]

Meanwhile, both population groups in the neighbouring countries of Azerbaijan and Armenia grew apart due to nationalism, populism and the harrowing stories of refugee compatriots about violence and atrocities in the years 1988-1994. Mixed marriages of Azerbaijanis and Armenians, of which there were many before 1988, chose to emigrate to Russia, Central Asia and elsewhere. In the decades after 1994, younger generations grew up in social bubbles, becoming ignorant of large minority societies. Since the advent of mobile telephony and the Internet, the current social media has only reinforced this trend.[23]

Both sides remained diametrically opposed. On the one hand, Armenia benefited from the status quo after 1994, she had consolidated her position in the seven conquered districts and Artsakh. In the long term, Armenia hoped for international recognition of Artsakh's independence. She also strove to keep the conquered districts as a buffer zone against Azerbaijan and possibly use them as negotiating cards for concessions from the Azerbaijani side. Azerbaijan, on the other hand, continued the negotiations, but carefully considered a military option to reverse the adverse strategic situation in the long run.[24]

14.2.3. Military, diplomatic and political developments – the arms race

The revenues from the oil and gas industry and its exports, the major pillar of the Azerbaijani economy, enabled Azerbaijan to expend far more on defence than Armenia. These expenditures amounted to up to three times that of her nemesis. In addition, Azerbaijan's larger population also meant that it could deploy more ready and reserve troops. Azerbaijan bought much Russian military equipment, such as artillery, vehicles, tanks and fighter planes, but certainly also Israeli armaments, i.e. artillery and drones. Azerbaijan also obtained from Israel the licenses to build certain drones. Israel became Azerbaijan's main arms supplier. In recent years, Turkey has become a top supplier too, with a crucial delivery in July 2020 of TB-2 Bayraktar drones, as will become apparent.[25]

In contrast, Russia remained Armenia's main supplier delivering a wide range of arms and munitions, such as artillery howitzers, vehicles, tanks, mortars, and fighter planes, most of them at very affordable prices or even for free. This enabled Armenia, despite her smaller economy and defence budget, to keep abreast of its arms race with Azerbaijan. As a member of the CSTO, Armenia could count on Russia's support in the event of an attack on the territory of the State of Armenia. In order to strengthen her military ties, Russia founded a base in Armenia, Gyumri.[26] However, Artsakh and the captured districts fell outside the CSTO's assistance clause. After all, Russia made efforts through the OSCE Minsk Group to find a solution through diplomatic channels, and also repeatedly called on the UN Security Council and the Madrid protocols under the Minsk Group, co-chaired by Russia, the United States and France, to hand over the conquered districts to Azerbaijan.[27]

14.2.4. Skirmishes and the planning for operation Iron Fist

The war in September-October 2020 was preceded by two major clashes between Azerbaijan and Armenia. In April 2016, a four-day skirmish broke out between the two belligerents over Nagorno-Karabakh. The clash left twenty dead and dozens injured on both sides. In July 2020, another skirmish broke out with several dead and injured. Tensions were heightened by the statement of the Artsakh government to move its seat of government to Shusha, which Artsakh accomplished in June 2020. Meanwhile, large-scale exercises of the Azerbaijani together with Turkish armed forces in the summer of 2020 were underway, which ended in September 2020.[28]

Aggravated by Artsakh's move and the ensuing heightened tensions within the Azerbaijan parliament and population, President Ilham Aliyev decided to carry out Operation Iron Fist. In previous years, Aliyev had abstained from a major offensive, because a failed operation would almost certainly jeopardise the very survival of his government. It appears that Aliyev was heading for the best-prepared, timely-placed operation. The operation's objectives were to liberate the Armenian captured districts around Artsakh, and to reach and cut off the Main Supply Route across Lachin between Yerevan and Artsakh. This would pose serious problems for the continuation of the Armenian defence in the entire region.[29]

Fig. 14.1. Armenia and Azerbaijan (line of control prior to the 2020 war). Maps by kind permission of the NLD Inst of Military History.

In terms of terrain, the North and Central part of Artsakh and the seven districts are very mountainous with few roads crossing the area. Amidst the mountainous terrain are the cities of Stepanakert and Shusha. Stepanakert formed for a long time the administrative centre of the Armenian government in Nagorno-Karabakh. Shusha was for both belligerents the region's most important city historically, culturally, religiously[30]. For both sides, Shusha had thus acquired a status comparable to that of the city of Jerusalem. The entire Eastern border plus the Southern part of the Armenian held territories, largely consisting of the occupied districts, is relatively flat and lends itself to manoeuvring. The focal point of the operation was a main attack over two approach roads from Southeast to Southwest along the Iranian border and then a bypass up North to reach the main road from the Armenian capital Yerevan to Lachin and Stepanakert.[31] The attack would be accompanied by attacks from the North and East to tie up as many Armenian units as possible there. The Armenian defence of Artsakh and the seven districts was arranged in successive defensive lines and fortifications by troops from Artsakh and Armenia supported by command posts, artillery and reserves.[32]

Fig. 14.2. Nagorno-Karabakh Conflict Zone (line of control prior to the 2020 war).

It is important to mention that to October 2020 there have only been short articles and analyses of operation Iron Fist and the defence of Artsakh. However, the majority of these focus on the remarkable performance of the Turkish and Israeli drones during the campaign. An overall view of the military actions of the Azerbaijani attackers and Armenian defenders has yet to be published, but fortunately some useful articles and reports from Amirkhanyan, Gressel, Postma, Rubin and Terzic provide us with first impressions of the Azerbaijani and Armenian conduct of the war.[33]

14.2.5. The conduct of Operation Iron Fist

Operation Iron Fist started at 08:10 hours on September 27. According to the Azerbaijanis, they responded to an Armenian attack, and according to the Armenian armed forces, the Azerbaijani attack came immediately. For De Waal it is certain that the Azerbaijanis started it. After all, the Armenian defenders, in accordance with their political and military strategy, had no interest in initiating a large-scale war that could change the favourable status quo over the area and would jeopardise a long-term track of international recognition of Artsakh and perhaps the Armenian authority over the conquered districts too.[34]

Gressel gives the best, albeit succinct, insight into the course of the Azerbaijani offensive. In a first phase of ground and air operations, Azerbaijan was able to gain air superiority over the area of operations. In a few days, Turkish and Israeli-made drones disabled several Armenian air defence systems, i.e. S-300 Surface to Air Missile batteries. Azerbaijani drones, tasked to conduct Intelligence, Surveillance and Reconnaissance missions, continually tried to locate the defenders.[35] The Azerbaijanis reportedly converted eleven obsolete Antonov An-2 biplanes into remote-controlled aircraft that tricked the Armenian Surface to Air Defence into turning on their tracking and targeting radar systems. The TB-2 Bayraktar drones disabled these systems with their rockets. Israeli loitering drones, able to search and crash on their targets, annihilated them too.[36] Striking is the low activity of the air forces of both belligerents; they kept their expensive fighter planes (about 50 aircraft in total) at bay.[37] Reciprocal shelling with artillery and rockets did take place at a few nodal points and cities, which would increase in importance in the beginning of October, but seem to have had no direct impact on the results of the ground offensive or the defence against it.[38]

By gaining local air superiority with the help of drones, the Azerbaijanis created a breach in the Armenian Surface to Air Defence, after which in a second phase the drones could take the lead as flying artillery in detecting and taking out Armenian tanks, artillery, vehicles, mortar positions and other positions, but also attacking command posts and reserve units. In their role as artillery observers they assisted the Azerbaijani artillery in annihilating these targets as well.[39]

In the third phase of combat, Azerbaijani mobile light infantry tank teams with well-organised fire support (artillery, rockets, mortars) managed to penetrate the drone-observed weak spots in the Armenian lines, and presumably even to break through these lines.[40] Drones blocked the approach routes for reserves, command communications and fire support requests so that strong points could be taken by the mobile infantry tank units. Meticulously, the Azerbaijani land forces succeeded in taking line by line and area after area from Southeast to Southwest between 27 September until 13 October. Armenian reserve forces launched several large, concentrated counterattacks, but these were repulsed with the help of the Azerbaijani drones.[41] On the one hand, losses of personnel for both sides during the campaign were high: at least 2,100 Azerbaijanis killed against at least 2,700 Armenians. On the other hand, the Armenian losses in rolling stock were appalling: hundreds of vehicles, tanks and pieces of artillery and mortars were knocked out, often by the Turkish TB-2s and the Israeli loitering drones.[42]

Meanwhile, following the outbreak of hostilities, the UN and OSCE Minsk Group pushed for an immediate ceasefire. After the mediation of Russia, the United States and France, short-lived ceasefires were concluded on various dates in October, yet these were quickly broken, thus repeatedly led to the continuation of Azerbaijani offensive and subsequent Armenian losses of war materials and territory and demoralisation among the Armenian armed forces.[43]

14.2.6. The campaign's end

The Azerbaijani Land Forces managed to reach and cut off the main road in the region, the Lachin Corridor, around November 8-10. They also captured the main city of Shusha, on November 8, with the help of light infantry commandos who first occupied a predominant hill just outside the city.[44] It was expected that the capture of Stepanakert would take a few more days. However, on November 9 under the auspices of Russia, a ceasefire was reached between Azerbaijan and Armenia, with President Putin receiving and hosting both President Aliyev and Prime Minister Pashinyan.[45] Russia negotiated the rapid deployment of a peacekeeping force of 2,000 Russian troops along the ceasefire line, the positions held by both sides until 9 November, and to secure the Lachin Corridor between Armenia and Artsakh. The Azerbaijanis thus retained territories in the enclave they had conquered in the North, East and South including the very important city of Shusha. This peacekeeping force would also patrol the transport road across Armenian territory along the Iranian border between Azerbaijan and Nakhchivan in order to open the passage between the two Azerbaijani territories for goods and passenger traffic.[46] In addition, the Armenian troops had to completely evacuate Artsakh and the surrounding seven districts before the end of December 2020, including three districts that were still entirely in Armenian hands, which they did. Thus, the Russian peacekeeping force became the protector of the Armenian enclave including Stepanakert. It was clear that in terms of sovereign status, Nagorno-Karabakh remained de jure under Azerbaijan.[47]

Fig. 14.3. Cease-Fire Agreement of November 9, 2020.

14.2.7. Implications of the Forty-Four Days War

After the forty-four days of war, the Azerbaijani population celebrated the victory exuberantly. During the war, public opinion in Azerbaijan was influenced by the camera images on electronic billboards and on social media of drones taking out positions, vehicles, artillery and tanks. Social media played a major role in Azerbaijan alongside traditional newspapers, radio and television stations. During the war, the press was strongly guided by the Azerbaijani state institutions. The victory boosted hopes of return among Azerbaijani displaced persons after the first war of 1988-1994 to their former homes in the seven districts and Nagorno-Karabakh.[48] Public opinion in Armenia, as in Azerbaijan, was strongly influenced by positive images of the war in their media. The Armenian outcry was, however, fierce after the ceasefire on November 9, 2020. The rosy pictures during the defensive campaign did not reflect the actual course of the war for the Armenian armed forces. For example, the Azerbaijani conquest of Shusha on November 8 was masked from the Armenian population. However, the harsh reality of the city's fall came out on November 9. The ceasefire agreement and the Armenian defeat led to a storming and occupation of the parliament in Yerevan and calls for Prime Minister Pashinyan's resignation, which he was able to avoid though with great difficulty. At the same time, the desire among the Armenian population for protection from Russia increased.[49]

Russia decided not to intervene in the forty-four-day war to punish the Pashinyan government for the earlier velvet, Purple Revolution in 2018, which took over power from a more Russian-minded government. She approached the assistance to Armenia solely from the agreements within the CSTO, namely only applicable if Armenian territory or Russian units there were attacked. Pashinyan's overtures to the West and Turkey in the intervening years were not appreciated by Russia. By eventually sending a peacekeeping force of 2,000 men, Russia was able to act not only as a mediator, but also as a peacemaker between the warring parties and to fulfil the role of protector of the Armenians in Artsakh.[50]

Turkey was also alert not to intervene immediately, in order to avoid a direct conflict with Russia. After the intensive exercises in the summer of 2020, the Turkish air base with 6 F-16s therefore stayed out of the fray. Turkey was able to strengthen its position in the Caucasus by providing support to the host country with war material, in particular drones and military advisers in Azerbaijani command and control.[51] After the military successes of the Turkish TB-2s in operations in Syria and Libya, as already mentioned, the deployment of these drones around Nagorno-Karabakh aroused amazement and admiration worldwide. The opening of a transport road from Azerbaijan to Nakhichevan also marked long-term economic success for Turkey. Russia had, however, thwarted the Turkish attempt to participate in the peacekeeping force, and the Turkish contribution has been limited to co-manning a joint monitoring headquarters to monitor compliance with the ceasefire using drones.[52]

14.3. Different narratives: turning the kaleidoscope

Based on the analysis of official communiques, news reporting and social media, different narratives emerge on the military and political developments, shedding more light on how perceptions were shaped, and influence was exerted. Armenian, Azerbaijani, Turkish and Russian perspectives are discussed. To some extent narratives can be linked together: Armenia traditionally receives support from Russia, both members of the CSTO, whereas Azerbaijan is backed by Turkey. As mentioned, although Turkey is a NATO member its relationship with Russia is complicated. Its delivery of drones to Azerbaijan in 2020 countered Russian interests. While for example Turkey's acquisition of Russian S-400 surface-to-air defence systems in 2017 strengthened ties, despite concerns from other NATO members. Furthermore, France, Russia and the United States co-chair the OSCE Minsk Group, which includes both Armenia and Azerbaijan as well. Moreover, broader global issues such as the COVID-19 pandemic or the election of Joe Biden as American president on November 3, 2020, have an indirect effect on the spread and impact of narratives on the Nagorno-Karabakh conflict. This is particularly visible on social media.

For Armenia and Azerbaijan the narratives are generally consistent and illustrate strong polarisation, the two countries expressing the wish to remain peaceful, blaming the enemy for violations of agreements, encouraging patriotism, praising army successes.[53] In both countries the traditional media landscape is strongly dictated by official government statements, mostly by the Ministry of Defence, reproduced in national news reporting. Yet both heads of state also seek to reach their people directly, with Armenian Prime Minister Pashinyan communicating via Facebook livestreams regularly, and Azerbaijani President Aliyev frequently using Twitter.[54]

In the Armenian narrative a central theme was victimhood of violence, which historically related to the Armenian genocide that had taken place a century ago.[55] Nagorno-Karabakh constituted an essential part of modern post-Soviet Armenian history. One study shows how right after the start of hostilities 7,764 Twitter accounts were created, many with an artificial booster-like signature.[56] They made up a share of over 10% of relevant accounts at the time, mostly amplifying Armenian government officials and public figures. Legitimate and fake account postings on Armenian social media also appealed to Kim Kardashian and her partner Kanye West for support (which she gave), and petitions were launched to gain White House support and end US aid to Azerbaijan. Efforts to raise support and funds continued throughout the selected timeframe for this study.[57] In Armenian news and social media, pan-Turkism is framed as a threat to the region, and simultaneously Russia is portrayed as the sole protector. The latter aligns with statements by Prime Minister Pashinyan.[58]

A major change occurred on November 10, 2020. With an official joint statement Pashinyan agreed to the ceasefire, accepting the new status quo in Nagorno-Karabakh, and the withdrawal of troops and 'returning' the regions Kelbajar and Lachim to Azerbaijan.[59] As mentioned, this led to large scale protests in the Armenian capital as people called for his resignation. On social media Prime Minister Pashinyan was accused of accepting bribes to accede to Azerbaijan, selling out Armenians. However, at that time, an apparent effort was also observed to regain public trust and pro-Kremlin sentiments among the Armenian population.[60]

In Azerbaijani news reporting the question was also raised whether Pashinyan would be able to keep his job. The overall official government narrative in Azerbaijan had mainly focused on the success of military advancements, largely consisting of Ministry of Defence 'operational updates' reproduced by news media. The Azerbaijani President spoke of a clear and phased plan for Nagorno-Karabakh, and the two adjacent regions Lachin and Kelbajar.[61] The overall theme was the 'sacred duty' of ending the 'occupation' of Nagorno-Karabakh and surrounding territories under Armenian control.[62] Mutual coexistence of Armenian and Azerbaijani communities was the desired end state, although 'the consequences of ethnic cleansing must be eliminated and all our internally displaced persons must return to their own homes' the President stated.[63] A focal point was the 'liberation' of the symbolic Karabakh city of Shusha. Reporting in Turkish international news media in which Azerbaijani citizens were interviewed, emphasised the symbolic significance of Shusha, deemed 'the diamond of Azerbaijan'.[64] On November 9 video imagery was trending on social media with the Azerbaijani President proclaiming Shusha's liberation and victory in battle dress.[65]

Another statement by the Foreign Ministry of Azerbaijan later that day is significant for its contrasting tone: Azerbaijan sincerely apologised for shooting down a Russian Mi-24 helicopter over Armenian territory.[66] It was carefully explained as a 'tragic incident' and 'accident'. The act was 'not directed against the Russian side' and Azerbaijan offered apologies and 'sincere condolences', affirming its readiness 'to pay appropriate compensation'. It shifted focus from the Azerbaijani-Turkish military operation to Russia's interests and involvement. Only seven hours later, in the early morning of November 10, was the final ceasefire agreement declared in a joint statement by the Azerbaijani, Armenian and Russian leaders. In an address to the nation President Aliyev then emphasised decisiveness and completeness of the victory, a tone reflected in Azerbaijani social and state media as well.

> "I gave them an ultimatum. I said – you must leave, if you don't, I will go to the end. Until the end! On November 8, I said in the Alley of Martyrs that I would go to the end, and no force could stop me. We won this victory on the battlefield."[67]

Because the provided Turkish drones were crucial for the military success of Operation Iron Fist, Azerbaijan's victory was partly also a Turkish one. The international media showed pictures of the victory parade with the TB-2 Bayraktar drones and their pilot crews in Baku. Furthermore, President Tayyip Erdogan was keen to express how 'the joy of our Azerbaijani brothers and sisters, who have been liberating step by step their occupied cities and Karabakh' was also 'our joy'.[68]

He further emphasised the strength of the connection, referring to 'the ancient Turkish city of Shusha', and the two countries acting internationally as 'two states, one nation'.[69] Implicitly this referenced the broader idea of the 'Great Turan', emphasising the unity of all Turks. Against the broader background of Turkey's regional position and ambition to increase its influence, also as a protagonist against 'islamophobia' in the world, the narrative on the Nagorno-Karabakh war emphasises how Turkey matters and takes responsibility. Foregrounding this in a phone call with Russian President Putin, Erdogan stated how 'Turkey and Russia have together contributed a lot' towards a lasting solution in Nagorno-Karabakh.[70] In addition he also suggested to further pursue cooperation in Syria. The creation of the Russian-Turkish Joint Centre in Azerbaijan coordinating the peacekeeping force signals an expansion of influence for both countries in the region.

Regarding Nagorno-Karabakh, Turkish news and social media reproduce the Turkish official narrative to a large extent. Of significance is a spike of Turkish language Tweets between November 8-10, around the date of the ceasefire. This is a clear difference from English, Armenian, Azerbaijani and Russian Tweets at the time. Most of these Turkish Tweets present a congratulatory tone in response to Azerbaijan's victory.[71] However, due to the tweet-retweet ratio and a six fold quantity above an average level of tweets, Kirdemir concludes (also in general) this included inauthentic activity, 'coordinated with highly active political accounts, troll accounts, and automated bot accounts'.[72] Of course, attribution of such activity is difficult. In any case, the celebratory tone and idea of a conclusive victory indicate an end-state being reached. An emphasis aligning also with Russian interests: to halt all military activities. Furthermore, another conclusion of the study was that in Russia-related Turkish-speaking channels, central themes were 'Russian-Turkish cooperation, anti-NATO sentiments, and threat perception towards the West'.[73]

This brings us to the Russian narrative, perhaps best described as consisting of a number of subvariants, tailored to various national and international audiences. Statements, speeches and reports on telephone conversations by President Putin published during the Nagorno-Karabakh war show a consistent effort to halt military action. Russia expressed continued support for Armenia throughout the conflict. Meanwhile, Russian and Armenian news and social media also enhance the notion of 'pan-Turkism' or 'neo-Ottomanism' as a threat and Russia as the sole protector for the country.[74]

In a series of consultations Putin mobilised support among OSCE Minsk group co-chairs and the UN Security Council to express grave concerns, 'condemn in the strongest terms the recent escalation of violence', deploring the loss of human life.[75] Fears were articulated that foreign fighters from Syria and Libya were joining the war against Armenia. Azerbaijan and Armenia were called upon to resume negotiations under the auspices of OSCE Minsk Group co-chairs Russia, France and the United States (thus not including Turkey as member). In an interview with Russian state media Putin emphasised military operations were 'not taking place in the Armenian territory', thus not activating Russian CSTO commitments.[76] On October 10 consultations between the Russian president and his Armenian and Azerbaijani counterparts resulted in a humanitarian ceasefire to allow for the exchange of bodies and prisoners. An initiative later repeated by France and the US. Overall, Putin's central concern was 'stopping the bloodshed'.[77] At this time, Putin expressed 'hope' towards Erdogan that Turkey would 'make a constructive contribution to de-escalating the conflict', emphasising their cooperation on Syrian and Libyan affairs.[78] The use of the term 'hope' signals Russia's limits to influence developments. However, in Russian news sources a more critical and aggressive tone also circulates. Turkey's influence in the Caucasus is a threat to Russia and it should answer for its support to Azerbaijan.[79]

Although less explicit, the shooting down of the Russian Mi-24 helicopter on November 9 does mark a change in Russian subnarratives as well. Although Russia formally accepted the Azerbaijani apology, the CSTO expressed deep concern over the incident. According to some, the symbolic value of the incident was used to Russia's advantage behind the scenes: Putin gave Azerbaijan an ultimatum that military operations must end after the recapturing of Shusha, or the Russian military would intervene.[80] Moreover, as mentioned, a peacekeeping force of 2,000 Russian soldiers was to be deployed on Azerbaijani held territory to monitor the cessation of hostilities and provide security for internally displaced persons and refugees to return home. Among ethnic Armenians in Nagorno-Karabakh the story went that Russia would also hand out passports.[81]

In the months after the November ceasefire agreement Russia was keen to emphasise its leading role to the world. Before international fora that include American and European countries such as the UN Security Council and the OSCE Minsk group, more cooperative language is used. While with summits of Commonwealth of Independent States leaders and BRICS countries, or the Shanghai Cooperation Organisation, a greater emphasis laid with Russia as sole initiator and facilitator of the negotiations. Regular Ministry of Defence press releases on the Russian peacekeeping force emphasised the providing of security, demining activities and assistance with the return of refugees.[82] The Russian state media generally reproduced the official statements and reports. After turning the kaleidoscope several times, a number of developments were identified regarding the Armenian, Azerbaijani and Turkish perspectives that fit the Russian agenda.

14.4. Russian efforts – a reflection

In early 2015, the Chief of the Main Directorate Operations of the Russian General Staff, Colonel-General[83] Andrey Kartapolov held a presentation at the Russian Academy of Military Science that accurately reflected the key elements of New Type Warfare (NTW), the Russian approach to hybrid warfare. In his speech, Kartapolov stated that achieving goals in NTW is a combination of non-military means and the employment of armed forces. He distinguished several steps in NTW: it starts with indirect and non-kinetic methods such as intensifying diplomatic pressure and information operations, and if necessary, it can also shift to classical methods of waging war, using various types of weapons in combination with large-scale information effects.[84]

This reflection shows how important information operations have become for Russian authorities during conflicts, whether it concerns the dissemination of disinformation, the manipulations of perceptions or the deployment of forces under the guise of conducting a peacekeeping operation. Most Russian authorities speak of information confrontation instead of information warfare, which includes a wide range of activities.[85] In their doctrine, the Russian government makes a distinction between information-psychological and information-technical approaches.[86] Cyber operations can be seen as the technical representation of information; it is about how information can be disseminated alongside other types of information media such as print, broadcast and social media, and, of course, diplomacy.[87] Russian information-psychological operations have been around for a very long time and were widely used during the Cold War. The main aim is to deceive the opponent or others, although this is often difficult to achieve, as one then needs to make an extensive analysis of the opponent to find out among other things, what the opponent's personal preferences, values and attitudes are. The Russian authorities therefore prefer to confuse an opponent regarding their true intentions rather than to deliberately mislead him.[88]

In practice, Russian authorities create confusion by very one-sided information through their own media.[89] In most cases the Russian state-owned mainstream media disseminate the view of the Kremlin, which is taken over by social media causing a form of social amplification regarding public opinion. Moreover, the Russian authorities, especially President Putin, use so-called doublespeak to create ambiguity: saying A but doing or meaning B. It is actually a simple method of deception. Putin knows exactly how to distort and mirror the course of events, such as dubious activities carried out by Russian security services or agencies themselves in a conflict and subsequently condemning and attributing these activities to the opponent. In addition, he knows like no other how to frame activities favourably for himself. For instance, he labels acts of war as 'peacekeeping duties'. This way of deception calls up strong associations with the activities of the 'Ministry of Truth' in George Orwell's dystopian novel '1984'. In his novel, Orwell wanted to make clear that the use of language strongly influences people's thinking, which in turn strongly influences politics. Orwell recognised that people who make heavy use of doublespeak may also suffer from doublethink, a form of schizophrenia in which a fractured human mind simultaneously accepts two contradictory beliefs as the truth, with all the consequences this entails.[90] During the 2020 Nagorno-Karabakh war, Putin used doublespeak not only to deploy troops in the region to exert influence there under the guise of peacekeeping forces to observe the truce. In addition, another example during the 2020 Nagorno-Karabakh is that, on the one hand, he expressed Russian support for Armenia and warned, together with the Russian media, against pan-Turkism and 'neo-Ottomanism', but on the other he flattered Turkish President Erdogan by saying that he hoped that Turkey will contribute constructively to the de-escalation of the 2020 Nagorno-Karabakh. Putin was indulging in all sorts of language manipulations to keep Erdogan on his toes, but also to retain influence in the region. All in all, these contradictory statements have led to a confusingly complex situation that is hence best understood from multiple perspectives. Although it is difficult to technically attribute the creation of artificial accounts and the subsequent messaging and amplification of Turkish, Azerbaijani and Armenian narratives to Russia, these fabricated or inauthentic activities on social media provide a natural fit with Russian interests.

14.5. Postscript

The current war in Ukraine has added a new chapter to Russian NTW. The Russian narrative on Ukraine includes elements of a 2014 'coup d'etat' and a 'special military operation' to 'fight Nazism' and protect Ukrainian brethren.[91] More recently other themes have become more prominent, such as Western 'neo-colonialism' that needs to be stopped by a 'preventive war' to protect the unique Russian culture and geopolitical space ('*Russkii mir*'), and reestablishment of 'New Russia' after referenda in captured Ukrainian regions.[92] However, Russia also seems to have met an enemy that punches above its military weight, facilitated by the creation of strong narratives and Western political and military support. The parallel fronts of international diplomacy and information operations have allowed Ukrainian president Zelensky to perform at his best.[93] His rhetorical skills as a former actor and the availability of social media allowed for effective and speedy communication from the first day of the war. It added to a broader Ukrainian narrative that exploited key events to strengthen morale and influence the physical battlespace – such as the sinking of Russian flagship Moskva or resistance at the Azovstal factory in Mariupol.[94] Determining specific effects of narratives remains complex, also as some narratives – both true or fake – are not designed top-down but emerge as frames that resonate among audiences.[95]

New studies on Ukrainian and Russian interacting narratives are highly relevant. However, complementing such typical with atypical case studies provides additional insights on Russian information operations, positions and interests. Renewed fighting in September 2022 among Armenia and Azerbaijan over the Nagorno-Karabakh region triggered European, Russian and American diplomatic initiatives. Early October, Armenian Prime Minister Pashinyan and Azerbaijani President Aliyev met with French president Macron and president of the European Council Michel.[96] Later a trilateral summit was hosted in Sochi where they met President Putin.[97] At another meeting on November 8 in Washington, US Secretary of State Blinken hosted their foreign ministers.[98] As more visits and summits follow, a historic pattern continues that expresses commitment of OSCE Minsk Group chairs to re-establish regional stability and 'stop bloodshed'. Meanwhile, Turkish President Erdogan underlined his support for Azerbaijan with a visit on October 21, physically embracing Azerbaijani President Aliyev and together opening a new airport on captured Nagorno-Karabakh territory.[99] While the war in Ukraine has changed perceptions of Russia in the (Western) world, narratives on Nagorno-Karabakh might also offer a different perspective on US-EU-Turkey-Russia relations, and how perceptions are shaped to exert influence. Linking narratives can nuance divides or the idea of all out polarisation as interests or positions align. In general, more emphasis on the emergence and impact of parallel narratives provides a better understanding of all dimensions of conflict and warfare.

Notes

[1] The authors want to thank cartographer Erik van Oosten of the Netherlands Institute of Military History (NIMH) for drawing and contributing the three maps of the Nagorno-Karabakh region inserted in this chapter.

[2] Russian authorities do not use the term hybrid warfare (in Russian 'gibridnaya voyna', but rather prefer the term New Type Warfare (NTW) instead. Timothy Thomas, "The evolving nature of Russia's way of war," *Military Review*, July-August (2017): 34-42.

[3] For example: Paul Ducheine and Frans Osinga eds., *Winning Without Killing: The Strategic and Operational Utility of Non-Kinetic Capabilities in Crises*, NL ARMS (The Hague: Asser Press, 2017).

[4] Timothy Thomas, "Russia's 21st century information war: working to undermine and destabilize populations," *Defence Strategic Communications* 1, no. 1 (2015): 12; T. S. Allen and A. J. Moore, "Victory without casualties: Russia's information operations", *Parameters* 48. no. 1 (2018): 60.

[5] Sebastiaan Rietjens, "Unravelling disinformation: The case of Malaysia Airlines Flight MH17," *The International Journal of Intelligence, Security, and Public Affairs* 21, no. 3 (2019): 195-218.

[6]   Han Bouwmeester, *Krym Nash: An Analysis of Modern Russian Deception Warfare*, PhD dissertation. Utrecht: Utrecht University, 2020, 403.

[7]   Svante Cornell, *Small Nations and Great Powers: A Study of Ethnopolitical Conflict in the Caucasus* (Richmond, England: Curzon Press, 2001), 56, 57; Cory Welt and Andrew S. Bowen, *Azerbaijan and Armenia: The Nagorno-Karabakh Conflict*, Congressional Research Service R46651, Version 1 New, January 7, 2021, 2. Preceded by a brief period, between February and May 1918, when Armenia, Azerbaijan and Georgia formed the Transcaucasian Federation. James J. Coyle, *Russia's Interventions in Ethnic Conflicts: The Case of Armenia and Azerbaijan* (Cham: Palgrave Macmillan, 2021), 6-8.

[8]   In this period one million Armenians were killed during death marches into the Syrian desert. Armenian women and children were forcibly converted to Islam. Thomas de Waal, *Great Catastrophe. Armenians and Turks in the Shadow of Genocide* (Oxford: Oxford University Press, 2015); Cornell, *Small Nations*, 58, 59; Masha Udensiva-Brenner, "Black Garden. Armenia and Azerbaijan Through Peace and War. An Interview with Thomas de Waal," *Harriman Magazine* (Summer 2018): 30-43, 34, http://www.columbia.edu/cu/creative/epub/harriman/2018/summer/ Black_Garden_Armenia_and_Azerbaijan_Through_Peace_and_War.pdf, consulted June 16, 2022.

[9]   Several groups in the Armenian and Azerbaijani societies armed themselves. Thomas de Waal, *Black Garden. Armenia and Azerbaijan Through Peace and War* (New York: New York University Press, 2013); Cornell, *Small Nations and Great Powers*, 58-63. Udensiva-Brenner, "Black Garden," 33-34. Welt and Bowen, *Azerbaijan and Armenia*, 2.

[10] Ali Askerov, "The Nagorno-Karabakh Conflict. The Beginning of the Soviet End," in *Post-Soviet Conflicts: The Thirty Years' Crisis*, ed. Ali Askerov, Stefan Brooks, Lasha Tchantouridzé (Lanham: Rowman & Littlefield, 2020), 55-80, 57, 58; Coyle, *Russia's Interventions*, 10, 11; Udensiva-Brenner, "Black Garden," 35; Welt and Bowen, *Azerbaijan and Armenia*, 2.

[11] Coyle, *Russia's Interventions*, 13-15; Udensiva-Brenner, "Black Garden," 35, 36.

[12] Coyle, *Russia's Interventions*, 15-30.

[13] The referendum was organised parallel to the referenda in Armenia and Azerbaijan. Askerov, "Nagorno-Karabakh Conflict," 58-61; Coyle, *Russia's Interventions*, 30-32.

[14] Alec Rasizade, "Azerbaijan's prospects in Nagorno-Karabakh," *Mediterranean Quarterly* 22, no. 3 (Summer 2011): 72-94, 77-83, https://doi.org/10.1215/10474552-1384882; Cornell, *Small Nations*, 81-95; Udensiva-Brenner, "Black Garden," 38.

[15] Coyle, *Russia's Interventions*, 2; Welt and Bowen, *Azerbaijan and Armenia*, 3, 4.

[16] De Waal, *Black Garden*, 318; Rasizade. "Azerbaijan's Prospects," 83-89; Askerov, "Nagorno-Karabakh Conflict," 61-67.

[17] Coyle, *Russia's Interventions*, 132-151; Cornell, *Small Nations*, 112-125; Welt and Bowen, *Azerbaijan and Armenia*, 3, 4; Philip Remler, *The OSCE as Sisyphus: Mediation, Peace Operations, Human Rights*, IAI Papers 21/16. Instituto Affari Internazionali -IAI- 2021, 14, 15, https://www.iai.it/en/pubblicazioni-tag/ all/363 [consulted June 16, 2022].

[18] De Waal, *Black Garden*, 284; Udensiva-Brenner, "Black Garden," 37, 40.

[19] The so-called Ganja gap. Turan Gafarlı, *Operation Iron Fist: Lessons from the Second Karabakh War*, Discussion Paper, TRT World Research Centre, February 2021, 6, 7, https://researchcentre.trtworld. com/wp-content/uploads/2021/02/Operation-Iron-Fist-1.pdf [consulted June 16, 2022].

[20] Ibidem, 6, 7.

[21] De Waal, *Black Garden*, 307, 308.

[22] Coyle, *Russia's Interventions*, 15; Udensiva-Brenner, "Black Garden," 34.

[23] De Waal, *Black Garden*, 310-316; Udensiva-Brenner, "Black Garden," 37, 38.

[24] De Waal, *Black Garden*, 305-307; Coyle, *Russia's Interventions*, 132-151; Askerov, "Nagorno-Karabakh Conflict," 76-79; Udensiva-Brenner, "Black Garden," 39, 40.

[25] De Waal, *Black Garden*, 305-306; Udensiva-Brenner, "Black Garden," 40, 41; Anna Maria Dyner, and Arkadiusz Legiec, *The Military Dimension of the Conflict over Nagorno-Karabakh*, PISM No. 241 (1671) 26 November 2020, Polski Instytut Spraw Miedzynarodowych-The Polish Institute of International Affairs, https://pism.pl/publications/The_Military_Dimension__of_the_Conflict_over_ NagornoKarabakh, consulted June 16, 2022.

[26] Askerov, "Nagorno-Karabakh Conflict," 75, 76; Udensiva-Brenner, "Black Garden," 39; Zhirayr Amirkhanyan, "A failure to innovate: the Second Nagorno-Karabakh War," *Parameters* 52, no.1 (Spring 2022): 119-134, 123, https://doi:10.55540/0031-1723.3133.

[27] Askerov, "Nagorno-Karabakh Conflict," 55, 56.

[28] Ibid., 73-75; Uzi Rubin, *The Second Nagorno-Karabakh War: A Milestone in Military Affairs*, Mideast Security and Policy Studies No. 184, The Begin-Sadat Center for Strategic Studies, Bar-Ilan University, December 2020, 10, https://besacenter.org/wp-content/uploads/2020/12/184web-no-ital. pdf, consulted June 16, 2022; Udensiva-Brenner, "Black Garden," 41; Welt and Bowen, *Azerbaijan and Armenia*, 7, 8; Amirkhanyan, "Failure to Innovate", 129-130; Brenda Shaffer, "The Trigger for War: Energy in the 2020 Armenia-Azerbaijan War," in *The Karabakh Gambit: Responsibility for the Future*, ed.Turan Gafarlı, and Michael Arnold (Istanbul: TRT World Research Centre 2021), 100-113.

[29] De Waal, *Black Garden*, 319-326.

[30] Shusha contains major monasteries and churches, one of them the cradle of the Armenian script. It was formerly the capital of the Azerbaijani realm under Turkish and Persian suzerainty in the eighteenth and nineteenth centuries. International Crisis Group, *Improving Prospects for Peace after the Nagorno-Karabakh War*, Crisis Group Europe Briefing No. 91 Baku/Yerevan/Tbilisi/ Moscow/Brussels, December 22, 2020, 9, 10, https://www.crisisgroup.org/europe-central-asia/ caucasus/nagorno-karabakh-conflict/b91-improving-prospects-peace-after-nagorno-karabakh-war, consulted June 16, 2022; Thomas de Waal in: "War in the Caucasus: Europe/Heavy Metal", *The Economist*, October 10, 2020, 31-32, 31; Miloslav Terzic, "Critical Review of the Protection of Aircraft Defense Forces during the Conflict in Nagorno-Karabakh in 2020," *Small Wars and Insurgencies* (Published online 12 January 2022): 3, 4, https://doi.org/10.1080/09592318.2021.2025286.

[31] Terzic, "Critical Review", 2-6.

[32] Amirkhanyan, "Failure to Innovate," 124; Terzic, "Critical Review", 2-6; Gustav Gressel, "Military Lessons from Nagorno-Karabakh: Reason for Europe to Worry," European Council on Foreign Relations, November 24, 2020, https://ecfr.eu/article/military-lessons-from-nagorno-karabakh-reason-for-europe-to-worry/, consulted June 16, 2022.

[33] Ibid.; Rubin, *A Milestone*, 1-16; Terzic, "Critical Review," 1-15; Amirkhanyan, "Failure to Innovate," 119-134; Joël Postma, "Drones over Nagorno-Karabakh. a glimpse at the future of war?" *Atlantisch Perspectief* 45, no. 2 (2021): 15-20, https://www.jstor.org/stable/48638213.

[34] Thomas De Waal in: "War in the Caucasus," *The Economist*, 32. However, De Waal does ignore the design and execution of Armenian preemptive strikes with more limited targets, such as the attack on the Tovuz power plant in July 2020; Shaffer, "Trigger", 100-113.

[35] Rubin, *A Milestone*, 8, 9. Postma, "Drones."

[36] The Israeli drones could fly independently with Artificial Intelligence, although during this campaign they were remotely controlled; Rubin. *A Milestone*, 1-16; Gressel, "Military Lessons"; Terzic, "Critical Review," 10.

[37] Rubin, *A Milestone*, 9, 10, 12-15. Terzic, "Critical Review," 7.

[38] Rubin, *A Milestone*, 12.

[39] Gressel, "Military Lessons." Terzic, "Critical Review," 1-15.

[40] Gressel, "Military Lessons."

[41] Ibid.

[42] Rubin, *A Milestone,* 10, 11.

[43] Welt and Bowen, *Azerbaijan and Armenia,* 8-11.

[44] It provided them with an overwatch position to conduct further operations in the city itself. Welt and Bowen, *Azerbaijan and Armenia,* 8-11; Gressel, "Military Lessons"; Michael Kofman, "Armenia-Azerbaijan War: Military Dimensions of the Conflict," *Russia Matters,* October 2, 2020. https://www.russiamatters.org/analysis/armenia-azerbaijan-war-military-dimensions-conflict, consulted June 16, 2022.

[45] Welt and Bowen, *Azerbaijan and Armenia,* 13.

[46] Ibid.,13-15.

[47] International Crisis Group, *Improving Prospects,* 3-6.

[48] International Crisis Group, *Improving Prospects,* 6-8; "War in the Caucasus," *The Economist,* 32; Postma, "Drones," 17.

[49] Arshaluys Barseghyan, Lusine Grigoryan, Anna Pambukchyan, Artur Papyan and Elen Aghekyan, "Disinformation and Misinformation in Armenia. Confronting the Power of False Narratives," Freedom House, June 2021, 18-20, https://freedomhouse.org/sites/default/files/2021-06/Disinformation-in-Armenia_En-v3.pdf, consulted June 16, 2022; Welt and Bowen, *Azerbaijan and Armenia,*16.

[50] Ibid., 6. For the initial reaction of the UN and the EU, see: International Crisis Group, *Improving Prospects,* 13-16.

[51] Galip Dalay, *Turkish-Russian Relations in Light of Recent Conflicts. Syria, Libya, and Nagorno-Karabakh,* SWP Research Paper 5/2021, 19 (Berlin: Stiftung Wissenschaft und Politik – SWP – Deutsches Institut für Internationale Politik und Sicherheit. https://doi.org/10.18449/2021RP05; Daria Ischachenko, *Turkey-Russia Partnership in the War over Nagorno-Karabakh: Militarised Peacebuilding with Implications for Conflict Transformation.* SWP Comment No. 53/2020, 1-4, Stiftung Wissenschaft und Politik -SWP- Deutsches Institut für Internationale Politik und Sicherheit, https://doi.org/10.18449/2020C53.

[52] The loss of the Armenian districts between Nagorno-Karabakh and Iran has shortened the border between both countries, that moreover has come under Russian and Turkish surveillance. Thus, the threat of Armenia's economic isolation has increased, that she tries to prevent by strengthening the alliance with Russia and reconnecting with Turkey. Brenda Shaffer, "The Armenia-Azerbaijan War: Downgrading Iran's Regional Role," *Central-Asia-Caucasus Analyst,* November 25, 2020, http://cacianalyst.org/publications/analytical-articles/item/13650-the-armenia-azerbaijan-war-downgrading-iran's-regional-role.html, consulted June 16, 2022; Dalay, *Turkish-Russian Relations,* 20-22; International Crisis Group, *Improving Prospects,* 11-12. See also: Thomas de Waal, "The South Caucasus, Can New Trade Routes help overcome a Geography of Conflict"; Carnegie Europe, November 2021, https://carnegieeurope.eu/publications/85729, consulted June 16, 2022.

[53] Anvar M. Mamadaliev, "Military propaganda at the second stage of the 2020 Nagorno-Karabakh War (October 12 – November 10): official materials of Armenia and Azerbaijan defense ministries", *Propaganda in the World and Local Conflicts* 8, no. 2 (2021): 99-114.

[54] ERMES III, "Event report media and disinformation in the Nagorno-Karabakh conflict," *College of Europe,* December 17, 2020, https://www.coleurope.eu/news/ermes-iii-event-report-media-and-disinformation-nagorno-karabakh-conflict, consulted May 12, 2022.

[55] ARMENPRES, "Armenia 'will not tolerate a second genocide,' President Armen Sarkissian tells Kathimerini," *Kathimerini daily,* 9 November 2020, https://armenpress.am/eng/news/1034410.html, consulted May 12, 2022.

56   Elise Thomas and Albert Zhang, "Snapshot of a shadow war: a preliminary analysis of Twitter activity linked to the Azerbaijan–Armenia conflict," *Australian Strategic Policy Institute*, October 1, 2020, https://www.jstor.org/stable/resrep26984, consulted May 12, 2022.

57   Barış Kırdemir, "Karabakh War in Online News and Social Media: Representation, Confrontation and Maneuvers of Information," Centre for Economics and Foreign Policy Studies EDAM, January 15, 2021.

58   Sabrina Abubakirova, "Aliyev and Pashinyan mark 'red lines' in interviews with RIA Novosti," *OC Media*, October 16, 2020, https://oc-media.org/aliyev-and-pashinyan-mark-red-lines-in-interviews-with-ria-novosti/, consulted May 12, 2022.

59   The Prime Minister of the Republic of Armenia, "Statement by the Prime Minister of the Republic of Armenia, the President of the Republic of Azerbaijan and the President of the Russian Federation," November 10, 2020, https://www.primeminister.am/en/press-release/item/2020/11/10/Announcement/, consulted May 12, 2022.

60   Kırdemir, "Karabakh War in Online News and Social Media," January 15, 2021, 23.

61   Abubakirova, "Aliyev and Pashinyan mark 'red lines'," October 16, 2020.

62   Robin Forestier-Walker, "Nagorno-Karabakh: Information war and competing media narratives," *Al Jazeera*, September 28, 2020, https://www.aljazeera.com/news/2020/9/28/nagorno-karabakh-information-war-and-competing-media-narratives, consulted May 12, 2022.

63   Abubakirova, "Aliyev and Pashinyan mark 'red lines'," October 16, 2020.

64   TRT World, "Nagorno-Karabakh fighting nears famous city of Shusha," November 1, 2020, https://www.trtworld.com/video/news-videos/nagorno-karabakh-fighting-nears-famous-city-of-shusha/5f9e683246e7130017c18f2b, consulted May 12, 2022.

65   Azərbaycan Respublikası Müdafiə Nazirliyi, "Video footage of the liberated from occupation Shusha city," *YouTube*, November 9, 2020, https://www.youtube.com/watch?v=So7e9mRblfY, consulted May 12, 2022.

66   APA, "Azerbaijan`s Foreign Ministry makes statement on the downing of Russian military helicopter," November 9, 2020", https://apa.az/en/xeber/foreign-politics/Azerbaijans-Foreign-Ministry-makes-statement-on-the-downing-of-Russian-military-helicopter-335118, consulted May 12, 2022.

67   President of Azerbaijan Republic, "Ilham Aliyev addressed the nation," November 10, 2020, https://president.az/en/articles/view/45924, consulted May 12, 2022.

68   Presidency of the Republic of Türkiye, "We share the joy of our Azerbaijani brothers and sisters who have been liberating step by step their occupied cities and Karabakh," November 8, 2020, https://www.tccb.gov.tr/en/news/542/122706/-we-share-the-joy-of-our-azerbaijani-brothers-and-sisters-who-have-been-liberating-step-by-step-their-occupied-cities-and-karabakh-, consulted May 12, 2022.

69   Presidency of the Republic of Türkiye, "President Erdoğan's Message on Azerbaijan's 'Flag Day'," November 9, 2020, https://www.tccb.gov.tr/en/speeches-statements/558/122725/president-erdogan-s-message-on-azerbaijan-s-flag-day-, consulted May 12, 2022.

70   Presidency of the Republic of Türkiye, "Phone Call with President Vladimir Putin of Russia," November 10, 2020, https://www.tccb.gov.tr/en/speeches-statements/558/122748/phone-call-with-president-vladimir-putin-of-russia, consulted May 12, 2022.

71   Kırdemir, "Karabakh War in Online News and Social Media," January 15, 2021, 9.

72   Ibid.

73   Ibid., 23.

74   Ibid., 5, 23.

75 President of Russia, "Statement of the presidents of Russia, the United States and France on Nagorno-Karabakh," October 1, 2020, http://en.kremlin.ru/events/president/news/64133, consulted May 12, 2022.

76 President of Russia, "Interview with Rossiya TV Channel," October 7, 2020, http://en.kremlin.ru/events/president/news/64171, consulted May 12, 2022.

77 President of Russia, "Replies to media questions on developments in Nagorno-Karabakh," November 17, 2020, http://en.kremlin.ru/events/president/news/64431, consulted May 12, 2022.

78 President of Russia, "Telephone conversation with President of Turkey Recep Tayyip Erdogan," October 14, 2020, http://en.kremlin.ru/events/president/news/64204, consulted May 12, 2022.

79 For example RIA Novosti, "They took the ring: Why is Turkey's influence in the Caucasus dangerous?," October 7, 2020, https://ria.ru/20201007/turkey-ccaucasus-1578482919.html, consulted May 12, 2022; Kırdemir, "Karabakh War in Online News and Social Media," January 15, 2021, 4-6.

80 Anton Troianovski and Carlotta Gall, "In Nagorno-Karabakh Peace Deal, Putin Applied a Deft New Touch," *New York Times*, December 1, 2020.

81 Ibid.

82 For example, Russian Ministry of Defence, "More than 4,800 explosive devices were neutralized by servicemen of the International Mine Action Centre of the Russian Ministry of Defence in Nagorno-Karabakh," December 14, 2020, https://eng.mil.ru/en/news_page/country/more.htm?id=12330060@egNews, consulted May 12, 2022.

83 Colonel-General is comparable with a Lieutenant General.

84 Thomas, "The Evolving Nature of Russia's Way of War."

85 Keir Giles, *Handbook of Russian Information Warfare: A Fellowship Monograph* (Rome: NATO Defence College, Research Division, 2016), 6.

86 Timothy Thomas, "Information security thinking: a comparison of US, Russian and Chinese concepts," *The Science and Cultural Series*, (August 2001): 344-358.

87 Keir Giles, *The Next Phase of Russian Information Warfare* (Riga: NATO Strategic Communications Centre of Excellence, 2016).

88 Herbert Romerstein, "Disinformation as a KGB weapon in the Cold War," *The Journal of Intelligence History* 1, no. 1 (2001): 54-67.

89 Tony Selhorst, "Russia's perception warfare: the development of Gerasimov's doctrine in Estonia and Georgia and it's application in Ukraine," *Militaire Spectator* 184, no. 4 (2016): 148-164.

90 Mark Satta, "Putin's Brazen Manipulation of Language is a Perfect Example of Orwellian Doublespeak," *Big Think Website*, March 19, 2022, https://bigthink.com/the-present/putin-orwellian-doublespeak/, consulted April 23, 2022.

91 Ronald Suny, "The Ukraine conflict is a war of narratives – and Putin's is crumbling," *The Conversation*, October 27, 2022, https://theconversation.com/the-ukraine-conflict-is-a-war-of-narratives-and-putins-is-crumbling-192811, consulted November 17, 2022.

92 Ibid.

93 Ajit Maan and Paul Cobough, "How Ukrainian Narrative Identity Dominates the War of Influence," *Homeland Security Today*, August 25, 2022, https://www.hstoday.us/featured/how-ukrainian-narrative-identity-dominates-the-war-of-influence/, consulted November 17, 2022.

94 Sofia Romansky, Lotje Boswinkel and Michel Rademaker, "The parallel front: An analysis of the military use of information in the first seven months of the war in Ukraine," The Hague: The Hague Center for Strategic Studies, October 2022.

95 Ibid.

96 Euronews, "Azerbaijan and Armenia leaders meet at European summit," https://www.euronews.com/2022/10/07/azerbaijan-and-armenia-leaders-meet-at-european-summit, consulted November 17, 2022.

97 Kremlin, "Trilateral talks with President of Azerbaijan and Prime Minister of Armenia," October 31, 2022, http://en.kremlin.ru/events/president/news/69729, consulted November 17, 2022.

98 Reuters, "Blinken hosts Armenia and Azerbaijan ministers, praises 'courageous steps'," November 7, 2022, https://www.reuters.com/world/asia-pacific/blinken-says-armenia-azerbaijan-making-courageous-steps-toward-peace-2022-11-07/, consulted November 17, 2022.

99 Euronews, "Heads of Turkey and Azerbaijan meet in territory captured during Nagorno-Karabakh war," October 21, 2022, https://www.euronews.com/2022/10/20/heads-of-turkey-and-azerbaijan-meet-in-territory-captured-during-nagorno-karabakh-war, consulted November 17, 2022.

References

Abubakirova, Sabrina. "Aliyev and Pashinyan mark 'red lines' in interviews with RIA Novosti", *OC Media*, October 16, 2020. https://oc-media.org/aliyev-and-pashinyan-mark-red-lines-in-interviews-with-ria-novosti/. Consulted May 12, 2022.

Allen, T.S. and A. J. Moore. "Victory without casualties: Russia's information operations." *Parameters* 48. no. 1 (2018): 59-71.

Amirkhanyan, Zhirayr. "A failure to innovate: the Second Nagorno-Karabakh War." *Parameters* 52, no.1 (Spring 2022): 119-134. https://doi:10.55540/0031-1723.3133.

APA. "Azerbaijan's Foreign Ministry makes statement on the downing of Russian military helicopter." November 9, 2020." https://apa.az/en/xeber/foreign-politics/Azerbaijans-Foreign-Ministry-makes-statement-on-the-downing-of-Russian-military-helicopter-335118. Consulted May 12, 2022.

ARMENPRES. "Armenia 'will not tolerate a second genocide,' President Armen Sarkissian tells Kathimerini." *Kathimerini daily*, 9 November 2020. https://armenpress.am/eng/news/1034410.html. Consulted May 12, 2022.

Askerov, Ali. "The Nagorno-Karabakh Conflict. The Beginning of the Soviet End." In *Post-Soviet Conflicts: The Thirty Years' Crisis*, edited by Ali Askerov, Stefan Brooks, Lasha Tchantouridzé, 55-80. Lanham: Rowman & Littlefield, 2020.

Azərbaycan Respublikası Müdafiə Nazirliyi. "Video footage of the liberated from occupation Shusha city." *YouTube*, November 9, 2020. https://www.youtube.com/watch?v=So7e9mRblfY. Consulted May 12, 2022.

Barseghyan, Arshaluys, Lusine Grigoryan, Anna Pambukhyan, Artur Papyan and Elen Aghekyan. "Disinformation and Misinformation in Armenia. Confronting the Power of False Narratives", Freedom House, June 2021. https://freedomhouse.org/sites/default/files/2021-06/Disinformation-in-Armenia_En-v3.pdf. Consulted June 16, 2022.

Bouwmeester, Han. *Krym Nash: An Analysis of Modern Russian Deception Warfare*. PhD dissertation. Utrecht: Utrecht University, 2020.

Cornell, Svante. *Small Nations and Great Powers: A Study of Ethnopolitical Conflict in the Caucasus.* Richmond, England: Curzon Press, 2001.

Coyle, James J. *Russia's Interventions in Ethnic Conflicts: The Case of Armenia and Azerbaijan.* Cham: Palgrave Macmillan, 2021.

Dalay, Galip. *Turkish-Russian Relations in Light of Recent Conflicts. Syria, Libya, and Nagorno-Karabakh.* SWP Research Paper 5/2021. Berlin: Stiftung Wissenschaft und Politik – SWP – Deutsches Institut für Internationale Politik und Sicherheit. https://doi.org/10.18449/2021RP05.

De Waal, Thomas. "The South Caucasus, Can New Trade Routes help overcome a Geography of Conflict." Carnegie Europe, November 2021. https://carnegieeurope.eu/publications/85729. Consulted June 16, 2022.

De Waal, Thomas. *Black Garden: Armenia and Azerbaijan Through Peace and War.* New York: New York University Press, 2013.

De Waal, Thomas. *Great Catastrophe: Armenians and Turks in the Shadow of Genocide.* Oxford: Oxford University Press, 2015.

Ducheine, Paul and Frans Osinga (eds). *Winning Without Killing: The Strategic and Operational Utility of Non-Kinetic Capabilities in Crises.* NL ARMS. The Hague: Asser Press, 2017.

Dyner, Anna Maria, and Arkadiusz Legiec. *The Military Dimension of the Conflict over Nagorno-Karabakh.* PISM No. 241 (1671) 26 November 2020. Polski Instytut Spraw Miedzynarodowych-The Polish Institute of International Affairs. https://pism.pl/publications/The_Military_Dimension_of_the_Conflict_over_NagornoKarabakh. Consulted June 16, 2022.

ERMES III. "Event report media and disinformation in the Nagorno-Karabakh conflict." *College of Europe.* December 17, 2020. https://www.coleurope.eu/news/ermes-iii-event-report-media-and-disinformation-nagorno-karabakh-conflict. Consulted May 12, 2022.

Euronews. "Azerbaijan and Armenia leaders meet at European summit." https://www.euronews.com/2022/10/07/azerbaijan-and-armenia-leaders-meet-at-european-summit. Consulted November 17, 2022.

Euronews. "Heads of Turkey and Azerbaijan meet in territory captured during Nagorno-Karabakh war." October 21, 2022. https://www.euronews.com/2022/10/20/heads-of-turkey-and-azerbaijan-meet-in-territory-captured-during-nagorno-karabakh-war. Consulted November 17, 2022.

Forestier-Walker, Robin. "Nagorno-Karabakh: Information war and competing media narratives." *Al Jazeera.* September 28, 2020. https://www.aljazeera.com/news/2020/9/28/nagorno-karabakh-information-war-and-competing-media-narratives. Consulted May 12, 2022.

Gafarlı, Turan. *Operation Iron Fist: Lessons from the Second Karabakh War.* Discussion Paper. TRT World Research Centre, February 2021. https://researchcentre.trtworld.com/wp-content/uploads/2021/02/Operation-Iron-Fist-1.pdf. Consulted June 16, 2022.

Giles, Keir. *Handbook of Russian Information Warfare: A Fellowship Monograph.* Rome: NATO Defence College, Research Division, 2016.

Giles, Keir. *The Next Phase of Russian Information Warfare.* Riga: NATO Strategic Communications Centre of Excellence, 2016.

Gressel, Gustav. "Military Lessons from Nagorno-Karabakh: Reason for Europe to Worry." November 24, 2020. European Council on Foreign Relations. https://ecfr.eu/article/military-lessons-from-nagorno-karabakh-reason-for-europe-to-worry/. Consulted June 16, 2022.

International Crisis Group. *Improving Prospects for Peace after the Nagorno-Karabakh War.* Crisis Group Europe Briefing No. 91 Baku/Yerevan/Tbilisi/Moscow/Brussels, December 22, 2020. https://www.crisisgroup.org/europe-central-asia/caucasus/nagorno-karabakh-conflict/b91-improving-prospects-peace-after-nagorno-karabakh-war. Consulted June 16, 2022.

Ischachenko, Daria. *Turkey-Russia Partnership in the War over Nagorno-Karabakh: Militarised Peacebuilding with Implications for Conflict Transformation.* SWP Comment No. 53/2020. Stiftung Wissenschaft und Politik -SWP- Deutsches Institut für Internationale Politik und Sicherheit. https://doi.org/10.18449/2020C53.

Kırdemir, Barış. "Karabakh War in Online News and Social Media: Representation, Confrontation and Maneuvers of Information." Centre for Economics and Foreign Policy Studies EDAM. January 15, 2021.

Kofman, Michael. "Armenia-Azerbaijan War: Military Dimensions of the Conflict." *Russia Matters,* October 2, 2020. https://www.russiamatters.org/analysis/armenia-azerbaijan-war-military-dimensions-conflict. Consulted June 16, 2022.

Kremlin. "Trilateral talks with President of Azerbaijan and Prime Minister of Armenia." October 31, 2022. http://en.kremlin.ru/events/president/news/69729. Consulted November 17, 2022.

Mamadaliev, Anvar M. "Military propaganda at the second stage of the 2020 Nagorno-Karabakh War (October 12 – November 10): official materials of Armenia and Azerbaijan defense ministries." *Propaganda in the World and Local Conflicts* 8, no. 2 (2021): 99-114.

Maan, Ajit, and Paul Cobough. "How Ukrainian Narrative Identity Dominates the War of Influence." *Homeland Security Today,* August 25, 2022. https://www.hstoday.us/featured/how-ukrainian-narrative-identity-dominates-the-war-of-influence/. Consulted November 17, 2022

Postma, Joël. "Drones over Nagorno-Karabakh: a glimpse at the future of war?" *Atlantisch Perspectief* 45, no. 2 (2021): 15-20. https://www.jstor.org/stable/48638213.

Presidency of the Republic of Türkiye. "We share the joy of our Azerbaijani brothers and sisters who have been liberating step by step their occupied cities and Karabakh." November 8, 2020. https://www.tccb.gov.tr/en/news/542/122706/-we-share-the-joy-of-our-azerbaijani-brothers-and-sisters-who-have-been-liberating-step-by-step-their-occupied-cities-and-karabakh-. Consulted May 12, 2022.

Presidency of the Republic of Türkiye. "Phone Call with President Vladimir Putin of Russia." November 10, 2020. https://www.tccb.gov.tr/en/speeches-statements/558/122748/phone-call-with-president-vladimir-putin-of-russia. Consulted May 12, 2022.

Presidency of the Republic of Türkiye. "President Erdoğan's Message on Azerbaijan's 'Flag Day'." November 9, 2020. https://www.tccb.gov.tr/en/speeches-statements/558/122725/president-erdogan-s-message-on-azerbaijan-s-flag-day-. Consulted May 12, 2022.

President of Russia. "Interview with Rossiya TV Channel." October 7, 2020. http://en.kremlin.ru/events/president/news/64171. Consulted May 12, 2022.

President of Russia. "Replies to media questions on developments in Nagorno-Karabakh." November 17, 2020. http://en.kremlin.ru/events/president/news/64431. Consulted May 12, 2022.

President of Russia. "Statement of the presidents of Russia, the United States and France on Nagorno-Karabakh." October 1, 2020. http://en.kremlin.ru/events/president/news/64133. Consulted May 12, 2022.

President of Russia. "Telephone conversation with President of Turkey Recep Tayyip Erdogan." October 14, 2020. http://en.kremlin.ru/events/president/news/64204. Consulted May 12, 2022.

Rasizade, Alec. "Azerbaijan's prospects in Nagorno-Karabakh." *Mediterranean Quarterly* 22, no. 3 (Summer 2011): 72-94. https://doi.org/10.1215/10474552-1384882.

Remler, Philip. *The OSCE as Sisyphus: Mediation, Peace Operations, Human Rights.* IAI Papers 21/16. Instituto Affari Internazionali -IAI- 2021. https://www.iai.it/en/pubblicazioni-tag/all/363 [consulted June 16, 2022].

Reuters. "Blinken hosts Armenia and Azerbaijan ministers, praises 'courageous steps'." November 7, 2022. https://www.reuters.com/world/asia-pacific/blinken-says-armenia-azerbaijan-making-courageous-steps-toward-peace-2022-11-07/. Consulted November 17, 2022.

RIA Novosti. "They took the ring. Why is Turkey's influence in the Caucasus dangerous?" October 7, 2020. https://ria.ru/20201007/turkey-ccaucasus-1578482919.html. Consulted May 12, 2022.

Rietjens, Sebastiaan. "Unravelling disinformation: the case of Malaysia Airlines Flight MH17." *The International Journal of Intelligence, Security, and Public Affairs* 21, no. 3 (2019): 195-218.

Romansky, Sofia, Lotje Boswinkel and Michel Rademaker. "The parallel front: an analysis of the military use of information in the first seven months of the war in Ukraine." The Hague: The Hague Center for Strategic Studies, October 2022.

Romerstein, Herbert. "Disinformation as a KGB weapon in the Cold War." *The Journal of Intelligence History* 1, no. 1 (2001): 54-67.

Rubin, Uzi. *The Second Nagorno-Karabakh War: A Milestone in Military Affairs.* Mideast Security and Policy Studies No. 184. The Begin-Sadat Center for Strategic Studies, Bar-Ilan University, December 2020. https://besacenter.org/wp-content/uploads/2020/12/184web-no-ital.pdf. Consulted June 16, 2022.

Russian Ministry of Defence. "More than 4,800 explosive devices were neutralized by servicemen of the International Mine Action Centre of the Russian Ministry of Defence in Nagorno-Karabakh." December 14, 2020. https://eng.mil.ru/en/news_page/country/more.htm?id=12330060@egNews. Consulted May 12, 2022.

Satta, Mark. "Putin's Brazen Manipulation of Language is a Perfect Example of Orwellian Doublespeak." *Big Think Website*, March 19, 2022. https://bigthink.com/the-present/putin-orwellian-doublespeak/. Consulted April 23, 2022.

Selhorst, Tony. "Russia's perception warfare: the development of Gerasimov's doctrine in Estonia and Georgia and it's application in Ukraine." *Militaire Spectator* 184, no. 4 (2016): 148-164.

Shaffer, Brenda. "The Armenia-Azerbaijan War: Downgrading Iran's Regional Role." *Central-Asia-Caucasus Analyst*, November 25, 2020. http://cacianalyst.org/publications/analytical-articles/

item/13650-the-armenia-azerbaijan-war-downgrading-iran's-regional-role.html. Consulted June 16, 2022.

Shaffer, Brenda. "The Trigger for War: Energy in the 2020 Armenia-Azerbaijan War," in *The Karabakh Gambit: Responsibility for the Future,* edited by Gafarlı, Turan, and Michael Arnold. Istanbul: TRT World Research Centre 2021.

Suny, Ronald. "The Ukraine conflict is a war of narratives – and Putin's is crumbling." The Conversation, October 27, 2022. https://theconversation.com/the-ukraine-conflict-is-a-war-of-narratives-and-putins-is-crumbling-192811. Consulted November 17, 2022.

Terzic, Miloslav. "Critical Review of the Protection of Aircraft Defense Forces during the Conflict in Nagorno-Karabakh in 2020." *Small Wars and Insurgencies* (Published online 12 January 2022): 1-15. https://doi.org/10.1080/09592318.2021.2025286.

The Economist. "War in the Caucasus: Europe/Heavy Metal." October 10, 2020, 31-32.

The President of Azerbaijan Republic. "Ilham Aliyev addressed the nation." November 10, 2020. https://president.az/en/articles/view/45924. Consulted May 12, 2022.

The Prime Minister of the Republic of Armenia. "Statement by the Prime Minister of the Republic of Armenia, the President of the Republic of Azerbaijan and the President of the Russian Federation." November 10, 2020. https://www.primeminister.am/en/press-release/item/2020/11/10/Announcement/. Consulted May 12, 2022.

Thomas, Elise, and Albert Zhang. "Snapshot of a shadow war: a preliminary analysis of Twitter activity linked to the Azerbaijan–Armenia conflict." *Australian Strategic Policy Institute.* October 1, 2020. https://www.jstor.org/stable/resrep26984. Consulted May 12, 2022].

Thomas, Timothy. "Information security thinking: a comparison of US, Russian and Chinese Concepts." *The Science and Cultural Series* (August 2001): 344-358.

Thomas, Timothy. "Russia's 21st century Information War: working to undermine and destabilize populations." *Defence Strategic Communications* 1, no. 1 (2015): 10-25.

Thomas, Timothy. "The evolving nature of Russia's way of war." *Military Review.* July-August (2017): 34-42.

Troianovski, Anton, Carlotta Gall. "In Nagorno-Karabakh Peace Deal, Putin Applied a Deft New Touch." *New York Times.* December 1, 2020.

TRT World. "Nagorno-Karabakh fighting nears famous city of Shusha." November 1, 2020. https://www.trtworld.com/video/news-videos/nagorno-karabakh-fighting-nears-famous-city-of-shusha/5f9e683246e7130017c18f2b. Consulted May 12, 2022.

Udensiva-Brenner, Masha. "Black Garden. Armenia and Azerbaijan Through Peace and War. An Interview with Thomas de Waal." *Harriman Magazine* (Summer 2018): 30-43. http://www.columbia.edu/cu/creative/epub/harriman/2018/summer/Black_Garden_Armenia_and_Azerbaijan_Through_Peace_and_War.pdf. Consulted June 16, 2022.

Welt, Cory and Andrew S. Bowen. *Azerbaijan and Armenia: The Nagorno-Karabakh Conflict.* Congressional Research Service R46651, Version 1 New, January 7, 2021. https://crsreports.congress.gov/product/pdf/R/R46651. Consulted June 16, 2022.

Epilogue

CHAPTER 15

War by numbers

A "technocratic hubristic fable"?[1]

Henk de Jong & Floribert Baudet

Abstract

This chapter discusses the idea that only quantitative data can help manage a war, explain its characteristics and dynamics, and predict its course. To this end we elaborate on the case of Robert McNamara, since his 'technocratic' statistical approach has guided the US war effort during the Vietnam War. In spite of the fact the US lost that war, the underlying idea has had a lasting influence that can be traced both to the conduct of contemporary wars, but also to the analysis of past wars. In the second section of the chapter we show that such approaches often disregard the complexity and imponderabilia of unique historical wartime contexts. We argue that for this reason the integration of quantitative-generalising and qualitative-historicising approaches is critical, whether aimed at understanding contemporary warfare or warfare in the past.

Keywords: Operations research; Qualitative approach; Qualitative approach; Data analysis; the McNamara fallacy

15.1. Introduction

In our data-obsessed cyber-age, it is easy to forget that the collection and use of data is hardly a new development. Since time immemorial, communities have counted livestock and bushels of grain, which were an indication of the chances of survival of that community, or at the very least implied the legitimacy of those in power. This applied to the military domain as well. The Mycenaeans of Bronze Age Greece, for instance, registered the number of available wheels and chariot frames, and allotted a corresponding number of horses to them so that they could be used in battle.[2] Series of baked clay tablets found in the ruins of Pylos list numbers of men available for service in the coast guard, and rowers to be sent to the city of Pleuron, just across the Gulf of Corinth.[3] The collection of data for military intelligence can boast a long history too. Sun Tzu in his *Ping Fa* stressed the importance of intelligence for a military commander, while leaders such as Ottoman sultan

Süleyman the Magnificent (1494-1566) and Frederick II of Prussia (1712-1786) were avid consumers of intelligence, especially on their enemies' military capabilities.[4]

By the 19th century European governments for the first time had acquired the technological means and the modern bureaucracy to effectively control their territory. The French Revolution had stimulated a sense of urgency in managing popular sentiment. At the same time, the exact sciences made great strides, increasingly accurately predicting the occurrence of natural phenomena. These developments produced the idea that it would be possible to identify laws governing *human* development as well. Measuring, counting and applying statistical analysis in order to identify patterns and enable accurate predictions gradually became second nature to 'positivist' liberal and conservative elites.[5] But even before that, in the military domain, theorists and practitioners like Jomini, and Von Bülow, were debating 'military systems' and applied mathematical logic to warfare.[6] The organisation and implementation of conscription from the late 18th century onwards required keeping track of quantifiable indicators such as age, weight, height and level of education. After 1850 such numerical indicators increasingly underlay the planning of operations and campaigns, as in Moltke's campaigns against Austria and France in 1866 and 1870, respectively.[7] Coal and steel production figures, the capacity of the railways, the potential of the logistics and meticulous calculations about the effectiveness of weapons became critical factors. Even studies in moral and psychological aspects of warfare were replete with measurement, quantification and calculations.[8] As each of the Great Powers, with the possible exception of Britain, had reasons to strike at the earliest possible moment rather than postpone what was believed to be inevitable, it may likewise be argued that the decision to go to war in 1914 was to a large degree inspired by prognoses of the number of troops their opponents could field, the efficiency of troop transports by train and the exact amount of firepower.[9]

During the World Wars the quantifying approach was also applied to military operations and Operational Surveys and Operational Analysis soon came of age.[10] Air Forces in particular were interested in hard data on the number of sorties, bombs dropped and their effects.[11] In this way, the contribution of the air force could be measured, adjusted, and, if necessary, justified. But they were hardly alone in this: the Battle of the Atlantic, that, ultimately, came down to enforcing a favourable ratio between the destruction of Allied tonnages of cargo or ships and the elimination of German submarines was also fought with statistics.[12] The same was true of course for the 'measurable' effects of the blockade of Japan and the ratio between Japanese and allied carriers, aircraft and pilots.[13] In the Cold War, both nuclear and conventional strategies were underpinned by endless sets of data on combat ratios and calculations on possible future wear and tear of enemy and friendly troops.[14] Its logic underlies both the peacetime and wartime direction of the military to this day.

While the importance of data for the conduct and the study of warfare can therefore not be denied, this chapter questions the idea that only quantitative data can help manage a war, explain its characteristics and dynamics, and predict its course. It *is* useful, necessary even, but there are a number of pitfalls involved. To highlight these, we elaborate on the case of Robert McNamara, since his 'technocratic' statistical approach has guided the US war effort during the Vietnam War. If there has been one person who has pushed the quantification method and its underlying idea that data can help predict the future, to extremes, it is McNamara. In spite of the fact the US lost that war, this idea has had a lasting influence that can be traced both to the conduct of contemporary wars, but also to the analysis of past wars. In the second section of the chapter we show that such approaches often disregard the complexity and *imponderabilia* of unique historical wartime contexts. We argue that for this reason the integration of quantitative-generalising and qualitative-historicising approaches is critical, whether aimed at understanding contemporary warfare or warfare in the past.

15.2. "Every quantitative measurement (...) shows that we are winning the war".[15]

A brilliant student of economics, statistics and management, in the months following the Japanese attack on Pearl Harbor Robert Strange McNamara (1916-2009) became involved in a US Army Air Forces program to teach their officers 'analytical approaches'.[16] In the context of this program, 'analytical' was synonymous with the quantifying method. Over time, he transferred to the Statistical Control Team at the Pentagon that was dominated by statisticians and RAND economists who held that the military had thus far been virtually 'blind; they had hardly any idea about numbers, types and dislocation of aircraft and spare parts. These 'Whiz Kids' advocated data-driven decision making. McNamara and his colleagues mapped out the stocks, the numbers and location of spare parts and in so doing were able to make operations much more efficient and cost effective.[17] He was subsequently made co-responsible for the analysis of the deployment of the US bombers in the Pacific Theatre and the B-29s in particular. McNamara now applied his mathematical and statistical approach to Operations Research and was able to improve both their effectiveness and efficiency.[18]

Returning to civilian life after the war, McNamara embarked on a career at the Ford car company, where he put his approach to good use. Developing a modern planning, organisation and management control system at Ford, integrating the crucial figures into trend graphs and using early computer modelling in decision making, he helped lay the foundation of an innovative and very influential management style.[19]

As Secretary of Defence in the Kennedy and Johnson Administrations, McNamara continued this approach. From the outset he demanded 'scientific' data from the armed forces about just about anything. McNamara advocated 'statistical rigor' among his officials and military personnel: each complex military problem was translated into numerical indicators and patterns on the basis of which he could make policy and strategic decisions. McNamara and his (civilian) staff used all kinds of numbers, paperwork and management tools for the *Planning, Programming and Budgeting System* (PPBS), the *Future Years Defense programs* (FYDP) and the *Draft Presidential Memorandum* (DPM), to analyse military conflicts, fund specific programs, cut budgets, as well as reorganise the ministry and military apparatus.[20]

Figures were not only compiled for analysis which brought about an increase in efficiency, but they also served as a political weapon, for in the process McNamara reduced the military apparatus to a supplier of data to the *Systems Analysis Office* he had created, when it, and the Joint Chiefs of Staff in particular, had been decision makers before.[21] As Sapolsky wrote: "Eisenhower fought political battles with his generals' stars; McNamara fought them with heaps of statistics."[22]

These innovations in Department of Defence decision-making and its justification, the introduction of modern management methods in order to enhance efficiency and effectiveness, and the establishment of civilian control over the military apparatus, undeniably were of profound importance and they should be characterised as McNamara's lasting legacy.[23] In his conduct of the Vietnam War, McNamara applied this approach as well; its drawbacks were quick to emerge, however.[24]

During his first visit to South-Vietnam, in May 1962, McNamara announced that "every quantitative measurement (...) shows that we are winning the war." His obsessive belief in graphs and charts was immediately exploited by South-Vietnamese leader Diem who presented him prefabricated graphs that only led to one conclusion: the war would soon be won indeed. At the end of 1962, McNamara entertained the thought of ordering the repatriation of the American advisors from South Vietnam, since his calculations indicated the war would certainly be won in 1964.[25] Nevertheless, already in December 1963, a still confident McNamara had to report to President Lyndon Johnson that 'the figures' that had been obtained probably did not correctly reflect reality: instead, the situation was "very disturbing" because "current trends" pointed to a possible Communist victory. He had to admit that earlier computer models and statistics were "grossly in error."[26]

Fascinatingly, this admission did not lead him to question the quantitative approach to warfare itself. Wrong figures merely had to be replaced by correct ones. His belief in the fundamental soundness of his approach was unshattered and in 1967 while admitting that "not every conceivable complex human situation can be fully reduced to the lines on a graph", he stated that "not to quantify what can be quantified is only to be content with something less than the full range of reason."[27]

Critics who expressed doubts about the quality of the South Vietnamese Army or warned about the corruption of the South Vietnamese regime, were swiftly and decisively brushed aside. They did not understand the numbers.[28]

This line of reasoning was based on the so-called 'systems analysis'. The point of departure for McNamara and his personal staff was that reality could be approached as a set of complex systems, and that these systems could best be understood by methodically quantifying and analysing their constituent components and their mutual coherence. Data from this complex world had to be "delineated, denoted, demarcated and quantified", after which it became possible to understand, tame and transform the world according to one's own will. By means of a rational analysis of the data, the world could be controlled and improved, or at least made more effective and efficient. The truth was in the data; not in reality.[29]

Building on this line of thought data-driven analysis, planning and control was intensified during the Vietnam War. In fact, almost every aspect of the war was quantified: the tonnage of goods arriving or destroyed in ports in Vietnam, the number of ships intercepted, the movements on the Hô Chí Minh Trail, the number of bombs on North Vietnam, the number of troops, the effects of brown-water operations, the percentage of land controlled by the US or the South Vietnamese Army, rice yields, population data, personnel damage assessments, etc. This obsession with data is one of the explanations for the escalation of the war, fixated as its 'managing directors' became on systematically increasing the number of GI's, US firepower and the bombing of targets in North Vietnam. By 1968 over half a million US soldiers were fighting in Vietnam. If the number of GI's multiplied, so would US firepower and this would ultimately tip the balance, or so the reasoning went. Likewise, in the same period operation Rolling Thunder, the continuous bombing of targets in North Vietnam, was significantly stepped up.

15.3. "If it's dead and Vietnamese, it's VC": Body count as indicator of success

The focus on data also translated into the conviction that the number of Viet Minh and Viet Cong dead, either from the air, by maritime blockade or Search and Destroy Missions, somehow was an indication of progress. McNamara and US military commander General William Westmoreland's military strategy had become a strategy of (measurable) attrition. The number of sorties or the tonnage of bombs dropped, and the ratio between one's own casualty rate and that of the enemy in particular, became their most important indicator of progress and success.[30] This latter indicator, the so-called body count, was at the core of McNamara's and Westmoreland's minds.[31] It soon became one of the most notorious manifestations of the quantitative approaches to war, as the unintended outcome of this focus was that killing rather than winning became central to the war effort.

But, as John Lewis Gaddis points out, "Statistics of enemies killed said (...) nothing about Vietnamese willingness to absorb those losses in pursuit of their political goal: expelling American troops and uniting the country under communist rule."[32] Not only did the figures not take into account the feelings of the Vietnamese population, their motives and motivation, the corruption of the South Vietnamese regime and its impact and consequences, the quality of the South Vietnamese troops, the morale of the GIs and the support of the home front, to name but a few relevant factors, but they were also largely unreliable. As a matter of fact, the figures were consistently inflated, in part because the measuring system ensured that civilians and enemy combatant were heaped together.[33] Worse still, indiscriminate violence in order to get the highest possible body count bred support for the VC and the North Vietnamese:[34]

> *Victory was a high body count, defeat a low kill ratio, war a matter of arithmetic. The pressure on unit commanders to produce enemy corpses was intense, and they in turn communicated it to their troops. This led to such practices as counting civilians as Viet Cong. "If it's dead and Vietnamese, it's VC" was our rule of thumb in the bush. It is not surprising, therefore, that some men acquired a contempt for human life and predilection for taking it.[35]*

Nonetheless, McNamara's stress on the statistical inevitability of the US victory became ever more outspoken and persistent. With the proper data, victory was in reach and all official indicators suggested that the US was winning the Vietnam War.[36] His obsession with data caused critics of his approach to see McNamara as a "human IBM machine" who cared more for computerised statistical logic than for human judgments.[37]

However, in spite of his outward assurances to the contrary, in private McNamara gradually opened up to the possibility that US objectives were not feasible.[38] He had always paid special attention to Air Force data. In his view, the effects of Operation Rolling Thunder, for example, were easier to measure than those of land operations, but by the Summer of 1967 he had come to realise that destroying 80% of North Vietnam's industry meant rather little, since that economy was only 1% dependent on industry. Two years of relentless bombing had caused about $300 million in damage but it had cost $900 million in bombers.[39] In addition, despite the ongoing aerial bombardment, the North Vietnamese were still able to send large numbers of people and equipment through the Hô Chí Minh Trail to the South. In other words, the numbers had turned against McNamara. Testifying to the Senate Armed Forces Committee in August 1967, he had to admit that he lacked "any confidence that they (North Vietnam) can be bombed to the negotiating table."[40]

Fascinatingly, McNamara's pessimism was buttressed by the same sort of statistics he had used to predict approaching victory earlier. His 1967 plea to Johnson to end the war – behind closed doors of course – was based on numbers and graphs just like his earlier predictions of a swift victory had been. At the end of 1967, on his last working day, the overworked McNamara shouted at his colleagues: "This goddamned bombing campaign, it's worth nothing, it's done nothing, they dropped more bombs than on all of Europe in all of World War II and it hasn't done a fucking thing!"[41] Even his despair was substantiated with numbers!

The Secretary of Defence only once considered an alternative approach. This was several months before, when he had already become convinced the war effort was in great trouble. McNamara then commissioned a major investigation into the way the US had become involved in Vietnam; an effort that can only be characterised as qualitative *historical* research. Tellingly, its gloomy conclusions were rated *top secret / sensitive* and shelved in some deep Pentagon drawer. The report stayed there until 1971 when it was leaked to the *New York Times* and became known as the *Pentagon Papers*.[42]

15.4. "What can't be easily measured really doesn't exist." The 'McNamara fallacy'

Sociologist Daniel Yankelovich was probably the first to term McNamara's obsession with numbers, graphs and models to study and control a complex phenomenon such as warfare, 'the McNamara fallacy'. He summarised it as follows:

> *The first step is to measure whatever can be easily measured. This is ok as far as it goes. The second step is to disregard that which can't be easily measured or to give it an arbitrary quantitative value. This is artificial and misleading. The third step is to presume that what can't be measured easily really isn't important. This is blindness. The fourth step is to say what can't be easily measured really doesn't exist. This is suicide.[43]*

The McNamara fallacy can be characterised as a tunnel vision. Experts on the Vietnam War and on statistical or quantitative management methods have highlighted additional methodological problems.[44] Kenneth Cukier and Viktor Mayer-Schönberger for example stress the poor quality of the data McNamara and his team used. The data failed to "capture what it purports to quantify", or was "used to frame desirable outcomes."[45] Jonathan Salem Baskin shows that McNamara and his team

> *spent over a decade looking within their model for revelations about underlying reality, and saw metrics that reaffirmed their hypothesis, while frequent and often loud external data points [that] challenged it were ignored.[46]*

Matthew Fay argues the approach itself was flawed because for military organisations it is hardly ever possible "to judge competing approaches in terms of their efficiency" in producing "output". What is more, McNamara's econometric approach failed to get to grips with the political aspects of the war. Not everything that he deemed measurable could actually be measured and when it could, he failed to see that the usefulness to do so depended on "recognising the real issues at stake", the real issues ultimately being political choices. In Fay's words, "Economics is a useful tool. But the essence of national defense is still political".[47] These critics also point out that McNamara and his staff assumed that the adversary reasoned along the same mathematical and apparently rational lines as they themselves did. This 'mirror imaging' was obviously a major miscalculation.

McNamara's former military subordinates felt the same way. After the war Brigadier General Douglas Kinnard opened the cesspool in his book *The War Managers.* In response to a questionnaire that he prepared, most of the top brass dismissed the tactics to win the war as ill-conceived: "The U.S. was committed to a military solution, without firm military objective – the policy was attrition – killing VC – this offered no solution – it was senseless." Sixty-seven percent of the generals that had served in the Vietnam War filled in Kinnard's form and the great majority was utterly dismissive of the disproportionate quantification of the war, and the body count in particular. In fact, only two percent of the respondents indicated they considered the body count a valid means to measure progress. The great majority responded that figures had been grossly inflated and that the model itself was nonsensical.[48]

There is ample evidence that lower ranked officers subscribed to this assessment.[49] Not only did the quantitative approach stimulate extensive fraud, it also fully disregarded the human dimension, Vietnam's social fabric and cultural-historical background, and the devastating effect of the South Vietnamese regime's political incompetence and corruption.[50] Worse still, McNamara's fixation on the body count was counterproductive because it engendered more resistance than it helped to suppress.[51]

15.5. "Combat Power = force strength x variables"?

The Vietnam experience could have led to discrediting the quantifying approach. But this did not happen. In a sense this approach was reinvigorated by Colonel Trevor N. Dupuy (1916-1995), US Army, who in the 1970s and 1980s wrote a number of books on the subject. Dupuy was convinced that studying past wars was crucial in preparation for future ones. By the end of the 1970s, he had developed a radically quantitative method to compute combat power. From a study of German and Allied troops in the Second World War he concluded that the Germans had been better soldiers than the Americans and he could provide 'hard evidence' to prove it.[52]

His analysis was based on a natural-science-like method that he had developed and which he called the *Quantified Judgment Model*.[53] This, he claimed could objectively determine the combat strengths of units and individuals. By attributing fixed figures to weapons and weapons systems and units and correcting them for what he termed 'variables' such as weather, terrain, etcetera, he was able to determine a so-called 'Score Effectiveness'. He even identified scientifically verifiable patterns that provided lessons from the past that could be applied in future wars. Dupuy's formula ran like this:

$$SE\left(f\right) = \frac{\sqrt{Case(e)}}{0.001 \times S(f) \times V(f) \times U(f)}$$

in which SE(f) stood for 'Score Effectiveness', Case(e) for 'the number of casualties inflicted on the enemy'; S(f) for 'friendly force strength'; V(f) for 'friendly force vulnerability factor'; and U(F) for the friendly posture factor. In principle, these were all easy to measure, Dupuy argued. In fact, his model promised an accuracy of up to two decimal places.

Dupuy's conclusions were criticised by John Sloan Brown of the Department of History of the United States Military Academy, West Point.[54] Dupuy's case selection was flawed, according to Brown, as he had selected cases that saw the Americans attacking and Germans defending. In addition, he had compared the best German troops, Panzer units, with average regular US army infantry. Furthermore, he did not take into account all kinds of factors that were also critical around 1944 such as the staggering lack of manpower on the German side, and logistics and strategical errors. Some of these factors, notably morale, motivation and the fog of war were also very difficult to quantify, to say the least. Brown argued that in Dupuy's simplified formula, *Combat Power = Force Strength x variables*, it was precisely these variables that determined the outcome. However, Dupuy presented these variables as a minor component. Brown reasoned that he could have reached the opposite conclusion with the same set of data, i.e., that the American soldiers were better fighters than the Germans and concluded that the model 'implie(d) a precision its subject does not warrant'.[55]

These were the opening salvos of what became known as the Dupuy-Brown Controversy. Unfortunately, Dupuy was not willing to engage Brown's criticisms, for in subsequent pieces he merely reiterated his claims and stressed that his model had been peer reviewed. Eventually, Brown was overawed and retracted some of his criticisms: the model itself *was* correct, he only had intended to question the *score effectiveness*. This outcome was unfortunate, because Brown's initial critique that Dupuy's formula multiplied arbitrary quantifications by non-quantifiable elements and that he ignored or downplayed the impact of factors that were hard to quantify or not quantifiable at all, was a warning that should have been heeded.

Instead, Dupuy's clinical numerical approach, like McNamara's before him, has remained attractive: we saw its logic repeated in military operations in the 1990s (the Kosovo War), and 2000s (Iraq and Afghanistan). In a sense, it promises a tidy and manageable model of warfare in which pawns can be moved at will to achieve a certain guaranteed effect. However, especially at the tactical level it would be utterly misleading to think that war is about moving pawns or about dots and arrows of the kind strategists and policy makers like to place on maps. But even at the strategic or policy level it would be wise to keep in mind that no matter how much data you collect, you only win when the opponent concedes defeat; war is an unpredictable clash of wills, and a dirty and chaotic one at that.

In spite of this, time and again claims are made to the contrary. A recent example is Ingo Piepers' PhD dissertation *Dynamics and Development of the International System: A Complexity Perspective.*[56] Piepers argued that the history of 'the international system' is best explained by applying the second law of thermodynamics. In the natural world, the build-up of pressure causes a phase transition, like ice that is heated will melt. The international system behaves in the same way, according to Piepers; for transitions to materialise a 'systemic war' is necessary. This will lead to a new system that will then function for a while until the inherent build-up of tensions inevitably leads to a new systemic war that will cause a transition to a new order. On the basis of what he perceived as historical trends in system-changes, Piepers identified four 'systemic wars', i.e., the Thirty Years' War (1618–1648), the French Revolutionary and Napoleonic Wars (1792–1815), the First World War (1914–1918), and the Second World War (1939–1945) and went on to predict the outbreak of a new one in 2020.[57]

A number of methodological flaws should be pointed out. A large-scale war, the Seven Year's War (1756–1763) was fought on at least three continents and involved all great powers and numerous smaller ones. It is treated as an anomaly, while the fact that the 'system' returned in more or less the same shape should have caused Piepers and his academic supervisors to rethink the model, as should the fact that the model dictated the treatment of post-World War II Europe as a single entity, when, in reality it was divided into two dynamic blocs with a number of states being members of neither. Secondly, it is far from obvious that the behaviour and interaction between human beings (which is what politics is about) is guided by laws derived from physics. Thirdly, Pieper's conclusions about the dynamics of inter-state interaction are built on rather modest evidence. Pieper's central premise that wars occur in fixed cycles was not (and could not be) substantiated if only because none of the central concepts such as 'international order', 'tensions', and even 'systemic war', was properly defined. While it no doubt is true that war breeds war, there is no simple connection between them and the development of the international system. To think otherwise is to ignore the complexity of the interplay of a very large number of variables, several of which cannot be quantified at all. Even in a simulator the outcome is never the same. Lastly, variables do not carry inherent numerical value or meaning. It is people that attribute meaning and act upon this attribution. Early 20th Century German demographics and steel production worried the French not because their numerical value was bigger than France's, or because both states were part of an inherently unstable system that was bound to invite a systemic war, but because they were Germany's. In 1871 France had lost territory to Germany and the only way to reclaim it was to defeat Germany. Morale was believed to compensate for numbers, hence the focus on the offensive spirit that nearly spelled the end for France in its opening campaign in 1914.

15.6. Statistics and the historical approach: a reappraisal

At this point, the reader might be tempted to infer that we propose to anathematise the quantitative approach. This would be premature. A modern army cannot do without data and figures; the complexities of the battlefield, logistics, and joint and combined operations simply require a metric approach. Likewise, at the operational and tactical levels, concepts such as the *Effects Based Approach to Operations* with its focus on clear objectives and achievable goals, demand an array of quantitative indicators that possess a certain predictive power. In addition, the political and strategic levels often require such figures to be able to justify (or 'sell') a certain course of action. Figures serve their purpose in establishing a measure of cost-effectiveness in the peace-time management of the armed forces. In short, many aspects and facets of war and warfare actually lend themselves very well to this approach.

But – and this is key – the quantitative and the qualitative approaches ought to benefit from one another, as transpires from the study of *past* wars. The historiography of warfare has long been dominated by the descriptive, so-called 'Rankean' tradition. In this tradition, there was a focus on *national* history, and on 'great captains'. It was characterised by a strong emphasis on what is historically unique and specific of the time and the place. From the 1950s however, a number of academic counter-movements gained steam among historians. Several of these made ample use of quantitative analysis. Their aim was not to glorify individuals and provide anecdotal evidence of illustrious battles, but to 'scientifically' discern historical patterns and laws. Since then, it has been shown many times that a 'hard', data-driven, quantitative approach and a more historical-interpretative approach to the military and to the field of military history are not mutually exclusive. On the contrary, they are quite compatible, and it could be argued that the one cannot do without the other.

Examples include Addington's *The Blitzkrieg Era and the German General Staff, 1865-1941*, and Van Creveld's *Fighting Power: German and U.S. Army Performance 1939-1945*.[58] Another great example is offered by Richard Overy's studies on the history of *Airpower*.[59] Overy frequently used quantitative methods, processing the numbers of aircraft, sorties, bombings, losses and effects measurements. But he did not squeeze the numbers into a formula or model or let them speak 'for themselves'. For example, the fact that virtually every city in Germany during the Second World War was reduced to ashes might justify the conclusion that the allied strategic offensive mathematically speaking had been effective. But Overy concluded that precisely because of this wanton destruction the main goal of breaking the morale of the German population had not been achieved. At the same time, he pointed to many unexpected and unintended but crucial effects of the bombing campaigns, such as refugee flows, disruption of logistics networks, etc. He elaborated on this again in *Why the Allies Won*.[60] The three-volume *Military Effectiveness*, edited by Alan Millet and Williamson Murray, is exemplary as well. Like Dupuy, they emphasised studying combat, but unlike him, they included unique contexts and paid due attention not only to firepower, but also to the political domain, the strategic setting, the operational level and tactics. In fact, the most fascinating insights in their study on *military effectiveness* arose from the interaction between the two domains and the two different methods used. In other words, the combination of quantitative and qualitative analysis led to a better understanding.[61]

This is not only true in academia. During the Second World War for instance, it was noted that planes that returned from bombing missions over Europe had similar patterns of damage. Quantitative analysis suggested that it was wise to reinforce those parts, but to little effect. It took considerable time before anyone dared suggest that the planes that displayed this particular pattern were the ones that had made it back to the United Kingdom; there were no returning planes with different hit patterns, because these had been downed by enemy fire. It was only when the Allies, rather counterintuitively, started reinforcing the parts that had initially seemed protected enough that their losses were reduced.[62]

15.7 Conclusion: "Numbers don't talk. People do."[63]

Earlier theorists aimed to equip the conduct of war with principles, rules, or even systems, and thus considered only factors that could be mathematically calculated (e.g., numerical superiority; supply; the base; interior lines). All these attempts are objectionable, however, because they aim at fixed values. In war everything is uncertain and variable, intertwined with psychological forces and effects, and the product of a continuous interaction of opposites.[64]

Whether one studies McNamara's obsession with data to understand and manage war effectively and efficiently, or critically examines past scientific research into combat power, it becomes clear 'the figures' in themselves do not provide an objective representation of (historical) reality. Rather, they form a set of pre-selected parameters that refer to partial aspects of the war situation. They must be made to speak. There is no clear self-evident correlation between the numerical representation of reality and that reality itself. A model is *not* reality and war is *not* a laboratory setup that you can use to make measurements. It is a complex and very chaotic whole, "intertwined with psychological forces and effect, and the product of a continuous interaction of opposites," as Clausewitz wrote.

Hopefully, this understanding will contribute to a certain restraint with regard to our present obsession with data, and to a realisation that the quantitative-generalising and qualitative-historicising approaches need each other. Both approaches have their strengths and weaknesses, which can be balanced only by recognising the necessary interaction of the two. The qualitative method can take into account complex variables that are difficult to measure, such as the motives of the actors involved and other *imponderabilia*. It can relate to time- and place-specific contexts. Its weakness though lies exactly in its focus on the unique and contextual, which makes comparison and generalisation often almost impossible. Whereas the quantitative approach may suggest pathways to success that do not exist in reality, the qualitative approach may easily degenerate into casual talk and anecdotal evidence.

Conversely, the mathematical, quantifying approach has obvious advantages, especially when it comes to data-driven modern warfare. You will have to measure, organise and model in order to properly interpret modern military planning, decision-making and actions. However, (methodological) problems arise when we limit ourselves to this approach; what should we do with the unmeasurable or the poorly measurable? What do the variables and data(sets) exactly say? There is obviously no such thing as 'pure data', especially when it comes to managing and understanding the complex phenomenon that is war(fare). After all, its collection reflects what *we* think important.[65] Worse still, data may "lure us to commit the sin of McNamara: to become so fixated on the data, and so obsessed with the power and promise it offers, that we fail to appreciate its inherent ability to mislead."[66] In short, data should not just be collected and modelled, they should be interpreted and contextualised. If, as Clausewitz wrote, war is like a chameleon that changes its colours all the time, we would do well to realise that when we come face to face with one, statistics alone are not enough to subdue the beast.

Notes

1   The phrase 'a technocratic hubristic fable', a characterisation of Robert McNamara's obsession with data, was taken from Paul Hendrickson, *The Living and the Dead: Robert McNamara and Five Lives of a Lost War.* Sydney: Vintage, 2000, 356.

2   Kim S. Shelton, "Foot Soldiers and Cannon Fodder: The Underrepresented Majority of Mycenaean Civilization," *EPOS: Reconsidering Greek Epic and Aegean Bronze Age Archaeology,* ed. Sarah P. Morris and Robert Laffineur (Liège: Université de Liège, 2007), 169–176.

3   Thomas Palaima, "Maritime Matters in the Linear B Tablets," *Thalassa: l'Égée Préhistorique et la Mer, Actes de la Troisième Rencontre Égéenne Internationale de l'Université de Liège,* ed. Robert Laffineur and Lucien Basch. Liège: Université de Liège, 1991, 273–310, at 308.

4   Sun-Tzu, *The Art of Warfare. Translated by Roger T. Ames.* New York: Ballatine Books, 2010, 113; Nathaniel.D. Bastian, "Information warfare and its 18th and 19th century roots," *The Cyber Defense Review* 4, no. 2 (2019): 31-36; Elisabeth Krimmer and Patricia Simpson, ed., *Enlightened War: German Theories and Cultures of Warfare from Frederick the Great to Clausewitz.* Rochester and New York: Camden House, 2011, 258–278.

5   Rom Harré, "Positivist Thought in the Nineteenth Century," in *The Cambridge History of Philosophy 1870-1945,* ed. Thomas Baldwin. Cambridge: Cambridge University Press, 2003, 7–26; Walter M. Simon, *European Positivism in the Nineteenth Century.* New York: Cornell University Press, 1963.

6   See, for instance, Antoine-Henri de Jomini, *Précis de l'art de la guerre.* 1838; Paris: Éditions Ivréa, 1994 and Heinrich Dietrich Adam von Bülow, *Geist des neuern Kriegssystems hergeleitet aus dem Grundsatze einer Basis der Operationen auch für Laien in der Kriegskunst faßlich vorgetragen von einem ehemaligen Preußischen Offizier.* Hamburg: Benjamin Gottlieb Hofmann, 1799.

7   See for example: Larry Addington, *The Blitzkrieg Era and the German General Staff, 1865–1941.* New Brunswick, NJ: Rutgers University Press, 1971.

8   Charles Ardant du Picq, *Études sur le Combat.* Paris: Hachette et Dumaine, 1880; Stéphane Audoin-Rouzeau, "Vers une Anthropologie Historique de la Violence de Combat au XIXe Siècle: Relire Ardant du Picq?," *Revue d'Histoire du XXIXe Siècle* 30 (2005) at https://doi.org/10.4000/rh19.1015 accessed 14 March, 2022.

9   See Christopher Clark, *The Sleepwalkers: How Europe Went to War in 1914.* London: Penguin Books, 2012; Alan John Perceval Taylor, *War by Timetable: How the First World War began.* London: MacDonald, 1969; David Stevenson, "War by Timetable? The Railway Race before 1914," *Past & Present* 162 (Feb. 1999): 163–194.

10 Erik P. Rau, "Combat science: the emergence of Operational Research in World War II," at https://doi.org/10.1016/j.endeavour.2005.10.002 accessed 12 March, 2022; Joseph McCloskey, "British Operational Research in World War II," *Operations Research* 35, no. 3 (May–June 1987): 453–470.

11 On this, see Richard Overy, *The Air War: 1939–1945.* London: Taylor & Francis, 1980; Idem, *Bomber Command, 1939–45.* London: Harper Collins, 1997; Idem, *The Bombers and the Bombed: Allied Air War Over Europe 1940-1945.* London: Viking Press, 2014; Idem, *The Bombing War: Europe 1939-1945.* London: Penguin Books, 2013; Maurice Kirby and Rebecca Capey, "The area bombing of Germany in World War II: an operational research perspective," *Journal of the Operational Research Society* 48 (1997): 661–677.

12 Jonathan Dimbleby, *The Battle of The Atlantic: How the Allies Won the War.* New York: Viking Press, 2015; David White, *Bitter Ocean: The Battle of the Atlantic, 1939–1945.* New York: Simon & Schuster, 2008.

13 Max Hastings, *Retribution: The Battle for Japan, 1944-45*. New York: Alfred A. Knopf, 2008; Eric M. Bergerud, *Fire in the Sky: The Air War in the South Pacific*. Boulder, CO: Westview Press, 2000; Clay Blair, Jr., *Silent Victory: The U.S. Submarine War against Japan*. Philadelphia: Lippincott, 1975.

14 John Lewis Gaddis, *The Cold War: A New History*. London; Penguin Books, 2005.

15 Stanley Karnow, *Vietnam. A History*. New York: Viking Press, 1983, 267.

16 For an introduction see Deborah Sharpley, *Promise and Power: The Life and Times of Robert McNamara*. Boston: Little, Brown, 1993.

17 As early as 1943, 3.6 billion dollars was saved through a better allocation of resources. See on this: George M. Watson, Jr. and Herman S. Wolk, "'Whiz Kid': Robert S. McNamara's World War II Service," *Air Power History* 50 no. 4 (2003): 4 –15.

18 On this: Sharpley, *Promise and Power*; Watson and Wolk, "Whiz Kid," and Errol Morris' documentary, *The Fog of War. Eleven Lessons from the Life of Robert S. McNamara* (2003).

19 Phil Rosenweig, "Robert S. McNamara and the evolution of modern management," *Harvard Business Review* 91 (2010): 87–93; Leo McCann, "'Management is the gate' – but to where? Rethinking Robert McNamara's 'career lessons'," *Management and Organizational History* 11 no. 2 (2016): 166–188.

20 Lawrence S. Kaplan, Ronald Dean Landa and Edward Drea, *The McNamara Ascendancy, 1961-1965* (Washington: Historical Office, Office of the Secretary of Defense, 2006).

21 Peter T. Tarpgaard, "McNamara and the rise of analysis in defense planning: a retrospective," *Naval War College Review* 48 no. 4 (Fall 1995): 67–87; Rosenweig, "Robert S. McNamara and the evolution of modern management"; Herbert R. McMaster, *Dereliction of duty: Johnson, McNamara, the Joint Chiefs of Staff, and the lies that led to Vietnam*. New York, Harper Collins,1998.

22 Harvey Sapolsky, "Rationalizing McNamara's Legacy," https://warontherocks.com/2016/08/rationalizing-mcnamaras-legacy/ accessed 20 March, 2022.

23 Abhijnan Rej, "The other legacy of Robert McNamara," https://warontherocks.com/2016/06/the-other-legacy-of-robert-mcnamara/ accessed 22 February, 2022; Phil Rosenweig, "Robert S. McNamara and the evolution of modern management," *Harvard Business Review* 91 (2010): 87–93; Alain C. Enthoven and K. Wayne Smith, *How Much Is Enough?: Shaping the Defense Program, 1961-1969*. Santa Monica: Rand Corporation, 2005; Tarpgaard, "Rise of Analysis in Defense planning," 67–87.

24 On Vietnam in general and McNamara's role in the war: Max Hastings, *Vietnam: An Epic History of a Tragic War 1945-1975*. London: William Collins, 2018; Geoffrey C. Ward and Ken Burns, *The Vietnam War: An Intimate History*. London: Penguin Books, 2017; Karnow, *Vietnam. A History*; Michele Chwastiak, "Rationality, performance measures and representations of reality: planning, programming and budgeting and the Vietnam War", *Critical Perspectives on Accounting* 17, no. 1 (2006): 29–55.

25 Karnow, *Vietnam*, 267.

26 Karnow, *Vietnam*, 325.

27 Kenneth Cukier and Viktor Mayer-Schönberger, "The dictatorship of data," *MIT Technology Review* 31, no. 05 (2013), at https://www.technologyreview.com/2013/05/31/178263/the-dictatorship-of-data/ accessed 29 March, 2022.

28 McNamara hated critics such as Lieutenant Colonel John Paul Vann. See: Neil Sheehan, *A Bright Shining Lie: John Paul Vann and America in Vietnam*. New York: Random House, 1988.

29 Tarpgaard, "Rise of Analysis in Defense planning", 67–87; Clark A. Murdock, "McNamara, systems analysis and the systems analysis office," *Journal of Political & Military Sociology* 2, no. 1 (1974) 89–104; Cukier and Mayer-Schönberger, "The dictatorship of data." (quote); Rej, "The other legacy of Robert McNamara"; Matthew Fay, "Rationalizing McNamara's Legacy," at https://warontherocks.com/2016/08/rationalizing-mcnamaras-legacy/ accessed 22 February, 2022.

30 Idem.

[31]   "Body count" in *The Encyclopedia of the Vietnam War: A Political, Social and Military History*, ed. Spencer Tucker. Oxford: ABC-Clio, 2001, 42 ff.

[32]   John Lewis Gaddis in *Classic Strategies of Containment* quoted in Fay, "Rationalizing McNamara's legacy." See also Gregory A. Daddis, "The problem of metrics: assessing progress and effectiveness in the Vietnam War," *War in History* 19, no. 1 (2012): 73–98.

[33]   See for example Christian G. Appy, *Working-Class War: American Combat Soldiers and Vietnam*. Chapel Hill and London: University of North Carolina Press, 2000, 153–56; Günter Lewy, *America in Vietnam*. Oxford: Oxford University Press, 1978, 450–1; Jonathan Schell, "The Village of Ben Suc," *The New Yorker* (July 15, 1967).

[34]   Stewart O'Nan and Philip Caputo, *The Vietnam Reader: The Definitive Collection of American Fiction and Nonfiction on the War*. New York: Anchor Books 1998, 156. See also *Reporting Vietnam, Part 1: American Journalism, 1959-1969*. New York: Library of America, 1998; *Reporting Vietnam Part Two: American Journalism 1969-1975*. New York; Library of America 1998.

[35]   O'Nan and Caputo, *The Vietnam Reader*, 156.

[36]   Daddis, "The Problem of Metrics," 73–98; Jonathan Salem Baskin, "According to U.S. Big Data, We Won the Vietnam War," *Forbes* 25 July 2014 at https://www.forbes.com/sites/jonathansalembaskin/2014/07/25/according-to-big-data-we-won-the-vietnam-war/?sh=d8f83183f218 accessed 21 February, 2022.

[37]   On this see the opening of Errol Morris' brilliant documentary, *The Fog of War. Eleven Lessons from the Life of Robert S. McNamara* (2003). Compare Cukier and Mayer-Schönberger's "McNamara relied on the figures, fetishized them. With his perfectly combed-back hair and his flawlessly knotted tie, McNamara felt he could comprehend what was happening on the ground only by staring at a spreadsheet—at all those orderly rows and columns, calculations and charts, whose mastery seemed to bring him one standard deviation closer to God" (Cukier and Mayer-Schönberger, "The dictatorship of data.")

[38]   Aurélie Basha i Novosejt, *'I made Mistakes': Robert McNamara's Vietnam War Policy, 1960–1968*. Cambridge: Cambridge University Press, 2019. See also McNamara's own accounts of the Vietnam War: Robert McNamara, *In Retrospect: The Tragedy and Lessons of Vietnam*. New York: Vintage Books, 1995; Robert McNamara, *Argument without End: In Search of Answers to the Vietnam Tragedy*. New York: Public Affairs, 1999.

[39]   Karnow, *Vietnam*, 457.

[40]   Idem, 509.

[41]   Idem, 512.

[42]   For the The Pentagon Papers see *The Pentagon Papers: The Defense Department History of United States Decisionmaking on Vietnam* at https://www.archives.gov/research/pentagon-papers. See also: Neil Sheehan. *The Pentagon Papers*. New York: Quadrangle Books, 1971; Daniel Ellsberg, *Secrets: A Memoir of Vietnam and the Pentagon Papers*. New York: Penguin Books, 2002.

[43]   Daniel Yankelovich, *Corporate Priorities: A Continuing Study of the New Demands on Business*. Stamford, Conn.: Daniel Yankelovich, Inc., 1972.

[44]   See Michael Shafer, *Deadly Paradigms: The Failure of U.S. Counterinsurgency Policy*. Princeton: Princeton University Press, 1988; Cukier and Mayer-Schönberger, "The dictatorship of Data"; Fay, "Rationalizing McNamara's legacy"; Baskin, "U.S. Big Data"; Jerome Slater, "McNamara's failures—and ours: Vietnam's unlearned lessons: a review," *Security Studies* 6, no. 1 (1996) 153–195.

[45]   Cukier and Mayer-Schönberger, "The Dictatorship of Data".

[46]   Baskin, "U.S. Big Data".

[47]   Fay, "Rationalizing McNamara's legacy".

48 Douglas Kinnard, *The War Managers*. Burlington; University of Vermont, 1977. See also Slater, "Vietnam's Unlearned Lessons", 25 (quote), 153–195.

49 'Cincinnatus', *Self Destruction: The Disintegration and Decay of the United States Army During the Vietnam Era*. New York: Norton, 1981.

50 Gregory H. Murry, "The On-Going Battle for the Soul of the Army" in: *Small Wars Journal* 12, no. 10 (2016) at https://smallwarsjournal.com/jrnl/art/the-on-going-battle-for-the-soul-of-the-army accessed 10 March 2022; Appy, *Working-Class War*, 153–56.

51 Idem.

52 Trevor N. Dupuy, *Numbers, Predictions and War: Using History to Evaluate Combat Factors and Predict the Outcome of Battles*. New York: Bobbs-Merrill, 1979; Idem, *Attrition: Forecasting Battle Casualties and Equipment Losses In Modern War*. Fairfax, VA: Hero Books, 1990.

53 Dupuy, *Numbers, Predictions and War*.

54 John Sloan Brown, "Colonel N. Dupuy and the mythos of Wehrmacht superiority: a reconsideration," *Military Affairs* 50, no. 1 (1986): 16–20. Their quasi-debate is in T.N. Dupuy, "Mythos or verity? The quantified judgment model and German combat effectiveness," *Military Affairs* 50, no. 4 (1986): 204–210; John Sloan Brown, "The Wehrmacht mythos revisited: a challenge for Colonel Trevor N. Dupuy," *Military Affairs* 51 no. 3 (1987): 146–147, and Trevor N. Dupuy, "A response to 'The Wehrmacht mythos revisited'," *Military Affairs* 51 no. 4 (1987) 96–197.

55 Brown, "Wehrmacht Superiority," 19.

56 Ingo Piepers, *Dynamiek en Ontwikkeling van het Internationale Systeem: een Complexiteitsperspectief*. Amsterdam: Universiteit van Amsterdam, 2006.

57 Ingo Piepers, *2020: WARning. Social Integration and Expansion in Anarchistic Systems: How Connectivity and Our Urge to Survive Determine and Shape the War Dynamics and Development of the System*. Amsterdam: Conijn Advies, 2016; Wil Thijssen, "De Derde Wereldoorlog kan elk moment beginnen, en ik voorspel het al jaren", *de Volkskrant* (10 January 2020).

58 Larry Addington, *The Blitzkrieg Era and the German General Staff, 1865–1941*. New Brunswick, NJ: Rutgers University Press, 1971; Martin van Creveld, *Fighting Power: German and U.S. Army Performance 1939–1945*. Westport, CN, Greenwood Press, 1982.

59 Richard Overy, *The Air War: 1939–1945*. London: Taylor & Francis, 1980. *Bomber Command, 1939–45*. London: Harper Collins, 1997; *The Bombers and the Bombed: Allied Air War Over Europe 1940-1945*; London: Viking Press, 2014, and *The Bombing War: Europe 1939-1945*. London: Penguin Books, 2013.

60 Richard Overy, *Why the Allies Won: Explaining Victory in World War II*. New York: Norton, 1995, see in particular pp. 101–33.

61 Allan Millet and Williamson Murray, ed., *Military Effectiveness. Vols. 1-3*. Boston: Allen & Unwin, 2012. See also Zoltan Jobbagy, "The efficiency aspect of military effectiveness," *Militaire Spectator* 178, no. 10 (2009): 506–514.

62 See M. Syed, *Black Box Thinking*. New York: Random House, 2015, 33–37.

63 Baskin, "According to U.S. Big Data".

64 Carl von Clausewitz, *On War*, ed. Michael Eliot Howard and Peter Paret. Princeton, NJ: Princeton University Press, 1976, 134.

65 For further reference, see James Mahoney and Gary Goertz, "A tale of two cultures: contrasting quantitative and qualitative research," *Political Analysis* 14, no. 3 (2006): 227–249; Marianne Franklin, *Understanding Research: Coping with the Quantitative-Qualitative Divide*. London: Routledge, 2012.Lisa M. Given, *The SAGE Encyclopedia of Qualitative Research Methods*. Los Angeles: SAGE, 2008; Steven J. Taylor, Robert Bogdan, *Introduction to Qualitative Research Methods*. New York: Wiley, 1998; Martin Hammersley, *Questioning Qualitative Inquiry*. London: SAGE, 2008; Martin Hammersley, *What Is Qualitative Research?* London: SAGE, 2013; Maggi Savin-Baden and Claire

Howell Major, *Qualitative Research: The Essential Guide to Theory and Practice.* London: Routledge, 2013; Norman K. Denzin and Yvonna S. Lincoln, ed., *Handbook of Qualitative Research* (2nd ed.). Thousand Oaks, CA: SAGE, 2000; John Creswell, *Research Design: Qualitative, Quantitative, and Mixed Method Approaches.* Thousand Oaks, CA: SAGE, 2003; Alan Bryman, *Social Research Methods* (4th ed.). Oxford: Oxford University Press. 2012.

[66]   Cukier and Mayer-Schönberger, "The dictatorship of data."

References

'Cincinnatus'. *Self-Destruction: The Disintegration and Decay of the United States Army During the Vietnam Era.* New York: Norton, 1981.

"Body Count." In *The Encyclopedia of the Vietnam War. A Political, Social and Military History*, edited by Spencer Tucker, 44 ff, Oxford: ABC-Clio, 2001.

Addington, Larry. *The Blitzkrieg Era and the German General Staff, 1865–1941.* New Brunswick, NJ: Rutgers University Press, 1971.

Apply, Christian G. *Working-Class War: American Combat Soldiers and Vietnam.* Chapel Hill and London: University of North Carolina Press, 2000.

Ardant du Picq, Charles. *Études sur le Combat.* Paris: Hachette et Dumaine, 1880.

Audoin-Rouzeau, Stéphane. "Vers une Anthropologie Historique de la Violence de Combat au XIXe Siècle: Relire Ardant du Picq?" *Revue d'Histoire du XXIXe Siècle* 30 (2005) https://doi.org/10.4000/rh19.1015

Basha i Novosejt, Aurélie. *'I Made Mistakes': Robert McNamara's Vietnam War Policy, 1960–1968.* Cambridge: Cambridge University Press, 2019.

Baskin, Jonathan Salem. "According to U.S. Big Data, We Won the Vietnam War." *Forbes*, 25 July 2014. https://www.forbes.com/sites/jonathansalembaskin/2014/07/25/according-to-big-data-we-won-the-vietnam-war/?sh=d8f83183f218

Bastian, Nathaniel. D. "Information warfare and its 18th and 19th century roots." *The Cyber Defense Review* 4, no. 2 (2019): 31-36.

Bergerud, Eric M. *Fire in the Sky: The Air War in the South Pacific.* Boulder, CO: Westview Press, 2000.

Blair, jr., Clay. *Silent Victory: The U.S. Submarine War against Japan.* Philadelphia: Lippincott, 1975.

Brown, John Sloan. "Colonel N. Dupuy and the mythos of Wehrmacht superiority: a reconsideration." *Military Affairs* 50, no. 1 (1986): 16–20.

Brown, John Sloan. "The Wehrmacht mythos revisited: a challenge for Colonel Trevor N. Dupuy." *Military Affairs* 51, no. 3 (1987): 146–147.

Bryman, Alan. *Social Research Methods* (4th ed.). Oxford: Oxford University Press. 2012.

Bülow, Heinrich Dietrich Adam von. *Geist des neuern Kriegssystems hergeleitet aus dem Grundsatze einer Basis der Operationen auch für Laien in der Kriegskunst faßlich vorgetragen von einem ehemaligen Preußischen Offizier.* Hamburg: Benjamin Gottlieb Hofmann, 1799.

Chwastiak, Michele. "Rationality, performance measures and representations of reality: planning, programming and budgeting and the Vietnam War." *Critical Perspectives on Accounting* 17, no. 1 (2006): 29–55.

Clark, Christopher. *The Sleepwalkers: How Europe Went to War in 1914*. London: Penguin Books, 2012.

Clausewitz, Carl von. *On War*, edited by Michael Eliot Howard and Peter Paret. Princeton, NJ: Princeton University Press, 1976.

Creswell, John. *Research Design: Qualitative, Quantitative, and Mixed Method Approaches*. Thousand Oaks, CA: SAGE, 2003.

Creveld, Martin van. *Fighting Power: German and U.S. Army Performance 1939–1945*. Westport, CN, Greenwood Press, 1982.

Cukier, Kenneth, and Viktor Mayer-Schönberger. "The dictatorship of data," *MIT Technology Review* 31, no. 05 (2013). https://www.technologyreview.com/2013/05/31/178263/the-dictatorship-of-data/

Daddis, Gregory A. "The problem of metrics: assessing progress and effectiveness in the Vietnam War." *War in History* 19, no. 1 (2012): 73–98.

Denzin, Norman K., and Yvonna S. Lincoln, ed. *Handbook of Qualitative Research* (2nd ed.). Thousand Oaks, CA: SAGE, 2000.

Dimbleby, Jonathan. *The Battle Of The Atlantic: How the Allies Won the War*. New York: Viking Press, 2015.

Dupuy, Trevor N. "A response to 'The Wehrmacht mythos revisited'." *Military Affairs* 51, no. 4 (1987): 96–197.

Dupuy, Trevor N. "Mythos or verity? The quantified judgment model and German combat effectiveness." *Military Affairs* 50, no. 4 (1986): 204–210.

Dupuy, Trevor N. *Numbers, Predictions and War: Using History to Evaluate Combat Factors and Predict the Outcome of Battles*. New York: Bobbs-Merrill, 1979.

Dupuy, Trevor, *Attrition: Forecasting Battle Casualties And Equipment Losses In Modern War*. Fairfax, VA: Hero Books, 1990.

Ellsberg, Daniel. *Secrets: A Memoir of Vietnam and the Pentagon Papers*. New York: Penguin Books, 2002.

Enthoven, Alain C., and K. Wayne Smith. *How Much Is Enough?: Shaping the Defense Program, 1961-1969*. Santa Monica: Rand Corporation, 2005.

Franklin, Marianne. *Understanding Research: Coping with the Quantitative-Qualitative Divide*. London: Routledge, 2012.

Gaddis, John Lewis. *The Cold War: A New History*. London. Penguin Books, 2005.

Given, Lisa M. *The SAGE Encyclopedia of Qualitative Research Methods*. Los Angeles: SAGE, 2008.

Günter Lewy, *America in Vietnam*. Oxford: Oxford University Press, 1978.

Hammersley, Martin. *Questioning Qualitative Inquiry*. London: SAGE, 2008.

Hammersley, Martin. *What is Qualitative Research?* London: SAGE, 2013.

Harré, Rom. "Positivist Thought in the Nineteenth Century." In *The Cambridge History of Philosophy 1870-1945*, edited by Thomas Baldwin, 7-26. Cambridge: Cambridge University Press, 2003.

Hastings, Max. *Retribution: The Battle for Japan, 1944-45*. New York: Alfred A. Knopf, 2008.

Hastings, Max. *Vietnam: An Epic History of a Tragic War 1945-1975*. London: William Collins, 2018.

Hendrickson, Paul. *The Living and the Dead: Robert McNamara and Five Lives of a Lost War* Sydney: Vintage, 2000.

Jobbagy, Zoltan. "The efficiency aspect of military effectiveness." *Militaire Spectator* 178, no. 10 (2009): 506–514.

Jomini, Antoine-Henri de. *Précis de l'art de la guerre*. 1838. Paris: Éditions Ivréa, 1994.

Kaplan, Lawrence S., Ronald Dean Landa and Edward Drea. *The McNamara Ascendancy, 1961-1965*. Washington: Historical Office, Office of the Secretary of Defense, 2006.

Karnow, Stanley. *Vietnam: A History*. New York: Viking Press, 1983.

Kinnard, Douglas. *The War Managers*. Burlington. University of Vermont, 1977.

Kirby, Maurice, and Rebecca Capey, "The area bombing of Germany in World War II: an operational research perspective." *Journal of the Operational Research Society* 48 (1997): 661–677.

Krimmer, Elisabeth, and Patricia Simpson, ed. *Enlightened War: German Theories and Cultures of Warfare from Frederick the Great to Clausewitz*. Rochester and New York: Camden House, 2011.

Mahoney, James, and Gary Goertz. "A tale of two cultures: contrasting quantitative and qualitative research." *Political Analysis* 14, no. 3 (2006): 227–249.

McCann, Leo. "'Management is the gate' – but to where? Rethinking Robert McNamara's 'career lessons'." *Management and Organizational History* 11, no. 2 (2016): 166–188.

McCloskey, Joseph. "British operational research in World War II." *Operations Research* 35, no. 3 (May–June 1987): 453–470.

McMaster, Herbert R. *Dereliction of Duty: Johnson, McNamara, the Joint Chiefs of Staff, and the Lies that Led to Vietnam*. New York, Harper Collins,1998.

McNamara, Robert. *In Retrospect: The Tragedy and Lessons of Vietnam* (New York: Vintage Books, 1995).

McNamara, Robert. *Argument Without End: In Search of Answers to the Vietnam Tragedy*. New York: Public Affairs, 1999.

Millet, Allan, and Williamson Murray, ed. *Military Effectiveness. Vols. 1-3*. Boston: Allen & Unwin, 2012.

Morris, Errol. *The Fog of War: Eleven Lessons from the Life of Robert S. McNamara*. documentary film, 2003.

Murdock, Clark A. "McNamara, systems analysis and the systems analysis office." *Journal of Political & Military Sociology* 2, no. 1 (1974): 89–104.

Murry, Gregory H. "The On-Going Battle for the Soul of the Army. " *Small Wars Journal* 12, no. 10 (2016) https://smallwarsjournal.com/jrnl/art/the-on-going-battle-for-the-soul-of-the-army.

O'Nan, Stewart, and Philip Caputo. *The Vietnam Reader: The Definitive Collection of American Fiction and Nonfiction on the War*. New York: Anchor Books 1998.

Overy, Richard. *Bomber Command, 1939–45*. London: Harper Collins, 1997.

Overy, Richard. *The Air War: 1939–1945*. London: Taylor & Francis, 1980.

Overy, Richard. *The Bombers and the Bombed: Allied Air War Over Europe 1940-1945*. London: Viking Press, 2014.

Overy, Richard. *The Bombing War: Europe 1939-1945*. London: Penguin Books, 2013.

Overy, Richard. *Why the Allies Won: Explaining Victory in World War II*. New York: Norton, 1995.

Palaima, Thomas. "Maritime Matters in the Linear B Tablets," *Thalassa: l'Égée Préhistorique et la Mer, Actes de la Troisième Rencontre Égéenne Internationale de l'Université de Liège,* edited by Robert Laffineur and Lucien Basch. 273-310. Liège: Université de Liège, 1991.

Phil Rosenweig, "Robert S. McNamara and the evolution of modern management." *Harvard Business Review* 91 (2010): 87–93.

Piepers, Ingo. *2020: WARning. Social Integration and Expansion in Anarchistic Systems: How Connectivity and Our Urge to Survive Determine and Shape the War Dynamics and Development of the System.* Amsterdam: Conijn Advies, 2016.

Piepers, Ingo. *Dynamiek en Ontwikkeling van het Internationale Systeem: een Complexiteitsperspectief.* Amsterdam: Universiteit van Amsterdam, 2006.

Rau, Erik P. "Combat science: the emergence of operational research in World War II." https://doi. org/10.1016/j.endeavour.2005.10.002

Rej, Abhijnan. "The other legacy of Robert McNamara." https://warontherocks.com/2016/06/the-other-legacy-of-robert-mcnamara/

Reporting Vietnam Part Two: American Journalism 1969-1975. New York. Library of America, 1998.

Reporting Vietnam, Part One: American Journalism, 1959-1969. New York. Library of America, 1998.

Sapolsky, Harvey. "Rationalizing McNamara's Legacy."https://warontherocks.com/2016/08/rationalizing-mcnamaras-legacy/.

Savin-Baden, Maggi, and Claire Howell Major, *Qualitative Research: The Essential Guide to Theory and Practice.* London: Routledge, 2013.

Schell, Jonathan, "The Village of Ben Suc." *The New Yorker* (July 15, 1967).

Shafer, Michael. *Deadly Paradigms: The Failure of U.S. Counterinsurgency Policy.* Princeton: Princeton University Press, 1988.

Sharpley, Deborah. *Promise and Power: The Life and Times of Robert McNamara.* Boston: Little, Brown, 1993.

Sheehan, Neil. *A Bright Shining Lie: John Paul Vann and America in Vietnam.* New York: Random House, 1988.

Sheehan, Neil. *The Pentagon Papers.* New York: Quadrangle Books, 1971.

Shelton, Kim S. "Foot Soldiers and Cannon Fodder: The Underrepresented Majority of Mycenaean Civilization." In *EPOS: Reconsidering Greek Epic and Aegean Bronze Age Archaeology,* edited by Sarah P. Morris and Robert Laffineur, 169–176. Liège: Université de Liège, 2007.

Simon, Walter. *European Positivism in the Nineteenth Century.* New York: Cornell University Press, 1963.

Slater, Jerome. "McNamara's failures—and ours: Vietnam's unlearned lessons: a review." *Security Studies* 6, no. 1 (1996): 153–195.

Stevenson, David. "War by timetable? The railway race before 1914." *Past & Present* 162 (Feb. 1999): 163–194.

Sun-Tzu, *The Art of Warfare.* Translated by Roger T. Ames. New York: Ballatine Books, 2010.

Syed, Matthew. *Black Box Thinking.* New York: Random House, 2015.

Tarpgaard, Peter T. "McNamara and the rise of analysis in defense planning: a retrospective." *Naval War College Review* 48, no. 4 (Fall 1995): 67–87.

Taylor, Alan John Perceval. *War by Timetable: How the First World War Began.* London: MacDonald, 1969.

Taylor, Steven J., and Robert Bogdan. *Introduction to Qualitative Research Methods.* New York: Wiley, 1998.

The Pentagon Papers: The Defense Department History of United States Decision-making on Vietnam https://www.archives.gov/research/pentagon-papers.

Thijssen, Wil. "De Derde Wereldoorlog kan elk moment beginnen, en ik voorspel het al jaren." *de Volkskrant* (10 January 2020).

Ward, Geoffrey C., and Ken Burns. *The Vietnam War: An Intimate History.* London: Penguin Books, 2017.

Watson, Jr., George M., and Herman S. Wolk, "'Whiz kid': Robert S. McNamara's World War II service." *Air Power History* 50, no. 4 (2003): 4 –15.

White, David. *Bitter Ocean: The Battle of the Atlantic, 1939–1945.* New York: Simon & Schuster, 2008.

Yankelovich, Daniel. *Corporate Priorities: A Continuing Study of the New Demands on Business.* Stamford, Conn.: Daniel Yankelovich, Inc., 1972.

About the editors

Robert Beeres holds a PhD in administrative sciences from Radboud University Nijmegen, The Netherlands. Currently he is a professor of Defence Economics at the Faculty of Military Sciences, Netherlands Defence Academy. His research interests include economics of arms export controls, defence capabilities, performance management and burden sharing within the EU and NATO. He has published numerous articles in peer-reviewed journals and books and co-edited a number of books.

Peter B.M.J. Pijpers, PhD is an Army Colonel (GS). He is Associate Professor of Cyber Operations at the Netherlands Defence Academy and researcher at the Amsterdam Centre for International Law (ACIL), University of Amsterdam. Dr Pijpers has (co-)authored numerous articles related to the legal and cognitive dimension of influence operations in cyberspace, below the threshold of the use of force.

Mark Voskuijl is professor of Weapon- and Aviation Systems at the Faculty of Military Sciences at the Netherlands Defence Academy. His research is focused on the performance and flying qualities of fighter aircraft, helicopters and unmanned aerial vehicles.

About the authors

Dr. F.H. (Floribert) Baudet is associate professor of Strategy at the Faculty of Military Sciences, Netherlands Defence Academy. He obtained his PhD in the history of international relations from Utrecht University. He has published widely on, among other things, Dutch foreign and defence policy, historical methodology in military sciences, the former Yugoslavia, and strategic communication.

Colonel Dr Han Bouwmeester is Associate Professor of Military Strategy and Land Warfare at the Faculty of Military Science, Netherlands Defence Academy. He is both an artillery officer and a scientist. Among other things, he commanded an artillery battalion and the Provincial Reconstruction Team in the Afghan province of Uruzgan. He also studied at the US School of Advanced Military Studies at Ft Leavenworth (KS), and obtained his PhD degree at Utrecht University. He publishes regularly on Russian military operations, deception warfare and various ways of influencing others.

Dr. Tessa op den Buijs is an assistant professor of Psychology at the Department of Military Behavioural Sciences and Philosophy of the Netherlands Defence Academy (NLDA). She lectures in Human Resource Management and Leadership, Research Strategies and Military Stress. She supervises bachelor, master and PhD students in writing their thesis and dissertation and is also a lecturer and thesis coordinator in the Frontex European Joint Master of Strategic Border Management (EJMSBM). Presently, she is engaged in several research projects: care needs of veterans, work-life balance, diversity and retention of military personnel.

Brigadier-General dr. Paul A.L. Ducheine is the Professor for Cyber Operations and Cyber Security at the Netherlands Defence Academy, and a Legal Advisor (Netherlands Army Legal Service). At the University of Amsterdam, he was named professor by special appointment of Law of Military Cyber Operations.

Martijn van Ee is assistant professor of mathematics and operations research at the Faculty of Military Sciences of the Netherlands Defence Academy. His research focuses on combinatorial optimisation and military operations research.

Paul C. van Fenema (PhD RSM) is a professor of Military Logistics at Netherlands Defence Academy. His research focuses on institutional change of interorganisational relationships and digital transformation. He has been published in journals such as Industrial Marketing Management, Organization Science, MIS Quarterly and Joint Forces Quarterly. He has worked at RSM, Tilburg University and Florida International University.

Gerold de Gooijer participated in various operations in Europe and the Middle East and is a recently retired army Colonel (MSc SCM). As a logistic expert he also was appointed as ERP project leader for several years. Currently he is affiliated with the Netherlands Defence Academy to finalise his PhD research on inter-organisational relationships in the field of "Total Force."

Axel Homborg holds a PhD in Materials Science from Delft University of Technology, The Netherlands. Currently he is an associate professor in Maintenance Technology at the Faculty of Military Sciences, Netherlands Defence Academy. His research interests include corrosion management, corrosion monitoring with an emphasis on passive classification and corrosion inhibition with a special interest in the replacement of chromium-6 in aerospace. He has published numerous articles in peer-reviewed journals.

André J. Hoogstrate received his doctorate in econometrics from Maastricht University (1998). After having worked as a forensic scientist for 20 years, he moved (2016) to the Netherlands Defence Academy. Research interests are forecasting, alignment of models and data with strategy and operations and the interpretation, validation, and verification of data-driven procedures.

Tess Horlings is a PhD researcher at the Netherlands Defence Academy. She has an academic background in Middle Eastern Studies and Military Strategic Studies, and professional experience in the national security domain. Her research focuses on the challenges and opportunities for intelligence organisations in dealing with data in the current Information Age. The research is multidisciplinary and uses a mixed methods approach, with a main focus on contributing empirical research findings to the Intelligence Studies domain.

Dr. H. (Henk) de Jong holds a PhD in history from the University of Amsterdam. He is an Assistant Professor of Military History at the Faculty of Military Sciences, Netherlands Defence Academy. His research focuses on cultural aspects of war and warfare.

Michiel de Jong holds a PhD in history from Leiden University, The Netherlands. Currently, he is an Assistant Professor of Military History at the Faculty of Military Sciences, Netherlands Defence Academy. His research interests include counterterrorism in France, Italy, Spain and Latin-America and economic, maritime and military history of the Netherlands and Europe in general. He is author and co-author of several books and has contributed articles to peer-reviewed journals and books.

Dr. Relinde Jurrius works as Assistant Professor of Mathematics & Operations Research at the Netherlands Defence Academy. Her research and teaching focuses on the application of new techniques in mathematics and computer science to relevant military problems, as well as the theory behind it.

Ton van Kampen developed himself during his long military career as a military expert on Operational Logistics. In 2002 he became Master of Transport and Logistic Management after graduating from Tilburg University. Today he is assistant professor of Military Logistics at the Netherlands Defence Academy.

Eric-Hans Kramer is Professor in Military Management and Organisation at Faculty of Military Science, Netherlands Defence Academy. Before that, he was associate professor of Human Factors and Systems Safety at the same organisation. His research focuses on contemporary organisational challenges of the military organisation in general and taskforces in particular. Furthermore, he is involved in developing interdisciplinary research, educational programs and in supervision of PhD projects.

Geraldo Mulato de Lima Filho is a Lieutenant Colonel at the Brazilian Air Force (FAB), where he has been working as a pilot for over 20 years. His research interests include decision support systems, MAV/UAV cooperative engagement, computational optimisation techniques, and applications of artificial intelligence methods.

Roy Lindelauf is an assistant professor of Quantitative Intelligence Analysis at the Faculty of Military Sciences and he coordinates the Data Science Center of Excellence of the Dutch MOD. In addition, he is a Research Fellow at Delft Institute of Applied Mathematics, TU-Delft. His research focuses on mathematical modelling of defence and security-related topics, data science and AI. He is project leader of a National Technology Project on developing algorithms for target discovery in terrorist and insurgent networks for the Joint Sigint Cyber Unit, together with Delft University of Technology.

Herman Monsuur is professor of Military Operations Research at the Faculty of Military Sciences of the Netherlands Defence Academy. His research focuses on combinatorial optimisation and military operations research.

Guido van Os holds a PhD in Public Administration from Erasmus University Rotterdam. Currently, he is assistant professor of Organisational Science and Management at the Faculty of Military Sciences, Netherlands Defence Academy. His research interests include the governance of technological innovations, digital transformation, and digitisation and organisational change.

Peter B.M.J. Pijpers, PhD is an Army Colonel (GS). He is Associate Professor of Cyber Operations at the Netherlands Defence Academy and researcher at the Amsterdam Centre for International Law (ACIL), University of Amsterdam. The main focus of his research is related to the legal and cognitive dimension of influence operations in cyberspace, below the threshold of the use of force.

Eric H. Pouw Ph.D. is a senior military legal advisor within the Royal Netherlands Army (RNLA), research fellow at the NLDA and researcher at ACIL, University of Amsterdam. He was, at the time of co-writing this chapter, a legal advisor within the Directorate of Knowledge and Development of the RNLA HQ on legal aspects relating to the Army's adaptation to information-driven operations.

Steven van de Put holds an LLM in Human Rights Law from Queen's University Belfast. Currently, he is working towards his PhD at Maastricht University. He is a lecturer at the Netherlands Defence Academy, where he teaches military law. His research interests include accountability for violations of international law and the use of international law in armed conflict.

Sebastiaan Rietjens (PhD) is a full professor of Intelligence & Security at the Netherlands Defence Academy and holds the special chair of Intelligence in War & Conflict at Leiden University. He has done extensive fieldwork in military exercises and operations and has published widely in international journals and books. His main research focus is on intelligence during military operations, peacekeeping intelligence, warning for hybrid threats and future developments that confront intelligence organisations.

Chris Rijsdijk holds a PhD from the University of Twente. Currently he is an assistant professor in platform systems at the Faculty of Military Sciences, Netherlands Defense Academy. His research interest includes data-driven decision support, for which he performs action research in numerous case studies within the Dutch MoD.

Gert Schijvenaars is an assistant professor in military logistics and information management at the Netherlands Defence Academy and is a retired army colonel. He got his master's degree in mechanical engineering and studied maintenance management at Delft University. He has been working within the NLMOD sustainment organisation on a large variety of assignments and at all different levels.

Prof. Theunissen has been active in the field of Avionics for over 30 years. The company ISD, that he founded in 1988, has developed prototype synthetic vision and DAA systems. In the U.S. he contributed to the development of the DAA Minimum Operational Performance Standards as part of RTCA Special Committee 228. In Europe, he contributed to the definition of requirements and capabilities for the Remain Well Clear function of a DAA system for UAVs in airspace classes D-G.

Dr. Wieger Tiddens has extensive experience and knowledge on the implementation and development of predictive maintenance. His passion is to achieve the maintenance of tomorrow. Within the Royal Netherlands Navy's Data for Maintenance group, he works on data-driven asset management and Predictive Maintenance for the current and future fleet.

Job Timmermans is an associate professor of Business Ethics at the Faculty of Military Sciences of the Netherlands Defence Academy. His interests cover the philosophical and sociological issues arising from the intersections of ethics, integrity and compliance within organisations. Previously, he has worked on several projects on Responsible Research Innovation (RRI) at Wageningen University, De Montfort University (UK), and Delft University of Technology.

Tiedo Tinga obtained his PhD in Mechanics of Materials from Eindhoven University of Technology, The Netherlands. Currently he is a full professor of Life Cycle Management at the Faculty of Military Sciences of the Netherlands Defence Academy. His research focuses on physics of failure, condition monitoring and predictive maintenance. He has supervised 17 PhD and PDEng students and published around 60 articles in peer-reviewed journals.

Mark Voskuijl is professor of Weapon- and Aviation Systems at the Faculty of Military Sciences at the Netherlands Defence Academy. His research is focused on the performance and flying qualities of fighter aircraft, helicopters and unmanned aerial vehicles.

Anna C. Vriend is an officer at the Royal Netherlands Navy and studies at the Royal Netherlands Naval Academy. She recently graduated from the Netherlands Defence Academy with a bachelor's degree in Military Systems and Technology.

Peter de Werd is Assistant Professor in intelligence and security at the Netherlands Defence Academy. He has an academic background in military sciences and political science, obtaining his PhD in intelligence studies at Utrecht University. His current research interests include contributing to reflexivist and critical theoretical debate in intelligence studies, narrative methodologies for intelligence analysis, and innovation of intelligence education.

Marten Zwanenburg is professor of military law at the Netherlands Defence Academy and the University of Amsterdam. Professor Zwanenburg previously worked as a senior legal counsel in the Legal Affairs Department of the Dutch Ministries of Foreign Affairs and Defence. Zwanenburg has published extensively in various international peer-reviewed journals and contributed to a number of the leading handbooks in the field of military law, including the Handbook of the International Law of Military Operations.